MANUAL OF VEGETABLE PLANT DISEASES

By

Charles Chupp

DPH

DISCOVERY PUBLISHING HOUSE PVT. LTD.
NEW DELHI-110 002

Reprinted - 2015

ISBN: 978-81-7141-251-8

Manual of Vegetable Plant Diseases

Published by:
DISCOVERY PUBLISHING HOUSE PVT. LTD.
4383/4B, Ansari Road, Darya Ganj
New Delhi-110 002 (India)
Phone: +91-11-23279245, 43596064-65
Fax: +91-11-23253475
E-mail: discoverypublishinghouse@gmail.com
sales@discoverypublishinggroup.com
web: www.discoverypublishinggroup.com

Printed at:
Infinity Imaging Systems
Delhi

MANUAL OF VEGETABLE PLANT DISEASES

PREFACE

SINCE the word vitamin has become a shibboleth for the home demonstration agent, the use of vegetables in the diet has received renewed attention. This added interest in truck crops is reflected by the increased number of recent published articles touching on the diseases of vegetables. Before 1889 there were only occasional pieces of literature dealing with maladies of the garden crops. Beginning at about that year, Halsted, Farlow, Humphrey, Selby, and other noted workers began their contributions to this phase of plant pathology. As the years went by, the number of investigators increased until at present between three hundred and four hundred contributions are published each year on vegetable diseases alone.

Notwithstanding the increase in the knowledge of the subject, there has been very little effort in bringing together all the material in such a form that it is available to the plant pathologists generally and to the growers and extension specialists in particular. The failure to make the compilation may be due to the fact that the task is a stupendous one, and as no person is intimately acquainted with every disease errors are sure to appear. Consequently the effort here presented may be a case of rushing in where wiser heads fear to enter. Even granting such a possibility, material has been collected from more than eight thousand citations in the preparation of this manual and is offered with the desire that it will aid not only in solving the gardeners' immediate problems, but will lay stress on many points that need further investigation. It is hoped that fellow pathologists will be interested enough to draw the attention of the author to errors that may be present.

When any particular plant disease is discussed before a

group of growers and is new to them, they at once want to know from where it came, whether it is something new, what it looks like, what causes it, what kind of weather and soil favor it, and how it can be controlled. With these questions in mind, nearly every disease of vegetables described in literature has been dealt with in some detail, paying special attention to symptoms and control. To make the discussions of greater value, especially to the student of plant pathology, many important citations have been added.

The nomenclature is one of the most difficult problems with which the pathologist is confronted. Many fungi have received several names and opinion frequently is divided in regard to their correct usage. In the main, the most common names have been retained. Additional scientific names have been included so that the table of contents will be a fairly complete host index for the extension specialist. In the nomenclature of bacteria, Smith's classification has been followed even though a more recent, and probably a better, classification is now available. The latter can hardly be used in a popular presentation, however, until the terms are applied generally by the investigators.

The author is indebted to a number of co-workers for suggestions and aid in revising the manuscript, and wishes to express his appreciation of the assistance given by Dr. M. F. Barrus, Dr. W. H. Burkholder, Dr. K. H. Fernow, Dr. E. F. Guba, Mr. E. E. Honey, Dr. L. M. Massey, Mr. A. G. Newhall, Dr. R. S. Kirby, Dr. H. E. Thomas, Dr. J. L. Weimer, Mr. D. S. Welch and Mrs. Charles Chupp. Mr. W. R. Fisher prepared most of the illustrations.

CHARLES CHUPP.

CORNELL UNIVERSITY, ITHACA, NEW YORK,
Oct. 1, 1924.

CONTENTS

CHAPTER I

CHAPTER II

CHAPTER III

CHAPTER IV

CHAPTER V

CHAPTER VI

CHAPTER VII

CHAPTER VIII

CHAPTER IX

CHAPTER X

CHAPTER XIII

CHAPTER XIV

CHAPTER XVII

CHAPTER XVIII

CHAPTER XIX

ILLUSTRATIONS

CHAPTER I

ASPARAGUS DISEASES

Asparagus is one of the oldest vegetables, having been in cultivation for over two thousand years. More than thirty thousand acres of it are now grown in the United States, with an annual value of over five million dollars. The states that produce most of the asparagus are California, New Jersey, Illinois, Massachusetts, South Carolina, Pennsylvania, and New York.

Fortunately the diseases of this palatable vegetable are either not destructive or have been overcome by the application of control measures. An excellent illustration of controlling one disease is the breeding of varieties resistant to the rust. There is no doubt that if Fusarium wilt or violet root-rot ever become destructive enough to threaten the cultivation of the crop, or even to reduce the yield materially over wide areas, satisfactory means of control will be developed for the two troubles. The confidence with which such a statement can be made shows the progress of recent years in all vegetable pathology, and expresses particularly the better understanding of control measures for vegetable diseases.

Asparagus Rust

Caused by *Puccinia asparagi* DC.

The rust on asparagus was discovered in Europe in 1805. On account of the various fungi that are parasitic on the rust organism, and because of resistant varieties of the host, the disease never appeared there in epidemic form. It was a different story, however, when the disease reached America

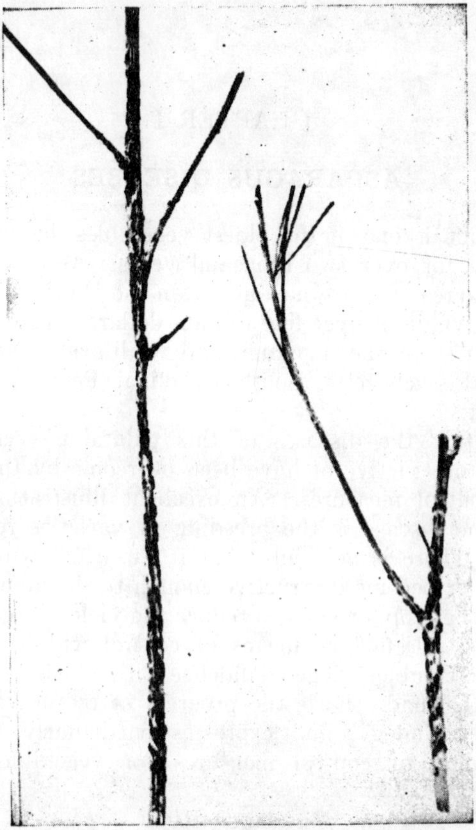

FIG. 1.—Asparagus stems showing rust pustules.

in 1896. With startling suddenness it swept the plantings about Boston, on Long Island, and in New Jersey, each season progressing farther south and west until California was invaded in 1902. Probably never has a disease swept over the country so rapidly and with such devastating results. It is now present in all parts of Europe and America where asparagus is grown, but has not been mentioned by pathologists in Asia, Africa, and Australia.

The greatest damage to the crop is not directly to the edible part, but, by attacking the tops several seasons in succession, the malady reduces the vitality of the root system so that the young shoots fail to form in the spring or are so poor that they must be sold as culls. At Concord, Massachusetts, the crop, which before 1896 was worth one hundred thousand dollars annually, was almost completely annihilated. On Long Island, where the yield had been fifteen to twenty-five hundred bunches an acre, it dwindled to seven or eight hundred. In California the loss was so heavy that a pathologist was hired to aid growers in solving this problem.

The rust attacks not only the common asparagus, known as *Asparagus officinalis,* but can also cause infection on onion and the *Asparagus* species *caspius*, *maritimus*, and *verticillatus*. It is not supposed to affect *A. scandens* and *plumosus*. The different varieties of the common asparagus vary in their susceptibility to the disease. Conover's Colossal is most severely affected while such varieties as Palmetto and Argenteuil show marked resistance. Within the last ten years still other improved resistant strains have been bred until asparagus rust has become almost a rarity in the very centers where previously it had been most destructive.

Symptoms.

Any one acquainted with the different stages of cereal rusts will have no difficulty in diagnosing the similar disease of asparagus. In an affected field the first symptom is the browning or reddening of a clump of the smaller twigs and needles (Fig. 1). The discolored area spreads rapidly in all directions until finally the entire planting appears as if the asparagus had ripened prematurely. If examined more closely, it will be seen that the red color is due to the stems and needles being covered with small pustules, which emit a dusty cloud when touched. As the season advances into autumn and winter, the red pustules gradually give place to dark-colored ones until the dead branches are black and prominent.

A different type of symptom may show earlier on asparagus that is not cut in the spring. This does not develop on the needles as do the red and black stages, but is confined to the stem and branches. The affected areas are oval-shaped and lighter green than the normal. Small circular pustules develop in concentric rings within the lesions. The whole is rather inconspicuous and usually not observed by one not familiar with this stage.

The rust pustules, bursting the epidermis under which they originate, often have a gray moldy appearance, at least as they become older. The moldy growth belongs to one or more fungi, which present the interesting anomaly of one micro-organism preying upon another that in turn is a serious pest on a higher type of plant. The most common one of the parasites is named *Darluca filum* Cast. Another one, and which is usually found on the spring stage of the rust, is *Tubercularia persicina* Ditt.

Cause of asparagus rust.

The disease is caused by *Puccinia asparagi*, a fungus closely related to the parasites causing rust on many cultivated plants and weeds. Its life history is composed of the four common rust stages, all of which are found on the one host; therefore, the fungus is designated as autoecious.

The first two spore stages take place together in the oval light green patches on the stems and branches. They occur where rust was present the year before, appearing on the first stalks that come through the ground. They are most abundant in April and May, although some have been found as late as July. The first or pycnial stage, in which the small globose pycniospores are borne in numerous minute spherical cavities, was once thought to represent the male element of the fungus, but it is now known to be functionless. In the second stage the spores are borne in closely packed chains in a cup-shaped fruit-body, the top of which breaks through the host epidermis. The fruit-body is known as an æcium and the spores as æciospores. They are relatively large and angular, are blown by wind, and in the presence of moisture germinate and enter the breathing pores of the asparagus plant.

When the mycelium has developed between the host cells until nearly every part of the tissue is invaded and the cells punctured with haustoria, a fruiting layer of the fungus is formed just below the epidermis of the host. The presence of the parasitic growth soon splits open the epidermis, and not only lays bare the spores for dissemination but also permits the moisture to escape from the stems and leaves, thus draining the vitality of the asparagus plant. In this stage of the life history the fruit-body is known as an uredinium and the spores as urediniospores or uredospores. They are brick-red in color, approaching a sphere in shape, and borne singly on short stalks. They are produced in such great numbers that when disseminated by the wind they cover the plants and ground with a reddish dust. They may even be found mixed with the cleaned seed and the roots that are removed for propagation. This has given rise to the belief that the fungus is carried long distances by means of the spores. Such a belief, however, is not well founded, since the urediniospores are short-lived, and, besides, none of the spores of the different stages would grow if buried in the ground with the roots or seeds. The urediniospores germinate at once in the presence of moisture and within twelve days after inoculation, may cause infection and yield a new generation of spores to threaten the crop.

Later in the season, usually when cooler weather has set in or following a prolonged drought, the urediniospores give place to large two-celled black spores, that may be borne in the same fruit-body as the former or in new fruit-bodies that may develop. The body is known as a telium, and the large spores as teliospores. They do not form a spore dust but cling to the sorus on the old stems until spring. On germination they put out a short, club-shaped, four-celled germ-tube known as a basidium or promycelium. Each of the cells produces a short thin stalk (sterigma) on which are borne singly thin hyaline spores, the basidiospores. The formation of basidiospores takes place at the time or soon after the young asparagus shoots are emerging from the ground in the spring. Infection is then produced on the young shoots by the entrance of mycelial threads, the aecial stage is formed, and the life cycle is repeated. If the young shoots are cut early for the market, the parasite will not have time to develop; but if a bed is left uncut or a few stray plants in a fence corner remain untouched, the fungus finds an ideal place for the production of inoculum that may infect any asparagus plants later in the spring.

There is a decided difference in the prevalence of the disease according to the moisture-retaining properties of the soil, the trouble being worse on the drier soils. In this case the host is less able to resist the attack as is shown when irrigated fields are given a less amount of water after the cutting season. On the other hand, soil-moisture increases dew which is the all-important factor in the occurrence of rust. Without dew there can be no rust. If a cheese-cloth tent is placed over asparagus or it is under the protection of a tree, there will be no disease. Again, in California where a trade-wind blows constantly from the west preventing the formation of dew, there is very little rust; but just outside of the path of these winds or wherever there is a windbreak, rust is plentiful. The dew is much more potent than rain, for the latter often washes the spores to the ground where they germinate and die. Fertilizers play no appreciable part in the prevalence of rust. Temperature has little influence since the parasite thrives under the same conditions that are necessary for the growth of the host. Cool weather may cause teliospore formation, but drought, slight nourishment, or a weakened condition of the host do so far more effectively.

Control of asparagus rust.

For those growers who still have susceptible varieties of asparagus, it is necessary to carry out rather complicated control measures. The rows should be planted in the direction of the prevailing winds, all windbreaks avoided, and the rows planted far enough apart to permit air to circulate freely among the branches, thus permitting the dew to dry quickly. Since the fungus obtains a foothold on uncut plants, all unused beds as well as stray specimens in fence corners and waste fields should be destroyed.

Spraying and dusting have been tried with a fair degree of success. In the East the practice has been to spray with bordeaux mixture to which a resin-fish-oil sticker has been added. In California sulfur dust is used. It is necessary to apply the latter when the dew is on the plants, so that the dust will adhere. In some of the fields which are too large to be dusted in the morning, the plants are first sprayed either with bordeaux mixture or a solution of whale-oil soap, and then while still wet are covered with the sulfur dust. The application of a fungicide is begun about three weeks after the cutting season and two additional applications made at twenty-day intervals.

It is no longer necessary to grow susceptible varieties of asparagus. Very desirable resistant strains are now on the market, the use of which will save the gardener the expense and labor of employing other control measures. By making selections from a large number of strains, Norton was able to obtain a resistant male plant which he named Washington. He also isolated two female plants, one of which he named Martha and the other Mary. Both of these when crossed with Washington gave strains that were not only resistant but also good commercial types. These new types were named Martha Washington anl Mary Washington.

If the grower is in the market for a resistant variety of asparagus and is not sure where to obtain it, he will do well to consult his county agricultural agent or his state plant pathologist regarding the matter. The gardeners' journals also print advertisements of reputable companies furnishing desirable resistant strains.

References

Anonymous. Washington asparagus: Information and suggestions for growers of new pedigreed rust-resistant strains. U. S. Dept. Agr. C. T. and F. C. D. Circ. 7: 1-8. 1919.

Halsted, B. D. The asparagus rust: its treatment and natural enemies. N. J. Agr. Exp. Sta. Bull. 129: 1-20. 1898.

Norton, J. B. Methods used in breeding asparagus for rust resistance. U. S. Dept. Agr. Bur. Plant Ind. Bull. 263: 1-60. 1913.

Sirrine, F. A. Spraying for asparagus rust. N. Y. (Geneva) Agr. Exp. Sta. Bull. 188: 233-276. 1900.

Smith, R. E. The water relation of *Puccinia asparagi*. Bot. Gaz. 38: 19-43. 1904.

Smith, R. E. Asparagus and asparagus rust in California. Calif. Agr. Exp. Sta. Bull. 165: 1-99. 1905.

Smith, R. E. Further experience in asparagus rust control. Calif. Agr. Exp. Sta. Bull. 172: 1-21. 1906.

Stone, G. E. and R. E. Smith. The asparagus rust in Massachusetts. Mass. (Hatch) Agr. Exp. Sta. Bull. 61: 1-20. 1899.

Walker, J. C. Rust of onion followed by a secondary parasite. Phytopath. 11: 87-90. 1921.

Fusarium Wilt of Asparagus

Caused by *Fusarium* sp.

In 1923 during the cutting season of asparagus in New York, a number of reports were received of diseased asparagus shoots, which reduced the salable yield appreciably. The shoots were wilted, dwarfed, and sometimes exhibited a dirty brown color. In certain cases the vascular system was also discolored. There seemed to be no infection in the older plants. When isolations were made, a Fusarium was always obtained. The cultures at first were white, but gradually changed to pink and red.

The same disease had previously been reported from New Jersey, Pennsylvania, Massachusetts, Illinois, California, and

Germany. In the last named place, the trouble was attributed to *Fusarium culmorum* Smith.

No control methods are available. When new plantings of asparagus are made, an effort should be put forth to procure only healthy plants, and these should be placed in uncontaminated soil. If the old planting becomes too badly diseased to be profitable, the stems and roots should be destroyed and a new bed started at some distance from the old infested one.

References

Cook, Mel T. A new disease of asparagus. Phytopath. **12:** 49. 1922.

Cook, Mel T. Dwarf asparagus. Phytopath. **13:** 284. 1923.

Müller, N. S. Bericht über die tätigkeit der agrikulturchemischen kontrollstation und der versuchstation für pflanzenkrankheiten der landwirtschaftskammer für die provinz Sachsen, Halle **1916-17:** 1-60. 1918.

Stone, G. E. and G. H. Chapman. Report of the botanists. Mass. Agr. Exp. Sta. Ann. Rept. **20:** 120-150. 1908.

Violet Root-Rot of Asparagus

Caused by *Rhizoctonia crocorum* var. *asparagi* (Pers.) DC.
(See Violet Root-rot of Potato for fuller discussion, p. 439.)

In some parts of England the violet-colored Rhizoctonia causes a serious rot of the roots and crowns of asparagus.

No definite control measures are known. Successful control demonstrations with bleaching powder have been reported when carrots were planted in heavily infested soil. From two to eight ounces of the powder for each square yard are harrowed into the soil in the early spring. It has been inferred that a similar treatment can be used for the control of the fungus when it attacks asparagus, although the latter does not seem to have been demonstrated.

References

Barker, B. T. P. and C. T. Gimingham. Experiments on the treatment of Rhizoctonia disease on asparagus. Univ. of Bristol, Agr. and Hort. Res. Sta. Ann. Rept. **1916:** 39-40. 1917.

Barker, B. T. P. and C. T. Gimingham. Further experiments on the Rhizoctonia disease of asparagus. Univ. of Bristol, Agr. and Hort. Res. Sta. Ann. Rept. 1917: 28-32. 1918.

Gimingham, C. T. and G. T. Spinks. Soil sterilization. Univ. of Bristol, Agr. and Hort. Res. Sta. Ann. Rept. 1919: 37-42. 1919.

Zopfia Root-Canker of Asparagus

Caused by *Zopfia rhizophila* Rabenh.

In Italy a root-canker of asparagus has caused considerable damage in certain plantings. The infection evidently takes place at the lower tip of the rhizome and gradually spreads forward until the whole root is included. When this last stage is reached, the plant dies. The canker spreads very slowly in the rhizome, and there is always a distinct line of demarcation between the healthy and diseased portions. This is not true of the roots, where the disease progresses very rapidly and the affected tissue is almost indistinguishable from the healthy. The roots are covered with small black fungous fruit-bodies.

The fungus was first found in Germany in 1874 on asparagus rhizomes which had been heeled-in for the winter, and under similar conditions was later observed in France. It was then not considered as a parasite and Arnaud even now insists that it is a saprophyte. The organism is closely related to the powdery-mildew fungi, differing in having colored mycelium and the perithecia not possessing real appendages. Apparently the organism lives in the soil, and when plants are weakened or wounded it is able to enter and cause a rot of the tissue.

Control measures that have been recommended are destruction by burning of all diseased roots, and disinfecting the place where the plant stood by applying to the soil six ounces of carbon bisulfide for each square yard.

References

Arnaud, A. Sur les genres Zopfia, Richonia et Caryospora. Bull. d. l. Soc. Mycol. France 29: 253-260. 1913.

Farneti, R. La cancrena delle zampe d' asparago. Riv. di Pat. Veg. **4**: 273-277. 1910.

Leptothyrium Canker of Asparagus

Caused by *Leptothyrium asparagi* Vogl.

In Italy a canker has been found on the roots, rhizomes and young shoots of asparagus. There are many small depressed spots about the withered base of the shoot, which may turn brown or almost black. The part of the rhizome next to the new stem growth may be similarly spotted and discolored. Dark areas on the roots are also present. The lesions bear some resemblance to those caused by Zopfia, but when isolations were made, a new species was found which was named *Leptothyrium asparagi.*

As the parasite is a soil organism, no means of control have been discovered.

References

Voglino, Pietro. Sopra una nuova infezione degli asparagi. Reale Accad. D'Agr. di Torino **56**: 176-180. 1917.

CHAPTER II

BEAN DISEASES

THE dry bean sections are confined mostly to California Michigan, and New York. Snap beans are grown not only on a commercial scale but as one of the most common vegetables in the home-garden. The value of the snap beans that are sold yearly is over eight million dollars while nearly eleven and a half million bushels of dry beans are placed on the market.

A very conservative estimate of losses compiled by the Plant Disease Survey Office gives about 4 per cent or nearly a half million bushels of beans lost on account of diseases. Notwithstanding the fact that beans frequently escape the numerous diseases to which they are susceptible, it is probable that a 10 per cent reduction in salable yield would not be an exaggeration.

Beans do not lend themselves readily to seed treatment, nor can they be sprayed successfully. Therefore, the one hope is that of obtaining strains which will be acceptable to the market and at the same time resistant to the most important diseases. Great strides have already been made in this direction, so that is is not an expression of over-confidence to state that in the future the growers of both snap and dry beans will be bothered very little with diseases, which in the past have been destructive.

REFERENCES

Plant Disease Bulletins volumes 1-6, and Supplements 12, 18, and 24. 1917-1922.

Fourteenth Census of the United States—1920 6: 1922.

Bean Anthracnose

Caused by *Colletotrichum lindemuthianum* (Sacc. and Magn.) Bri. and Cav.

Although bean anthracnose was not described until 1878, it is proved that the disease had been present many years before that, for pods collected in 1843 show the characteristic spotting. The first specimens were found in Europe, but the trouble is now known to occur wherever beans are grown. In years when it is epidemic it causes enormous losses, for not only are the seedlings killed, but the older plants are weakened, the leaves injured, the pods spotted, and the seeds withered and discolored. In growing dry beans the reduced stand and light weight of the seed are the chief sources of loss, while on snap beans the pod-spot is most to be dreaded. Such green beans often show no spotting when picked, but develop serious infection when placed in hampers and shipped long distances or held in storage for a few days.

Because of its prominence, the disease has been given a number of names. In most cases it is known as anthracnose, but many growers designate it also as pod-spot, canker, leaf-spot, and even as blight and rust. These last two names are objectionable since there is a bacterial blight and a true rust of beans, so that when the terms are used in connection with anthracnose they cause confusion.

The disease is found only on the common bean and some other closely related species. For example, it will attack lima-beans, cowpeas, kulthi bean, and scarlet runner. On none of these latter, however, has it ever threatened to become epidemic. The different varieties of the common bean vary in their susceptibility to anthracnose. The problem is complicated further by the fact that there are distinct races of the fungus that may react differently on the same variety or strain of bean. The tendency towards immunity is discussed at greater length in connection with control measures.

Symptoms.

A number of organisms may be able to discolor the bean seed, but the characteristic lesion produced by anthracnose can usually be distinguished by the decided blackening of the

FIG. 2.—Different varieties of bean seed affected by anthracnose.

affected parts (Fig. 2). This is not always true, however, for the spot may be brown or tan-colored, and even when the black is present the brownish tinge usually borders the area. The spot is more prominent on the white bean seed than on those which naturally are colored. The discoloration may include only a small area of the bean or cover half the surface, and may extend inward only through the seed-coat or may penetrate into the embryo. The badly affected beans

Fig. 3.—Anthracnose killing a bean seedling. Infection came from an infected seed.

often are shriveled, and sometimes may be recognized by their lightness in weight.

When planted the diseased bean may not germinate at all or bear a stem which drops the cotyledons. When the cotyledons do persist they are often marked with blackened cankers, the centers of which under moist conditions are lined with a pink gelatinous mass. The young stem, also, is affected either along its entire length or in one or more spots (Fig. 3). The surface tissue first becomes brown, then sinks and turns black

FIG. 4.—Anthracnose killing the veins on the lower surface of a bean leaf.

until cankers are formed which in appearance resemble these on the cotyledons. Occasionally the cankers become so deep or so numerous that the stem is no longer able to support its own weight. The part below ground is rarely affected.

A very young leaf may have almost its entire area invaded and appear as if injured by bacterial blight, but the more common types of lesions are the small angular dead areas on the upper surface and the killing and blackening of the veins of the lower side of the leaf (Fig. 4). The killed tissue may crack open giving the foliage a ragged appearance, or the petioles may be so weakened with cankers that the leaf droops and dies.

The pod-spot (Fig. 5) is the most noticeable symptom of the anthracnose. Starting as a small brown area, the lesion enlarges, becomes black in the center, and sunken as the host tissue collapses. In severe cases the canker passes entirely through the wall of the pod and affects the bean seed below. The spot may vary in size from that of a mere speck to one the width of the pod. A number of adjoining spots may coalesce, forming a continuous lesion the whole length of the pod. Aside from the reddish or brown margin and the charcoal-black center, the spot attracts further attention when, in the presence of moisture, its center becomes flesh-colored due to the formation of spores in a gelatinous matrix. In dry weather the brightly colored mass dries down into brown or black pimples.

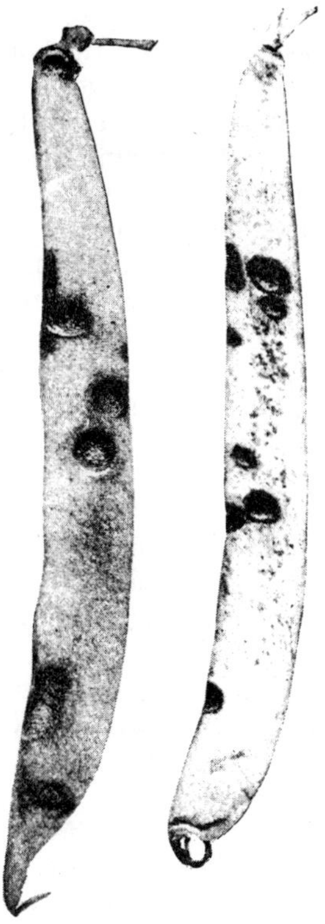

FIG. 5.—Anthracnose spots on bean pods.

Cause of bean anthracnose.

The parasite, *Colletotrichum lindemuthianum*, sometimes known as *Glœosporium lindemuthianum*, is one of a large group of fungi characterized by its shallow open fruit-body known as an acervulus. There may be from one to many acervuli in each lesion or canker. The acervulus is made up of a thick mass of short conidiophores, among which arise tall bristle-like stalks, the setæ. At the tip of each conidiophore a constriction takes place and a small unicellular spore is cut off. This process is repeated many times, so that it has been estimated that on one badly diseased pod a half million spores are produced.

When a diseased bean is planted the cotyledons, pushed above the ground, have cankers in which the spores are borne in large pink masses. They are embedded in a gelatinous substance, therefore are dependent on rain or dew for their dissemination. The water washes away the mucilaginous covering and the spores drop from the cotyledons to the ground or are washed down on the stem. Later they are washed in the same way from the leaves and stems to the pods.

After germination, the spore produces on the end of a germ-tube a secondary spore or appressorium. This in turn sends out a slender thread that penetrates directly through the cuticle of the bean. The actual time of incubation is about thirty-six hours, although the period between inoculation and the appearance of typical lesions is from four to five days. The mycelium passes directly through the walls from cell to cell, and grows so luxuriantly that often cells are filled with the hyphæ. The host tissue is killed, and the fungus soon begins to form acervuli that burst through the epidermis. Dissemination in the field is obtained by mechanical means. When wet beans are cultivated, the spores cling to the man or animal or to the tools and are carried to every plant which is touched. Insects also may play a part in the inoculation.

The fungus lives over winter in the bean seed, in diseased material left in the field, or in bean straw. The fungus does not live long in the soil when separated from the host. A sexual spore stage was found in some test-tube cultures and named *Glomerella lindemuthianum* Shear.

The organism is very susceptible to changes in temperature and humidity. The disease is most abundant in a cold wet soil, and disappears during the hot summer months of the southern states or in the western states where dry-farming is practiced. The optimum temperature for the fungus lies between 63° and 75° F. while the minimum and maximum are approximately 57° and 80° F. respectively. Since the spores are set free only in the presence of moisture, a rain or dew is essential. When diseased seeds are planted and receive no immediate rains, they may produce a healthy crop for the infected cotyledons drop off before the spores have a chance to disseminate. If, on the other hand, the temperature is low and the soil wet, the seedlings may become infected before they appear above the ground, thus insuring plenty of inoculum for the remainder of the season.

Control of bean anthracnose.

Certain general practices such as crop rotation, good soil drainage, and destruction of weeds are as applicable to bean anthracnose as to nearly every vegetable disease. An additional point is that of cultivating only when the plants are dry to prevent the carrying of the spores from one vine to another.

Seed treatment and spraying with bordeaux mixture have been recommended, but neither has proved wholly satisfactory. When the seed is treated in a solution, the skins are likely to loosen and slip off. Furthermore, no fungicide has been found that will penetrate the seed far enough to kill the parasite and will not at the same time injure the viability of the bean. Spraying has been only partially successful. Applications must be made every week from the time the plants show above the ground until picking time. Inasmuch as the disease does not appear every year and as the trouble is not controlled completely in an epidemic year even when bordeaux mixture is applied, the cost of spraying is not justified except under unusual circumstances.

It has been suggested that hand-picking the seeds and discarding all the ill-favored ones will eliminate infection in the field, but actual experience has discredited this method of eradicating the Colletotrichum. Even white beans that are infected will often pass as healthy, while colored beans are still more difficult to detect. As the disease is not systemic, perfectly healthy pods will contain seeds, all of which are free from the parasite. Consequently it is possible to select clean stock by hand-picking enough unblemished pods to furnish the desired amount of seed. Clean seed, however, is not enough to insure a healthy crop; the planting must be far enough from an infested field so that the spores will not be carried from it to the healthy plants.

FIG. 6.—Two plants inoculated with Colletotrichum spores. One was susceptible and the other resistant.

An attempt was made in the eastern states to obtain healthy bean seed by having native stock grown in the dry-farming districts of the West. The plan worked very well as far as the anthracnose was concerned, but the western beans were severely infected with bacterial blight, which is proving a more destructive disease than anthracnose.

The most desirable method of controlling any plant disease is that of procuring strains resistant to the pathogene (Fig. 6). For several years different workers have tried to discover or breed good commercial types of beans that are sufficiently resistant to anthracnose to be profitable when grown. There were two or three very grave obstacles to be overcome in attempting to obtain such strains. First, there are so many races of the fungus that a race isolated from beans collected from one locality may appear in culture or on the host exactly like that found in another section; but when both are inoculated in different varieties of beans, they may react wholly unlike. One may infect one strain of bean and prove harmless to another, while the second type may do just the reverse. There are now at least three known forms of *Colletotrichum lindemuthianum*, alpha, beta, and gamma. It has even been asserted that there are as many as eight such biologic forms.

Another difficulty encountered in breeding resistant beans is that the consumer is accustomed to a certain shape, size, and color of the bean, and does not readily accept alterations in these conventional types. Fortunately some varieties of beans have races which are as distinct as are the forms of Colletotrichum. Notable examples among the dry beans are the Wells Red Kidney, White Imperial, Perry Marrow obtained by breeding, and Nova Scotia Marrow, which are resistant to the alpha and beta strain but are susceptible to the more rare gamma strain; and the Michigan Robust Pea, susceptible to the alpha and resistant to the beta and gamma strains. The Yellow Eye and Nova Scotia Marrow are susceptible in their seedling stage, but when more nearly mature are quite resistant to the three strains of the fungus. With this array of desirable varieties, the problem of controlling anthracnose on dry beans has been fairly well solved. There is no doubt that similar strides will be made in improving snap beans. There are already a number of the wax and green-pod bush and pole beans resistant to one form of the Colletotrichum. Extensive breeding work is under way. The only difficulty to be overcome is that of keeping the strains unmixed, and of retaining a reliable source, to which the grower can turn for strains certain to be resistant to the disease.

References

Barrus, M. F. Variation of varieties of beans in their susceptibility to anthracnose. Phytopath. **1**: 190-195. 1911.

Barrus, M. F. An anthracnose-resistant red kidney bean. Phytopath. 5: 303-311. 1915.

Barrus, M. F. Varietal susceptibility of beans to strains of *Colletotrichum lindemuthianum* (Sacc. and Magn.) B. and C. Phytopath. 8: 589-614. 1918.

Barrus, M. F. Bean anthracnose. N. Y. (Cornell) Agr. Exp. Sta. Memoir 42: 97-215. 1921.

Burkholder, W. H. The production of an anthracnose-resistant white marrow bean. Phytopath. 8: 353-359. 1918.

Burkholder, W. H. The gamma strain of *Colletotrichum lindemuthianum* (Sacc. and Magn.) B. and C. Phytopath. 13: 316-323. 1923.

Edgerton, C. W. The bean anthracnose. La. Agr. Exp. Sta. Bull. 119: 1-55. 1910.

Lauritzen, J. I. The relation of temperature and humidity to infection by certain fungi. Phytopath. 9: 7-35. 1919.

Leach, J. G. The parasitism of *Colletotrichum lindemuthianum*. Minn. Agr. Exp. Sta. Tech. Bull. 14: 1-41. 1923.

McRostie, G. P. Inheritance of anthracnose resistance as indicated by a cross between a resistant and a susceptible bean. Phytopath. 9: 141-148. 1919.

Muncie, J. H. Experiments on the control of bean anthracnose, and bean blight. Mich. Agr. Exp. Sta. Tech. Bull. 38: 1-50. 1917.

Whetzel, H. H. Bean Anthracnose. N. Y. (Cornell) Agr. Exp. Sta. Bull. 255: 431-447. 1908.

Bacterial Blight of Beans

Caused by *Bacterium phaseoli* EFS.

The bacterial blight was first observed in 1886 in New Jersey, but was not destructive enough to receive serious attention until it was found in New York six years later. It is now widely distributed, having been reported from nearly every state and from France, England, Russia, South America, South Africa, Bermuda, Japan, and the Philippines. It evidently causes the greatest losses in the United States, where it is becoming the most troublesome disease of beans. The causal organism attacks only this host, but there are bacteria on closely related genera that are so nearly identical with the bean organism that they are classed as physiological races of the parasite.

Symptoms.

The symptoms of the bacterial blight on the bean foliage are much more conspicuous than are those of the anthracnose.

FIG. 7.—Bacterial blight on bean leaf.

The lesion begins as a small water-soaked or wilted area, which quickly enlarges, and as it does so the affected tissue dies and turns brown (Fig. 7). The dead tissue often becomes thin and parchment-like. Bordering each dead area is a zone of lighter green or yellow, the halo that characterizes so many bacterial lesions. The spots ordinarily are large in size, irregular in shape, and noticeable on both sides of the leaf. Part of the spot may have on its surface a thin white or yellow incrustation composed of the bacteria that ooze from the invaded cells and dry when they reach the surface. A few spots on one leaflet are sufficient to cause its death, or the whipping about by the wind and rain will break apart the affected portions and give the plant a ragged appearance. When infection is general in a field, the dead and fallen leaves and the wilted vines are strong proof that the name blight was rightly applied.

When the spots on the pods are small and numerous, and even when they grow larger they may be mistaken for sunscald (Fig. 8). These bacterial blotches are similar to those on the leaf, frequently starting as water-soaked lesions or as brownish centers, with red or yellow margins, with irregular shape, large size, and sometimes with bacterial incrustations. The tissue does not dry nor shrivel as it does in the leaf, and is never charcoal-black as is true in the case of anthracnose. The disease may penetrate the pod only slightly, or pass through the wall and into the seed below.

The affected bean seed may be much shriveled and discolored or show only the slightest tinge of yellow at the hilum with no apparent injury. The surface of the injured spot is yellowish, and often shiningly smooth as if polished. It is difficult to recognize the disease on colored beans unless the injury is severe enough to cause reduction in size or wrinkling of the surface.

One of the serious forms of the disease is the stem-girdle. At the node where the cotyledons were attached, or at the joint where the lowest branches unite, a red-bordered cleft partly or wholly encircles the stem. This girdle is very narrow, but deep enough to weaken the plant, so that in a heavy wind it falls over and wilts. Plants that show no injury

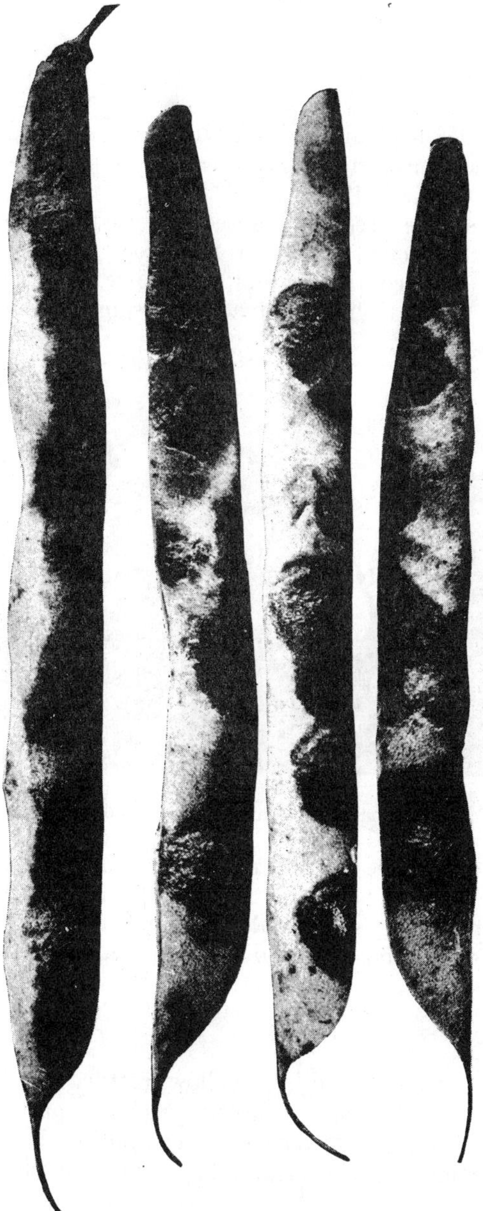

FIG. 8.—Bacterial blight on bean pods.

at the nodes may also exhibit wilted foliage, and both types may have tip-burn of the leaves before the wilting is manifest (Figs. 9, 10).

When affected seed is planted the seedling may not push through the ground at all, or if it does, it is deformed and dwarfed. The cotyledons and stems of the young plant may show the typical blight blotches. In some cases the growing

FIG. 9.—Bacterial blight causing defoliation of a bean plant.

tip is injured, resulting in what the growers designate as "snake-head." There are a number of other causes which may bring about the same malformation, so that "snake-head" within itself is not a sure diagnostic character of the disease.

Cause.

The blight is caused by a yellow, motile, rod-shaped organism, named *Bacterium phaseoli.* Another parasite on beans and described as *Bacterium flaccumfaciens* is either a very closely related species or probably a virulent race of the same species.

FIG. 10.—Bacterial stem-canker of bean.

The bacterial blight is of interest particularly because at times the organism is systemic in the plant, passing from one part to the other through the xylem vessels. In anthracnose the spores are scattered to different parts of the plant by outside agencies; but *Bacterium phaseoli* can pass from the inner part of the cotyledon through the water-ducts directly into the stem and from there to all parts of the host. Since even the seed in the pod is not exempt from this manner of infection, it is not possible to obtain healthy stock by selecting unblemished pods. The whole plant must be healthy or there can be no guarantee that the seeds are disease-free, no matter how fine they may appear.

The bacterium lives in diseased bean refuse and straw during the winter and is able to survive a dormant period of three years or more while within the living seed. When such seed is planted, the parasite enters the seedlings, causes lesions, and then oozes to the surface where it is spattered by the rain, or carried by insects, tools, men or horses to plants in other parts of the field. The bacteria enter the stomata when drops of dew or rain are present, and may cause a visible lesion within a few days or remain in the host for a number of days before any signs of invasion are evident. The nodal infection mentioned above may come directly from the cotyledons, or the angle where the stem and branch meet may catch the bacteria as they are washed down from the leaves and pods.

The disease is not severe in the field until the high temperature of midsummer is present. The bacterium is favored by the heat and is retarded by cool weather. It thrives best when there is plenty of dew or rain, but may sweep across a field in epidemic form when there is a relatively small rainfall as is evinced in the more arid regions of some of the western states. This is further brought out by work reported from South Dakota where seed beans were dipped in a liquid with a suspension of legume organisms. The wet seeds that were planted gave a badly diseased crop, while the progeny planted dry did not show nearly so much infection. When the same experiment was tried in Virginia where there was plenty of air- and soil-moisture, the dipping of the seed had no effect on the relative amount of disease in the crop.

Control of bacterial blight.

The part of the discussion on the control of bean anthracnose dealing with crop rotation, cultivation when plants are dry, eradication of weeds, seed treatment, and spraying with bordeaux mixture apply also to bacterial blight. Unfortunately no variety is nearly enough immune to the blight to serve as a basis for successful breeding work. Nevertheless, experiments are under way and it is probable that seed of resistant varieties will be on the market within a few years. In the meantime the grower must rely on less commendable control measures.

The first step in growing a healthy crop is the procuring of clean seed; and the only assurance of healthiness in any stock is seeing it grow in the field. There is no infallible method of distinguishing slight infection in the threshed crop. Therefore, it would be well if an inspection service could be established like that in operation for the certification of potatoes. An unbiased trained observer, with definite standards for his guidance, could readily determine whether the product of a grower's field were fit for planting. If the field met the requirements, a bill of certification could be issued, and which would serve as a guide for the grower who found it necessary to buy his bean seed.

The farmer, who expects to grow beans for certification, will find that many precautions are necessary to keep his stock healthy from year to year. An isolated bean plot planted with the best seed and cared for in the most approved manner is indispensable. The plot should be watched carefully for the slightest trace of disease and every affected plant removed from the field at once. The seed from this plot can be used the following year for planting in the regular field. It should be planted as late as the season will permit.

References

Burkholder, W. H. The bacterial blight of the bean: a systemic disease. Phytopath. 11: 61-69. 1921.

Burkholder, W. H. Varietal susceptibility among beans to the bacterial blight. Phytopath. 14: 1-7. 1924.

Edgerton, C. W. and C. C. Moreland. The bean blight and preservation and treatment of bean seed. La. Agr. Exp. Sta. Bull. 139: 1-43. 1913.

Hedges, Florence. A bacterial wilt of the bean caused by *Bacterium flaccumfaciens* nov. sp. Science 55: 433-434. 1922.

Hedges, Florence. Bacterial pustule of soy bean. Science 56: 111-112. 1922.

Leonard, L. T. An influence of moisture on bean wilt. Jour. Agr. Research 24: 749-752. 1923.

Muncie, J. H. Experiments on the control of bean anthracnose and bean blight. · Mich. Agr. Exp. Sta. Tech. Bull. 38: 1-50. 1917.

Rapp, C. W. Bacterial blight of beans. Okla. Agr. Exp. Sta. Bull. 131: 1-39. 1920.

Smith, E. F. The cultural characters of *Pseudomonas hyacinthi, P. campestris, P. phaseoli,* and *P. stewartii,* four one-flagellate, yellow bacteria parasitic on plants. U. S. Dept. Agr. Div. Veg. Phys. and Path. Bull. 28: 1-153. 1901.

Bean Mosaic

Cause undetermined

As soon as mosaic was recognized as a disease of vegetables, it was noticed on the bean plant. It is not known whether it is a trouble of recent origin or whether it has been present for many years and the affected plants considered merely as weak or off-type. Suffice to know that now it is distributed in all parts of the United States, in England, and probably in other countries where the host is grown. Certain varieties of beans are so seriously affected that some growers have temporarily been forced out of the business until satisfactory resistant strains can be obtained. This is true particularly in the pea bean section.

Symptoms.

The bean leaves affected with mosaic have irregular light yellow areas merging with dark green patches, producing the

Fig. 11.—Bean leaf showing mottling and puckering caused by mosaic.

characteristic mottling or mosaic effect (Fig. 11). The darker areas develop faster than do those having the yellow tinge with the consequence that the leaf is much puckered, especially along the midrib, and the edges are cupped downward. The affected plant may have a sickly yellow color; nevertheless it remains alive until the end of the season. If infection takes place late in the summer, the development of the pods is not checked; early infected plants, however, rarely bear any seed even though they may continue to blossom until autumn.

Cause.

So far as is known, the only place where the mosaic virus lives during the winter is in the bean seed. When such seed is planted, the progeny will also be diseased. The virus is carried from one plant to another in any manner in which plant sap may be conveyed. When a leaf on a healthy bean plant is rubbed with a leaf affected with mosaic until the tissue is broken, the healthy plant will become diseased within sixteen to eighteen days. The virus may also be carried by various insects. When the inoculum once enters a plant, it apparently spreads to all parts of the vine, including the pods, since a diseased individual usually bears infected seed.

Infection is favored by high temperature and humidity at the time of inoculation.

Control.

Attempts have been made to discover some treatment of the seed, and which would kill the stored virus; but the relation between the latter and the seed is evidently so intimate that one cannot be killed without destroying the other.

Where the very susceptible pea bean has been grown, the disease may be avoided by substituting the Robust pea bean, which is almost perfectly immune. Other varieties of beans resistant in varying degrees to mosaic are Red Marrow, White Marrow, Long Pod, Keeney Rustless, German Black Wax, Michigan White Wax, Refugee Wax, Prolific Black Wax, Scarlet Flagolet, Red Valentine, and probably many more. Breeding work will finally add still others to the list.

Until resistant strains are more easily obtainable, it is necessary to procure healthy seed either by growing disease-free stock in an isolated seed-plot or by buying from a known source. In connection with the latter alternative, an inspection and certification service as outlined for the control of bacterial blight of beans can be adapted for the elimination of mosaic.

References

Barss, H. P. Bean blight and bean mosaic. Crop Pest and Hort. Rept. Ore. **3**: 192-194. 1921.

McRostie, G. P. Inheritance of disease resistance in the common bean. Phytopath. **10**: 319. 1920.

McClintock, J. A. Lima bean mosaic. Phytopath. **7**: 60, 61. 1917.

Nelson, Ray. Transference of the bean mosaic virus by *Macrosiphum solanifolii*. Science **56**: 342-344. 1922.

Reddick, Donald. A hybrid bean resistant to anthracnose and to mosaic. Phytopath. **12**: 47. 1922.

Reddick, Donald and V. B. Stewart. Varieties of beans susceptible to mosaic. Phytopath. **8**: 530-534. 1918.

Reddick, Donald and V. B. Stewart. Additional varieties of beans susceptible to mosaic. Phytopath. **9**: 149-152. 1919.

Dry Root-Rot of the Bean

Caused by *Fusarium martii phaseoli* Burk.

Root-rots of beans have often been mentioned in literature, and reported from many different bean-growing areas. It seems, however, that no one has ever made an effort to prove the identity or diversity of the causes in each case. It, of course, is known that Thielavia produces a root-rot, yet this is so different from the ordinary root trouble that the two can never be confused. The dry root-rot has been observed in New York, Vermont, Michigan, and probably the same malady has caused loss in nearly every other state and in Canada. No data are at hand regarding the disease on other continents. Where it does occur, the losses are very heavy. In the dry shell bean section of New York many farms are no longer fit for growing this crop, and cereals or potatoes take its place.

The organism is able to infect every variety of the common bean, the tepary bean, the scarlet runner, the moth bean, the lima-bean, the adzuki bean and the black-eye cowpea. It does not affect the common pea, although there is another variety of the same pathogene that does injure the latter host.

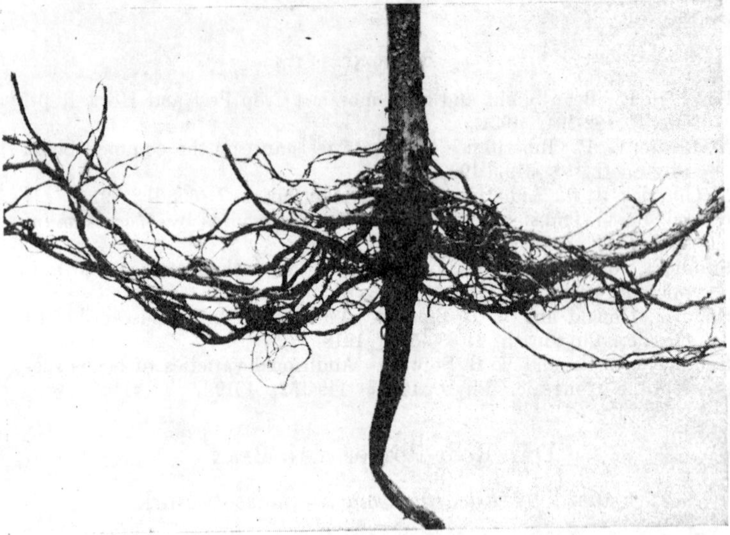

FIG. 12.—Dry root-rot of bean showing development of new roots above the canker.

Symptoms.

The invaded part of the plant may include any or all of the roots and sometimes the base of the stem. Infection may show one or two weeks after the seedling is out of the ground, or not until the plant has attained considerable size. The tap-root takes on a bright red color over its entire surface, or only in streaks, while its tissues remain firm and turgid. As the disease progresses the lateral roots as well as the end of the tap-root shrivel (Fig. 12). They may then be attacked by other fungi, bacteria, or by insects, and decay. If, during such an attack of the disease the plant is not killed outright, which it rarely is, new lateral roots often grow from the stem above the canker, thereby prolonging the life of the host.

The top of the plant is not invaded, yet the effects of the disease on it are very noticeable. In severe cases many plants are killed, leaving only a series of missing hills in the row. When the advance of the disease is slower, the vines exhibit the symptoms of slow growth, dwarfing, unhealthy color, few pods, and often with yellowing and dropping of the foliage. When the tap-root is killed, the lower part of the stem may become pithy or hollow. There is seldom any pronounced wilting of the plant.

Cause.

The root-rot fungus is morphologically identical with *Fusarium martii* but as this particular fungus does not infect beans, the organism on the bean root has been named *Fusarium martii phaseoli.* On culture media the fungus has tints of green or blue, while the spores in mass are frequently yellowish. The spores are of the usual fusarial type, and mostly three- or four-septate. They are seldom found in connection with the diseased plant in the field, evidently being produced after the host has been killed and left in the soil. The germ-tube of the spore and probably also mycelium growing in contaminated soil enter the host directly through the epidermis. Visible symptoms are present in a week after infection takes place. The mycelial threads instead of growing singly are often found in strands of a dozen or more both between the cells and over the outside of the root. It is the general practice in the dry bean section, after pulling and threshing the beans, to feed the roots and stems as bean-straw to the sheep and cattle. It is interesting to note in this connection that the most severely infested area in New York is co-extensive with the more intensive sheep industry. Evidently the chief method which the fungus has of living over the winter is in the bean-straw and in manure, and through these agents is distributed to previously healthy fields. In the snap bean section where the bean straw is not fed, the dissemination of the fungus proceeds much more slowly, and in no place has it become of very great economic importance. The fungus when it once gains admission into a field is able to live indefinitely in the soil even though it has no susceptible hosts on which to prey.

The parasite is not influenced much by moisture as it can live wherever the bean will survive. It, however, is affected by the temperature. It has been found that infection will occur at any point between 59° and 93° F., although the best growth of the fungus is at about 86° F. Thus in the northern tier of states it probably will cause most damage during an unusually warm summer.

Control of root-rot.

This is another of the soil organisms which is so difficult to combat. Such sanitary measures as long rotations and destruction of diseased straw undoubtedly aid in reducing the amount of loss, but will not meet all the requirements. If a crop has become diseased, a fair yield may sometimes be procured by not cultivating after the symptoms appear. The adventitious roots that are formed are so near the surface of the soil that they are destroyed by the cultivator teeth; the plants then have nothing to support them. If these roots are not injured they frequently are able to nourish the vine and produce an average set of pods. The only recommendation that can safely be made at present is that the grower plant beans until the "sick" soil no longer produces profitable yields, then change to other crops for five or more years. This will not starve out the parasite, but ordinarily will reduce it enough so that a fair yield can be obtained again, provided that diseased straw has not been used as a mulch in the meantime.

As an encouragement, it may be said that breeding for

resistant varieties is under way and may finally produce suitable types. Among the white marrow beans are occasional vines which mature late and have seeds that incline toward flatness, consequently the strain is designated as the flat marrow. It is of little value within itself as it is too long in maturing for northern states, and its pods do not ripen uniformly; nevertheless, it can be used as one parent in breeding resistant varieties. The work has already been undertaken, but because there are so many factors involved in immunity to root-rot, seed of resistant stock has not yet been put on the market. Such stock no doubt will be available later.

References

Burkholder, W. H. The dry root-rot of the bean. N. Y. (Cornell) Agr. Exp. Sta. Memoir **26**: 999-1033. 1919.

Burkholder, W. H. The effect of two soil temperatures on the yield and water relations of healthy and diseased bean plants. Ecology **1**: 113-123. 1920.

Burkholder, W. H. The effect of varying soil moistures on healthy bean plants and on those infected by a root parasite. Ecology **5**: 179-187. 1924.

McRostie, G. P. Inheritance of disease resistance in the common bean. Jour. Amer. Soc. Agron. **13**: 15-32. 1921.

Reddick, Donald. Effect of soil temperature on the growth of bean plants and on their susceptibility to a root parasite. Amer. Jour. Bot. **4**: 513-519. 1917.

Black Root-Rot of Bean

Caused by *Thielavia basicola* (B. and Br.) Zopf.

Occasionally in bean fields a second root-rot is found which in its effect on the vine resembles the dry root-rot. There is one outstanding difference which makes it easily recognizable from all the other troubles. The affected tap-root is roughened and charcoal-black in color. The darkening may extend even slightly above the surface of the soil. In a wet period a frosty appearance frequently covers the blackened area. The destruction of lateral roots and the dwarfing and yellowing of the vines are in all respects similar to those described for dry root-rot. It is seldom of economic importance on this host, although it is destructive to several other cultivated crops, as tobacco and violets, and occurs on almost a limitless number of plants. Among the vegetables are horse-radish, pea, and watermelon.

The fungus has long been known and has been described under various names. Since the ascerigerous stage was found, it is recognized as *Thielavia basicola.* It reproduces by means of three types of spores; hyaline spores borne within a narrow hyphal thread, thick-walled reddish-brown chalamydospores formed in chains of five to seven, and ascospores produced in globose completely closed fruit-bodies. The organism lives over in the soil and on different types of hosts. It usually is considered as a weak parasite, and is easily injured by high temperatures. Only when the soil has a temperature less than 75° F. does the trouble become at all serious. Since the soil becomes warmer than this in July and August even in the northern tier of states, the disease on beans is not serious enough to demand control measures.

References

Cooke, M. C. The root-rot fungus. Gard. Chron. ser. 3. **41**: 361. 1907.

Johnson, James. Host plants of *Thielavia basicola.* Jour. Agr. Research **7**: 289-300. 1916.

Johnson, James and R. E. Hartman. Influence of soil temperature on Thielavia root-rot. Phytopath. **8**: 77. 1918.

Zopf, W. Über die Wurzelbräune der Lupinen, eine neue Pilzkrankheit. Ztsch. f. Pflanzenkr. **1**: 72-76. 1891.

Rhizoctonia Stem- and Pod-Rot of Bean

Caused by *Rhizoctonia solani* Kühn.

Nearly all vegetables are affected by Rhizoctonia, the bean being no exception. On this host three types of injury are produced: damping-off of seedlings, stem- and root-rot, and pod-rot. The damping-off takes place only when the plants

FIG. 13.—Rhizoctonia stem-rot or wire-stem of beans.

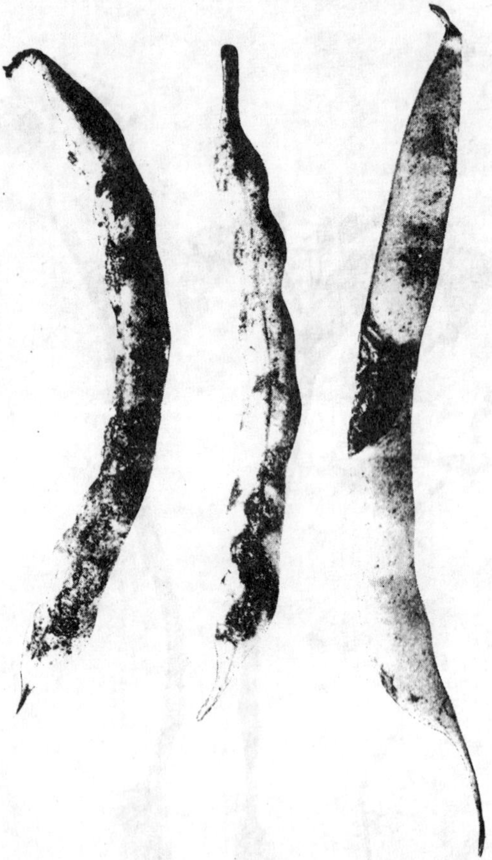

FIG. 14.—Rhizoctonia canker on bean pods.

are very small, and then only rarely, for in most cases the fungus forms a canker on the side of the stem, which has no resemblance to the real damping-off .

The most important injury occurs on the roots or stems (Fig. 13). The lesions are brick-red, irregular in shape, and sunken, somewhat like the Rhizoctonia cankers on potato stems. Similar spots occur on pods (Fig. 14), particularly those which touch the ground. The lesions are large irregular blotches that may cover the greater part of the pod epidermis. The discoloration may be only superficial or extend inward to the seed, causing it to rot if young, or merely discoloring it if more nearly mature.

The fungus is probably the same as the one on potato and whose perfect stage is *Corticium vagum* B. and C., although it may be possible that it is a special race. It lives over winter in the soil, in plant refuse, and in the living bean seed. The mycelium, when well developed in the host, forms minute sclerotia which aid the parasite in surviving periods of inactivity. This is so noticeable that a fungus on beans has been described as *Rhizoctonia microsclerotia* Matz.

The pathogene, evidently, is harmful only in wet soils, or in a rainy period when the temperature is above 54° or below 79° F.

Little can be done in controlling the disease. Long rotations with cereals and grasses, obtaining healthy seed, and the use of well-drained soils will no doubt restrain the parasite from doing any great damage in a given field. The trouble is usually not serious enough to demand particular attention.

References

Barrus, M. F. Rhizoctonia stem-rot of beans. Science, N. S. **31**: 796-797. 1910.

Fulton, H. R. Diseases of peppers and beans. La. Agr. Exp. Sta. Bull. 101: 1-21. 1908.

Hedgcok, G. G. A note on Rhizoctonia. Science, N. S. **19**: 268. 1904.

Matz, J. Una enfermedad dañina de la habichuela. Porto Rico Dept. Agr. and Labor Sta. Circ. **57**: 3-8. 1921.

McCready, S. B. Bean diseases. Ont. Agr. Coll. and Exp. Farm Ann. Rept. **36**: 44-47. 1910.

Peltier, G. R. Parasitic rhizoctonias in America. Ill. Agr. Exp. Sta. Bull. 189: 283-390. 1916.

Richards, B. L. Soil temperature as a factor affecting the pathogenicity of *Corticium vagum* on the pea and the bean. Jour. Agr. Research **25**: 431-450. 1923.

Bean Rust

Caused by *Uromyces appendiculatus* (Pers.) Lév.

The rusts and smuts of plants are so conspicuous that they were observed long before there was a science which dealt with plant diseases. Included among these long known troubles is the bean rust, the fungus of which was described in 1797. This rust is common in all the eastern and central parts of the United States and in the West has been reported from Colorado, New Mexico, California, and Oregon. Its distribution in foreign countries has not been recorded in sufficient detail to be sure that it is of economic importance except in the United States. In some of the southeastern states and in California it causes considerable loss, but in the northern states, especially in the dry bean sections, the rust comes too late in the season to be destructive.

The fungus is able to infect nearly all varieties of the common bean, and has been reported on the lima-bean, tepary bean, scarlet runner, cowpea, and a number of other closely related hosts. It is possible that some of these hosts are attacked by closely related rusts which have been mistaken for the one on beans. This is true particularly of the cowpea.

Symptoms.

The rust pustules may appear on any part of the plant above ground, being most numerous on the under sides of the leaves (Fig. 15), less abundant on the pods, and occurring sparingly on the stems. The pustules are small, distinct, circular, slightly raised sori that are typical of any true rust. In the beginning they are reddish-brown in color, later in the season turning black. As many as two thousand pustules may be found on one leaf, and as each is made up of a powdery mass of spores, they give a rusty color to anything which touches them, even covering the ground with a faintly brownish dust. The pustules are raised from the surface of the host more than are those of the cereal rust, and particularly is this true on the pods where the elevations may sometimes be almost an eighth of an inch. On some varieties of beans the sori are surrounded by a yellow halo. When the leaf has become thoroughly infected, it shrivels and falls from the plant. In the case of a rust epidemic, most of the vines in a field may be defoliated.

FIG. 15.—Rust on bean leaves.

Cause.

The organism, *Uromyces appendiculatus*, unlike many other rust fungi, develops entirely on the one host, it being autoecious and having all the spore stages of a rust. The æcial or cluster-cup stage is rare, while the uredinial and telial stages comprise the brown and black pustules, respectively, that cause the injury on the host. The large, globose, slightly roughened urediniospores are borne in the open sori or short stalks. Later in the season the one-celled, elliptical, thick-walled teliospores are borne in the same sori or in additional ones that may develop. Both types of spores are blown by the wind, or carried on clothing and tools to healthy plants. The urediniospores may germinate at once and cause infection, or may be able, at least in the southern states, to survive the winter and serve as inoculum in the spring. Teliospores do not germinate until after a rest period. They give rise to basidia or promycelia on each of which are borne four basidiospores. These, as well as the surviving urediniospores, are present in the spring to cause infection on the new crop.

There is some indication that there are at least two biologic forms of *Uromyces appendiculatus.*

Control.

Spraying or dusting for the control of bean rust has not proved satisfactory. The other alternative is that of planting resistant stock. There are many varieties listed in seed catalogues as rust-proof, but unfortunately this refers only to anthracnose or blight and is not related to the true rust. Many of the well known varieties, however, are resistant, as was shown by work done in Virginia. As a rule the pole beans are more susceptible than are the bush varieties, and the green pod strains less resistant than are those with wax pods. For example, the well known Kentucky Wonder pole bean is reported universally as severely affected. Among the pole beans that have proved somewhat resistant are Marblehead, Tennessee Wonder, Brockton, Indian Chief, Everbearing, and Mont d'Or; and among the bush beans are Hodson Green Pod, Refugee, Early Mohawk, May Queen, Improved Goddard, Low's Champion, Early Refugee, Mexican Red, Golden Eye, Detroit, Hodson Wax, New Pearl, Wardwell, Challenge, Crystal White, Flagolet, Webber, California, White Kidney, Well's Red Kidney, Yellow Eye, Dwarf Horticultural, and French's Horticultural. If rust is a factor in bean production, a grower should be able to find among such a large group a resistant variety of bean that will fulfill the demands of his market.

References

Beach, S. A. Some bean diseases. N. Y. (Geneva) Agr. Exp. Sta. Bull. 48: 308-333. 1892.

Fromme, F. D. The rust of cow peas. Phytopath. 14: 67-79. 1924

Fromme, F. D. and S. A. Wingard. Bean Rust. Va. Agr. Exp. Sta. Bull. 220. 1-18. 1918.

Fromme, F. D. and S. A. Wingard. Varietal susceptibility of beans to rust. Jour. Agr. Research **21**: 385-404. 1921.

Halsted, B. D. Bean diseases and their remedies. N. J. Agr. Exp. Sta. Bull. 151: 1-28. 1901.

Lakon, G. Uber die Empfänglichkeit von *Phaseolus vulgaris* L. and *P. multiflorus* Wild. für den bohnenrost und andere krankheiten. Ztschr. f. Pflanzenkr. **26**: 83-97. 1916.

Angular Leaf-Spot of Beans

Caused by *Isariopsis griseola* Sacc.

The angular leaf-spot of beans has such a characteristic appearance on the foliage, and the fungus is so distinctive in its fruiting that the trouble has long been recognized. It is present in many of the far eastern states, in many of the European countries, in Africa, and most severely in Porto Rico. It is seldom of great economic importance, although occasionally it may cause defoliation and killing of the pods. It has also been reported on peas.

Symptoms.

The lesions on the leaf are small angular brown spots, usually with no different colored border, and so numerous that they give a checkerboard appearance to the surface of the foliage. The fungus when fruiting forms a gray moldy covering over the dead area on the under surface of the leaf. Later, when it is dry, the fruiting bodies appear almost like small black stromata, or when only superficially examined might even be confused with immature pycnidia.

The spots on the pods are very conspicuous in being black with red or brown centers. The two zones are sharply divided from each other, as well as from the adjoining healthy tissue. The spots vary greatly in size, some being no more than mere specks while others are almost the whole width of the pod. It is not uncommon for several spots to unite forming a very large lesion, which, however, differs from the smaller dead areas in not having the reddish-brown centers. Small black fungal growths are scattered thickly over the killed tissue of the pod in the manner described for the under surface of the leaf. Lesions are rarely found on the stems.

Cause.

The fungus, *Isariopsis griseola,* is characterized by having its conidiophores united into a coremium, resembling a bound sheaf of wheat. On the tips of the stalks are borne large two- to four-celled spores, which are probably blown about by wind or splashed by rain. Almost nothing is known of the life history of the parasite. It has been suggested that it may live over winter and be disseminated on the seed.

Control.

No careful experiments in controlling the disease have been attempted. Seed treatment as recommended for the control of Diaporthe, and spraying the vines with bordeaux mixture 4-4-50, have been advised.

References

Halsted, B. D. Bean diseases and their remedies. N. J. Agr. Exp. Sta. Bull. 151: 1-28. 1901.

Plant Disease Survey Office. U. S. Dept. Agr. Plant Disease Survey bulletin. Supplement **1**: 77. 1919; **16**: 234. 1921; **22**: 349, 350. 1922; **26**: 96. 1923.

Schoevers, T. A. C. Nieuwe Ziekten, waarop gelet moet worden. Tijdschr. Plantenz. **26**: 208-211. 1920.

Sclerotinia Rot of Beans

Caused by *Sclerotinia libertiana* Fuckel

(See discussion of Lettuce Drop, page 243.)

The stems and pods are severely affected in occasional fields after a continued rainy period (Fig. 16). The tissue is reduced

FIG. 16.—Sclerotinia rot of bean stems and pods.

to a soft decayed mass. The disease is recognized by the mass of white fluffy mycelium over the affected area, and the large black sclerotia embedded in the fungous mat.

Keeping the weeds removed from the field and the soil well drained will aid the plants in drying, and therefore will reduce the amount of injury that might otherwise take place. Deep plowing to bury the sclerotia, rotations with cereals, and burning the old diseased vines will reduce the amount of inoculum for the following year.

References

McClintock, J. A. Sclerotinia blight, a serious disease of snap beans caused by *Sclerotinia libertiana* Fckl. Va. Truck Exp. Sta. Bull 20: 419-428. 1916.

Southern Sclerotium Rot of Beans

Caused by *Sclerotium rolfsii* Sacc.

The sclerotium rot is typically a trouble of plants in the southern states. It was once reported in Illinois, although it is doubtful whether the fungus can ever survive the winter so far north. The disease is extremely destructive wherever it occurs, and on most vegetable crops. It has been reported on beans, cabbage, young corn, eggplant, potato, radish, pepper, rhubarb, watermelon, sweet-potato, and tomatoes, besides other field crops and weeds. It may also cause injury on pumpkins, squash, cabbage, potatoes, and sweet-potatoes in storage.

When the bean seedling is affected, the whole plant may be invaded and soon die. The older vines are attacked near the surface of the soil where deep brown lesions are produced on the stems. The fungus gradually works downward into the roots. The leaves turn yellow, droop and fall, leaving bare stalks with possibly a few pods clinging to them. The fungus appears at the base of the stem as a white mycelial mass in which brown sclerotia are borne. In wet weather a white weft of hyphæ will grow from the stem over the surface of the ground.

The fungus, *Sclerotium rolfsii,* found many years ago in Florida, has procured its genus name from the fact that no spore stage has ever been discovered. It is composed of rather coarse white mycelium having clamp connections at the septa as do the Basidiomycetes. When the mycelium has developed well in or on the host, sclerotia are formed. They are irregularly globose, brown, and resemble in size and appearance a mustard seed. The shape and color of the sclerotia differentiate the organism from Sclerotinia or Botrytis, two common sclerotial forms on vegetables cultivated in the northern states.

The mycelium grows into the host tissue without the presence of wounds, and within a few days causes a visible lesion. The mycelium may persist all the year on living hosts, but when necessary the sclerotia aid in withstanding long rest periods. When opportunity again offers itself, the sclerotia send out hyphæ which grow into the soil and later cause infection.

The fungus is very susceptible to cold, therefore will never menace crops in the North. It reproduces most abundantly in sandy or sandy-loam soil. It requires much water and is destructive in proportion to the amount of moisture present. No definite control measures can be recommended. Burning diseased plant refuse, practicing rotations in which cereals or other crops resisting the disease are alternated with vegetables, and destruction of susceptible weeds, are general precautions that will assist in reducing the loss from sclerotium-rot.

References

Rolfs, P. H. The tomato and some of its diseases. Fla. Agr. Exp. Sta. Bull. 21: 25-36. 1893.

Taubenhaus, J. J. Recent studies on *Sclerotium rolfsii* Sacc. Jour. Agr. Research **18:** 127-138. 1919.

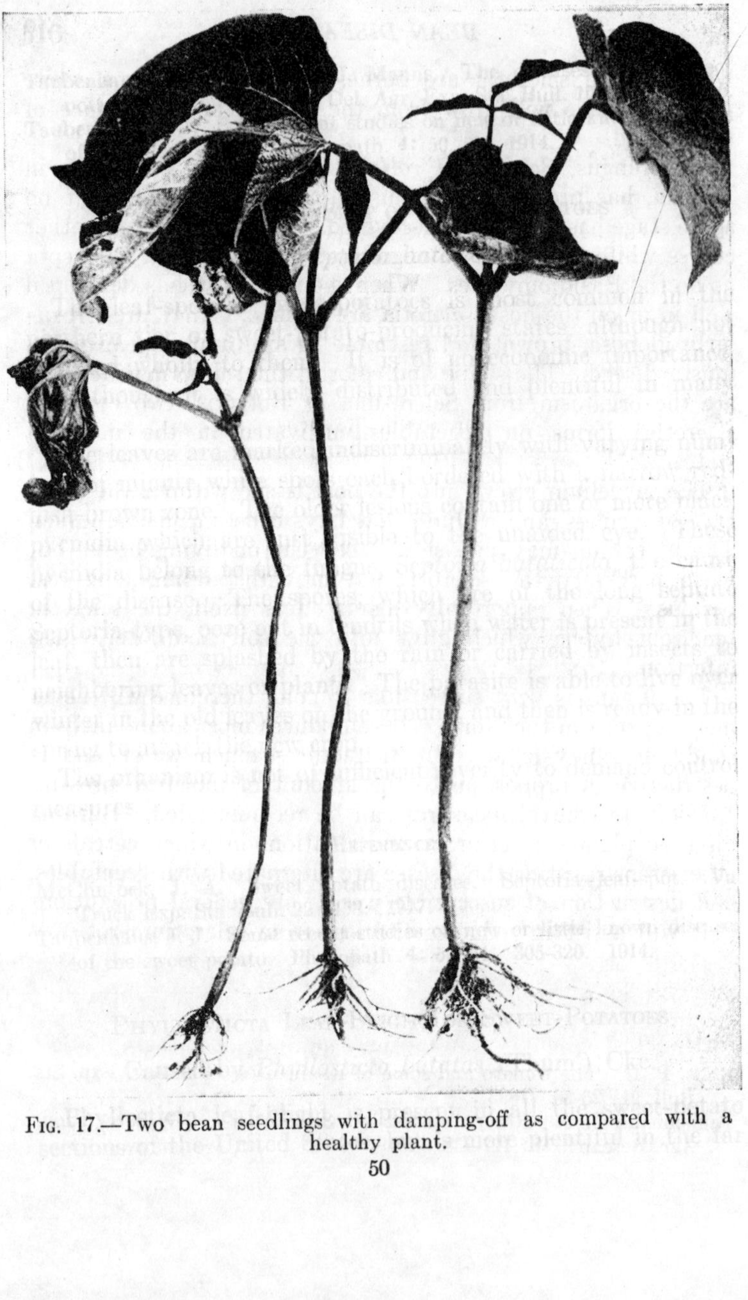

FIG. 17.—Two bean seedlings with damping-off as compared with a healthy plant.

Damping-off of Beans (Fig. 17)

Caused by various fungi

(See Damping-off of Tomatoes, p. 546)

Gray Mold-Rot of Beans (Fig. 18)

Caused by *Botrytis* sp.

(See Gray Mold-Rot of Lettuce, p. 253)

Leaf-Scorch and Sunscald of Beans

Cause physiological

The leaf-scorch and sunscald in many respects resemble the bacterial blight of beans. The leaves may be killed entirely or have only limited irregular dead areas, which are separated from the healthy portions by reddish borders. The leaf-scorch is essentially a tip-burn.

The sunscald appears on the pods and on the sides exposed to the sun. The injury begins as small brown or reddish spots which gradually lengthen into streaks or combine into large irregular brown areas covering almost all, if not all, of one side of the pod. It is rarely found on both sides. At times it is so severe that the discoloration extends through the pod wall to the bean seed itself, which is slightly stained.

The leaf-scorch is caused by drying winds on a warm day when the evaporation of water from the leaf is greater than that obtained from the roots. It is more prevalent during a dry windy period following one in which the vines have made very rapid succulent growth.

The sunscald is not produced by the heat as may be supposed, but by the shorter rays of light. In the high altitudes of the western states the humidity usually is less and the portion of ultra-violet light is much greater than in the lower levels, and it is in these higher altitudes where the scald is

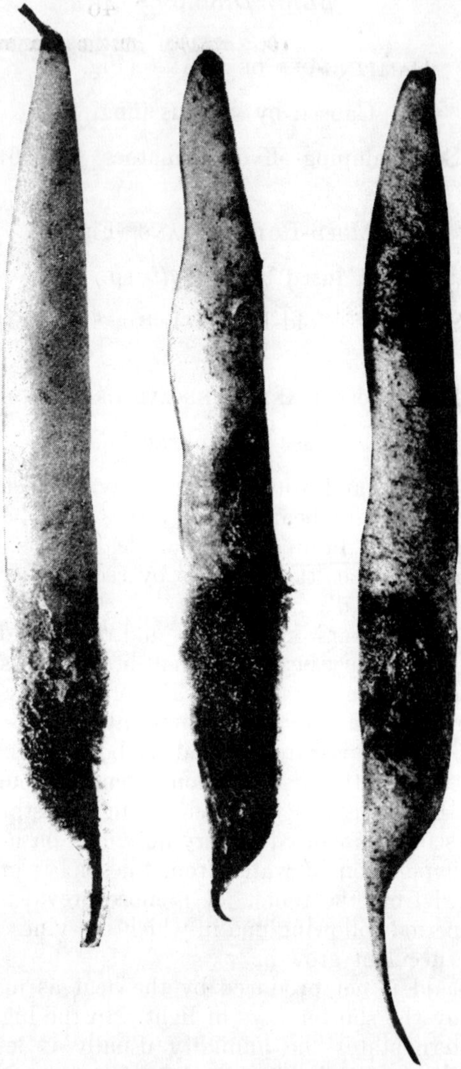

FIG. 18.—Botrytis rot on bean pods.

most severe. Here also, the leaves drop after irrigation ceases, thereby exposing the pods to the detrimental rays of light, and increasing sunscald.

No control measures are available.

References

Clinton, G. P. Report of the Botanist for 1906. Conn. Agr. Exp. Sta. Ann. Rept. 1906: 307-368. 1907.

MacMillan, H. G. Sunscald of beans. Jour. Agr. Research **13**: 647-650. 1918.

MacMillan, H. G. Cause of sunscald of beans. Phytopath. **13**: 376-380. 1923.

Heat Injury of Beans

In Colorado, where the summer heat of the soil and air sometimes becomes excessively high, the base of the bean plant at the surface of the soil has been observed to be shrunken and collapsed in such a manner that the vine falls over, and may occasionally appear wilted. The affected tissue does not change in color, and ordinarily the epidermis remains unbroken. The prostrated plants, however, do not revive. The roots show no injury.

Where the trouble occurred, the soil at a depth of one inch averaged 109.8° F. for ten days, rising as high as 113° F. on some days. The air registered 99.3° F. for the same length of time. It is possible that for a brief period during the hottest part of the day, the vines were exposed to temperatures as high as 130° F.

Reference

MacMillan, H. G. and L. P. Byars. Heat injury to beans in Colorado. Phytopath. 10: 365-367. 1920.

Powdery-Mildew of Beans

Caused by *Erysiphe polygoni* DC.

The powdery-mildew of the bean, no doubt, is present in many countries, but has been reported as destructive only in

Bermuda, where the first specimens were collected in 1889. In 1922, Whetzel found that the disease still caused losses there.

The fungus grows on the surface of all parts of the plant above ground, producing a talcum-like effect on the leaves, pods and stems. The leaves curl, take on an unhealthy color, and die prematurely. The pods are dwarfed and fewer in number.

The fungus is evidently a strain of *Erysiphe polygoni.* Salmon, in his monograph of the family, gives a long list of synonymous names, which have been applied to the parasite.

The talcum-like growth is made up of mycelium and numerous simple conidiophores on which the conidia are borne in chains. Rarely perithecia are formed later. Its detailed life history is not known.

Where the losses from the disease are great enough to warrant the effort, it may be possible to control the fungus by dusting the plants once or twice with finely ground sulfur, beginning as soon as the mildew appears.

References

Galloway, B. T. Powdery mildew of the bean. Jour. Myc. 5: 214. 1889.

Whetzel, H. H. Report of the Pathologist, Bermuda. Bermuda Board and Dept. Agr. Repts. **1921**: 41. 1922.

Downy-Mildew of Limá-Bean

Caused by *Phytophthora phaseoli* Thaxt.

Downy-mildew was observed first in Connecticut in 1889, and was later reported from New Jersey, Delaware, Long Island, Maryland, Pennsylvania, Virginia, West Virginia, Ohio, California, and from Russia. It is most destructive along the Atlantic shore, and gradually decreases in virulence farther inland. It affects both the bush and pole limas, but is not found on any other host. In years when the disease is abundant, from 50 to 90 per cent of the crop is destroyed.

FIG. 19.—Downy-mildew on lima-bean pods.

Such a severe epidemic, however, does not often occur two years in succession, and there may be several seasons when the mildew is almost wholly absent.

Symptoms.

The most conspicuous symptom is the white downy mold in patches or completely covering the pod (Fig. 19). This mold at first hides the affected host part so that the only visible change in the pod tissue may be a border of purple between the healthy and diseased portions. The fungus grows through the pod wall into the bean. The whole structure finally shrivels, dries, and becomes black. Such black dried pods are common in badly infested fields.

The fungus attacks also the young shoots, flowers, and leaves. The tender branches and flower-pedicels are distorted in shape and covered with the mold. The white mycelial weft appears very sparingly on the leaf, but the veins may be purplish in color, twisted or otherwise distorted.

Cause.

The genus, Phytophthora, to which the parasite belongs, is one of the most destructive genera of fungi that attack plants. *Phytophthora phaseoli* is no exception. The time of incubation is so short, the amount of inoculum so abundant, and the means for dissemination so nearly perfect, that within a few days it can sweep over a planting and leave nothing but blasted pods. Much of the early infection comes from the diseased seed. The affected bean does not produce a diseased seedling directly, but when the invaded seed is placed in the moist soil, the fungus threads grow outward and fruit on the surface of the seed. If weather conditions are favorable, the conidia may then be splashed to the plant above ground. The white mycelial growth that forms such a conspicuous diagnostic character on the host is a great mass of branched elongated conidiophores, which by their proliferation bear recurrent crops of lemon-shaped conidia. These may germinate directly by means of a germ-tube, but more often exude their contents in the form of flagellate zoospores. The latter in the presence of moisture germinate at once with a germ-tube, and when on the host are able to penetrate the epidermis on any part of the bean plant. The mycelium grows between the cells, and almost at once kills the cell contents on which it feeds. Within a few days new fruiting-bodies are formed on the surface of the pod, stems, or leaves, and the life history is repeated. The presence of conidiophores on the blossom, although not so plentiful as on the pod, is of great importance, since bees or insects collecting nectar carry the conidia from diseased to healthy blossoms. The wind, too, is an important factor in dissemination, as are flowing water and the splashing of rain.

Both the conidia and zoospores are short-lived and do not aid in keeping the fungus alive during the winter. The mycelium also is incapable of withstanding age unless it is growing within the bean seed, where it is able to survive until spring. The most common form in which the parasite withstands the winter months is as a sexual spore or oospore. This is a thick-walled, large, nearly spherical fruit-body found in the old diseased parts of the plants, particularly in the pods and the cotyledons of the seeds. The oospores are fairly abundant and serve as sources of inoculum for the new crop.

Cool weather and moisture are favorable for the development of the downy-mildew, as is attested by the fact that in a rainy season the fungus grows even after the first fall frosts. It is found most abundantly in the districts where the night temperatures are low and the midday temperatures relatively high.

Control of downy-mildew.

After studying the life history of the parasite, it is evident that long rotations, destruction of the diseased vines in the autumn, and the selection of seed beans from a healthy crop, are three important steps in controlling the mildew. When pole beans are planted, only two or three vines should be grown on each pole and the hills should be far enough apart so that the vines at the top of the poles will not intermingle, permitting the dew and rain on the foliage and pods to dry quickly. As a supplement to the above named control measures, spraying with bordeaux mixture 4-4-50 is suggested. There are two objections to the application of a fungicide: first, the disease occurs only during certain years, but as the epidemic cannot be foretold it is necessary to spray each year; secondly, the fungicide may spot the pods. Each grower must decide whether the loss of his lima-beans has been frequent enough to justify his use of bordeaux mixture. In the northern states applications seldom need to be made before the middle of July. Three or four additional sprayings are made at weekly or ten-day intervals.

References

Clinton, G. P. Downy mildew, *Phytophthora phaseoli* Thaxt., of lima beans. Conn. Agr. Exp. Sta. Ann. Rept. **1905**: 278-303. 1906.

Halsted, B. D. The downy mildew of lima beans. N. J. Agr. Exp. Sta. Bull. 151: 18-24. 1901.

Sturgis, W. C. The mildew of lima beans (*Phytophthora phaseoli* Thaxt.). Conn. Agr. Exp. Sta. Ann. Rept. **1897**: 159-166. 1898.

Sturgis, W. C. Mildew of lima beans. Conn. Agr. Exp Sta. Ann. Rept. **1898**: 236-241. 1899.

Thaxter, R. Report of the mycologist. Conn. Agr. Exp. Sta. Ann. Rept. **1889**: 167-171. 1890.

Diaporthe Pod-Blight of Lima-Bean

Caused by *Diaporthe phaseolorum* (Cooke and Ellis) Sacc.

The pod-blight has proved injurious only in the states bordering the Atlantic Ocean: Virginia, Maryland, Delaware, New Jersey, Long Island, and Connecticut. It, however, has been observed in West Virginia, Michigan, and possibly in England. Evidently it is indigenous to New Jersey where it was first found in 1891, fourteen yars before it was reported from any other district. It occurs more abundantly on the pole lima-beans than on the bush varieties. In a year favòrable for the fungus, it may cause the loss of a large percentage of the crop.

Symptoms.

The disease first appears on the leaves, then later spreads to the maturing pods. The spots on the foliage are large, brown, irregular, and often with a discolored border. As the tissue is dying or after its death, large black pycnidia form in concentric circles in the lesions. In a heavy wind the killed tissue may drop out, leaving ragged appearing leaves.

Very young pods are rarely affected. The lesions on the older pods may occur at any point and usually spread in all directions, leaving them black and wilted. A large number of black pycnidia are embedded in the epidermis and serve as an excellent diagnostic sign of the disease. If the seed has formed before infection takes place, it may shrivel, or if the pod is invaded while young, no seed is formed.

Cause.

The fungus is generally known as *Phoma subcircinata* E. and E. even though the perfect stage has been discovered and named *Diaporthe phaseolorum*. The parasite lives over winter in affected seeds and in bean refuse left in the field. In the pycnidia from each pod are produced an innumerable number of spores, which after germination enter through the stomata or wounds into the host tissue. The mycelium grows in every direction, finally forming clumps of hyphæ mostly in the stomatal cavities and from which new fruit-bodies are formed. The spores are disseminated by wind and by pickers. Perithecia may be produced during the winter and spring in the diseased tissue, and their ascospores together with the mycelium hibernating in the seed serve as inoculum for the infection of the new crop.

The spores in the pycnidia are sometimes two-celled, and in

this manner resemble those of Ascochyta. It, therefore, has been suggested that the fungus which has been described as *A. phaseolorum* Sacc. is merely a strain of the Phoma. This point needs further investigation.

There is a second spot on beans, lima-beans, and cowpeas, which may resemble that caused by Phoma, except that the lesions are smaller, more nearly circular, and have fewer and smaller pycnidia. The pathogene, *Phyllosticta phaseolina* Sacc., is in this case quite distinct from the Phoma.

Wet warm weather favors the pod-blight organism. The optimum temperature is between 80° and 87° F. Since the disease is confined almost wholly to the Atlantic coast, it evidently is stimulated by certain combinations of temperature and humidity that are peculiar to this district.

Control of pod-blight.

As the parasite is carried in the seed, the first consideration is that of obtaining clean seed from healthy plants. In order to make sure that no spores or mycelium are clinging to their surfaces, the seed may be dipped in corrosive sublimate (1 ounce in 7.5 gallons of water) for ten minutes, then removed, rinsed and dried. If, after these precautions have been taken, infection still takes place in the field, bordeaux mixture 4-4-50 may be used satisfactorily. The applications are begun when the vines are one to two feet tall, and continued at weekly intervals until danger of infection is past.

References

Clinton, G. P. Report of the botanist. Conn. Agr. Exp. Sta. Ann. Rept. 1905: 263-330. 1906.

Cook, Mel T. Report of the plant pathologist. N. J. Agr. Exp. Sta. Ann. Rept. 1912: 517-518. 1913.

Harter, L. L. Pod-blight of the lima bean caused by *Diaporthe phaseolorum*. Jour. Agr. Research 11: 473-504. 1917.

Smith, C. O. Leaf-spot of beans and cowpeas. Del. Agr. Exp. Sta. Bull. 70: 15-16. 1905.

Bacterial Spot of Lima-Bean

Cause by *Bacterium viridifaciens* T. and W.

The bacterial spot of lima-beans has evidently been present in the United States for many years, even though it was not fully recognized as a distinct disease until 1917. In that year

Fig. 20.—*Bacterium viridifaciens* causing spotting of lima-bean foliage.

the organism was isolated from material collected in Wisconsin and in New York. In a few cases it did considerable damage in home gardens. The following year it was severe in some war gardens at Ithaca, New York, and since then has been observed several times in Wisconsin. Judging from its former ravages, the pathogene will probably never do as much damage as does the bacterial blight.

Symptoms.

The spots on the leaves (Fig. 20) and pods are rarely so large or so numerous as are those caused by *Bacterium phase-*

oli. The foliar lesions are small, reddish-brown, irregularly circular spots, in which the tissue is thinner than that of the normal leaf. As the dead cells become dry, the center of the spot turns gray and may crack open or fall away. The killed areas may coalesce occasionally. Similar spots are produced on the stems and pods, except that they are elongated on the former.

Cause.

The pathogene, *Bacterium viridifaciens,* is a small, white, rod-shaped organism with the group number 211.2322133. It lives over winter in the seed and in bean refuse. In the spring it probably is splashed by rain from the cotyledons or the soil to the new leaves, where it enters through the stomata.

The disease is more abundant in wet weather, and spreads when the temperature is about 85° F.

Control.

No definite control measures have been demonstrated. Practicing three-year rotations of crops and obtaining healthy seed are suggestions based on knowledge of the life history of the organism.

REFERENCE

Tisdale, W. B. and Maude M. Williamson. Bacterial spot of lima bean. Jour. Agr. Research 25: 141-154. 1923.

CERCOSPORA LEAF-BLOTCH OF BEAN

Caused by *Cercospora cruenta* Sacc.

The leaf-blotch is fairly common in America and Europe, and has been observed in China, but is not of economic importance. It affects common beans, lima-beans and cowpeas, being most common on the two last hosts. Ordinarily only the leaf is attacked, and often only the older lower foliage. The brown or red dead spots are angular in shape and at times numerous enough to cause the invaded leaf to drop. The lesions are so nearly like those caused by one or two other fungi that it is necessary to make a microscopical examination for a true diagnosis.

The fungus is a Cercospora, and most of the reports in America suggest *Cercospora cruenta*. Among the other species of this genus which have been mentioned as occurring on beans are *C. canescens* E. and M., *C. phaseolorum* Cooke, and *C. olivascens* Sacc. It is not proved whether any of these or all are synonymous. All the species are characterized by the thin, long, septate spores, borne on short exposed condiophores. It is possible that the fungus is carried in the seed, at least it has been isolated from the discolored epidermis of cowpeas. Little else is known of its life history.

No attempts have been made to control the disease. If it should ever become bothersome, it no doubt could be eliminated by obtaining seed from healthy plants, and possibly by spraying with bordeaux mixture.

References

Halsted, B. D. Bean diseases and their remedies. N. J. Agr. Exp. Sta. Bull. 151: 1-28. 1901.

Jackson, H. S. Diseases of field crops in Delaware in 1907. Del. Agr. Exp. Sta. Bull. 83: 3-25. 1908.

Yeast-Spot of Lima-Beans

Caused by *Nematospora phaseoli* Wingard

In 1901 a yeast was discovered causing a disease of hazelnuts in Italy. Later a species of the same genus was found on tomatoes, and recently another was isolated from discolored lima-beans. The yeast occurs on the seed in the pod at any time in its development, and may be present in pods that show no outward symptoms. The lesions are dark brown, sunken, wrinkled areas. After infection takes place, the beans cease developing, and later may be found dwarfed or dead. The earlier the attack, the more injury the bean sustains.

The parasite is different from those on hazelnut and tomatoes and has, therefore, been named *Nematospora phaseoli*. It is disseminated with the seed but does not develop if the temperature is much below 86° F. The yeast cells are of various shapes and sizes even resembling short mycelial hyphæ. Asexual reproduction is by means of the usual yeast budding. There are also numerous fusions of two cells resulting in the formation of asci each with eight ascospores in two groups of fours. The details of its life history have not been recorded.

Planting disease-free seed probably suffices in controlling the trouble.

REFERENCES

Anderson, H. W. Notes on the Nematospora disease of lima beans. Phytopath. **14**: 31. 1924.

Wingard, S. A.. Yeast-spot of lima bean. Phytopath. **12**: 47, 525-532. 1922.

BACTERIAL STREAK OF BEANS

Caused by *Bacillus lathryi* Manns and Taub.

Streak is a serious disease of sweet-peas. The causal organism has also been isolated from cowpeas, end-rot of tomato, clover, and common and lima-beans. On the beans dark streaks are formed on the stems. Occasionally the lesions may be depressed. Both on the stems and on the pods the disease resembles that caused by bacterial blight. The trouble on the bean is of little economic importance.

The organism, *Bacillus lathryi,* is a rod-shaped yellow bacterium which liquefies gelatine very slowly. Its group number is 211.2222522.

No control methods are required.

REFERENCE

Manns, T. F. Some new bacterial diseases of legumes and the relationship of the organisms causing the same. Del. Agr. Exp. Sta. Bull. 108: 1-44. 1915.

CHAPTER III

DISEASES OF BEETS, CARROTS, AND CHICORY

THE very extended literature dealing with diseases of beets refers almost wholly to the sugar-beet, which is grown extensively in America and Europe. The garden beet is susceptible to most, if not all, of the maladies that affect sugar-beets, but the more limited quantities in which the former is grown, together with a greater probable resistance to disease, has left it almost free of the many fungous and bacterial troubles that are so destructive to sugar-beets. For this reason only a few of the more common diseases are discussed.

REFERENCE

Halsted, B. D. Some fungous diseases of beets. N. J. Agr. Exp. Sta. Bull. 107: 1-13. 1895.

CERCOSPORA LEAF-SPOT OF BEET

Caused by *Cercospora beticola* Sacc.

The leaf-spot has been known many years, the causal fungus having been described in 1873. The trouble is present wherever common beets are grown, and is generally destructive on sugar-beets, which are the most susceptible host. Other plants on which the leaf-spot is found are mangel-wurzel, swiss chard, and *Martynia louisiana*.

Symptoms.

The disease on the leaves is characterized by spots, which are at first small brown flecks with reddish-purple borders

(Fig. 21). When the spots reach a size of an eighth of an inch or more in diameter, they become ashen-gray in the center, the border still remaining a purple color. As the spots mature,

Fig. 21.—Cercospora leaf-spot of beet.

they often become so thin and brittle in the center that the dead tissue drops out leaving a ragged hole. Rarely the lesions may become as large as five centimeters in diameter; ordinarily they are very small and numerous. In severe cases the spots kill the foliage. When successive crops of leaves are killed, the crown of the beet root may have elongated and become roughened, suggesting the name "pine-apple disease," a term that has been used to designate the trouble. Lesions similar to those on the leaves occur on the leaf-petioles, flower-pedicels, seed-pods, and seeds, and have even been reported on the roots of the beets.

Cause.

The fungus, *Cercospora beticola,* forms an ashen-colored fruiting layer over the invaded tissue. The layer is composed of short conidiophores grouped in clumps, and on which are borne singly the thin, long, septate spores. The fungus is disseminated by means of splashing water, on insects, cultivating tools, workmen, and by wind. It survives the winter in the old diseased refuse or on the seed. In the spring the mycelium fruits on the leaves partly trampled into the soil and furnishes inoculum for the new crop. Infection takes place only through the stomata, and for this reason is most common on mature leaves. Pool and McKay have conducted some interesting experiments in which they have shown that the germ-tube enters only when the stomata are open, and that any influences, such as moisture, light, and temperature, that react on the guard-cells, determine the amount of disease present.

High humidity and temperature tend not only to stimulate the growth of the fungus, but also keep the stomata open, with the result that under such influences the disease progresses most rapidly. The optimum temperature is approximately 85° F. When the temperature reaches 100° or falls much below 60° F., conidial production is greatly reduced. Stomata remain open throughout the day if the humidity does not drop below 60, but probably close at any point much below 50.

Control.

Control of the parasite consists for the most part in field sanitation. As the fungus lives in the old leaves, they should be plowed under deeply in the fall so that they will disintegrate before spring. A two- or three-year rotation of crops and isolation from infested fields are also essential. In small gardens the affected foliage can be removed and either fed to cattle or burned. In large fields it may be convenient to harvest the tops and place them in a silo where the fermentation kills the mycelium.

If the disease is severe and the crop of sufficient value, spraying with bordeaux mixture 4-4-50 may be profitable. It has been shown that the fungus can be controlled when the fungicide is applied every two weeks or ten days, beginning as soon as the disease is evident in the field.

As the Cercospora is seed-borne, seed treatment will decrease the amount of early infection. The seed is dipped for seven minutes in formaldehyde (15 parts in 1000 parts of water), rinsed in water and planted at once or dried.

References

Duggar, B. M. Three important fungous diseases of the sugar beet. N. Y. (Cornell) Agr. Exp. Sta. Bull. 163: 352-359. 1899.

Halsted, B. D. Some fungous diseases of beets. N. J. Agr. Exp. Sta. Bull. 107: 7-11. 1895.

McKay, M. B. and Venus W. Pool. Field studies of *Cercospora beticola.* Phytopath, 8: 119-136. 1918.

Pammel, L. H. Spot disease of beets. Iowa Agr. Exp. Sta. Bull. 15: 238-243. 1891.

Pool, V. W. and M. B. McKay. The control of the sugar beet leaf-spot. U. S. Dept. Agr. Bur. Plant Ind. Circ. 121: 13-17. 1913.

Pool, V. W. and M. B. McKay. Relation of stomatal movement to infection by *Cercospora beticola.* Jour. Agr. Research 5: 1011-1038. 1916.

Pool, V. W. and M. B. McKay. Climatic conditions as related to *Cercospora beticola.* Jour. Agr. Research 6: 21-60. 1916.

Townsend, C. O. Leaf-spot, a disease of the sugar-beet. U. S. Dept. Agr. Farmers Bull. 618: 1-18. 1914.

Curly-Top of Beet

Cause not determined

According to reports which apparently are authentic, the curly-top of beet has been present in western United States for many years, although it caused no great losses until about 1900. The literature published during the succeeding fifteen years is very interesting to read as it depicts the gradual development of the knowledge regarding the cause of the trouble. The search for truth was like the childrens' game of hunting the thimble, each investigator becoming warmer

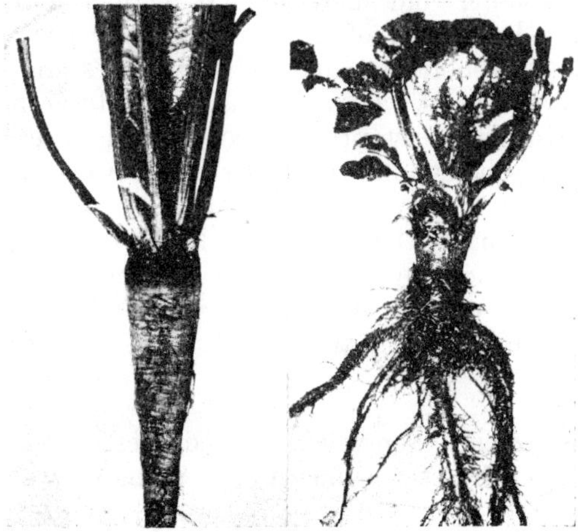

FIG. 22.—Curly-top of garden beet. At the left a healthy plant as compared with a diseased plant of the same age.

and warmer until finally the relation of the leaf-hopper and the dissemination of the virus was discovered.

The curly-top of the beet is now known to be present on both American continents, but does not seem to occur in the eastern hemisphere. The greatest injury is to sugar-beets grown in California and Utah, although the garden beets and mangel-wurzel may also be affected badly. A number of weed hosts such as chickweed, bur-clover, red-stem filaree, malva, lamb's quarter, and cultivated plants as buckwheat and spinach are susceptible, and aid in propagating the virus.

Symptoms.

Notwithstanding the fact that the curly-top is a virus disease disseminated by insects, it is distinct from the true mosaic of sugar-beets, which has been studied in Europe and America. The curly-top (Fig. 22) may appear at any time in the growth of the plant, but does most damage when the seedling is infected. The young plant is dwarfed, the leaves are smaller and more numerous than normally, the veins on the lower side of the leaf are more prominent, the leaf-petioles are shortened, and the leaf-blades are reduced in length and width as well as being decidedly crinkled and puckered. In many instances the roots of the diseased plants are exceedingly hairy, and the root tissue is woody and tough. Occasionally there may be a darkening in the crown of the root and a wilting of the plant, but these two symptoms may be absent or induced by other causes. A badly affected plant dies before the end of the season or remains stunted and unhealthy until the end of the summer.

Cause.

The cause of curly-top was a mooted question for a number of years. Step by step evidence was gathered which at last proved that the disease was produced by a virus carried by the beet leaf-hopper. This insect, as soon as it feeds on a diseased plant, becomes contaminated with the inoculum which it disseminates to healthy plants when feeding upon the latter. Not only is the leaf-hopper capable of carrying the virus as soon as it feeds on diseased foliage, but after the insect once becomes contaminated it can feed more than three months on non-susceptible hosts and still be able to transmit the virus to beets. After the beet harvest, the leaf-hoppers migrate to weeds, most of which are immune to curly-top, and feed there until the early winter rains cause the growth of red-stem filaree, a susceptible host. The leaf-hoppers reared on this weed and on the chickweed in the spring are supplied with the virus which they carry to the beet fields. There is no other natural method in which the curly-top is transmitted from one plant to another. It can be transmitted artificially by grafting as is true of ordinary mosaic.

Control of curly-top.

As yet no definite control measures are available, but as the garden beet is not often seriously affected, the gardener has no particular need for protective measures against this trouble. Destruction of susceptible weeds and eradicating the leaf-hopper by the use of proper insecticides no doubt will reduce the amount of disease.

References

Ball, E. D. The leaf-hoppers of the sugar beet and their relation to the "curly-leaf" condition. U. S. Dept. Agr. Bur. Ent. Bull. **66**: 33-52. 1911.

Boncquet, P. A. and C. F. Stahl. Wild vegetation as a source of curly-top infection of sugar beets. Jour. Econ. Ent. **10**: 392-397. 1917.

Carsner, E. Susceptibility of various plants to curly-top of sugar beet. Phytopath. **9**: 413-421. 1919.

Robbins, W. W. Mosaic disease of sugar-beets. Phytopath. **11**: 349-365. 1921.

Severin, H. H. P. Investigations of the beet leaf-hoppers (*Eutettix tenella* Baker) in California. Jour. Econ. Ent. **12**: 312-326. 1919.

Townsend, C. O. Curly-top, a disease of the sugar beat. U. S. Dept. Agr. Bur. Plant Ind. Bull. 122: 1-32. 1908.

Crown-Gall of Beet

Caused by *Bacterium tumefaciens* EFS. and Townsend

For nearly a hundred years tumors have been observed on beets, and commented on in literature. When first found, the galls were few in number and were considered merely as interesting monstrosities. As the disease became more common, and particularly since it was found more often near the head-lands than in the middle of the field, crown-gall was considered as resulting from wounds. It was not until about 1900 that the bacterial nature of the disease was demonstrated.

The parasite is able to infect a large number of unrelated hosts. Aside from beets, the disease may be injurious to roses, raspberries, grapes, peaches, tomatoes, and in fact to nearly every vegetable grown in infested soil.

Symptoms.

The galls (Fig. 23) ordinarily are not present on the very young beet, but begin to appear about the middle of the

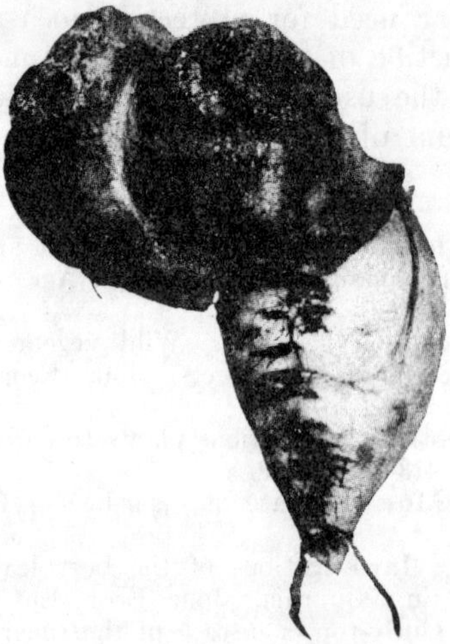

FIG. 23.—Crown-gall on beet.

summer and continue to increase in number and in size until the crop is harvested. At that time all sizes of the tumors may be present. The gall starts as a slight swelling, usually with a smooth surface, and frequently without a distinct line of demarcation between the excrescence and the beet proper. In other cases the excrescence lies close to the outer surface of the host, being connected only by a slender neck of tissue. There may be one to many galls on a single beet, and these appear on any part of the root. It is not uncommon to find the crown of the fleshy root completely covered by the tumor. The swellings may be of any size, but vary according to the succulence and rate of growth in the host, being larger when the beet is developing quickly.

Another type of swelling known as tuberculosis is found on sugar-beets. This is distinguished from the crown-gall in that its inner tissue shows brown water-soaked areas, the outer surface is much cracked and roughened, and the whole excrescence usually decays.

Cause.

The parasite, *Bacterium tumefaciens*, is a white rod-shaped organism with the group number 212.2322023. There is so much variation among the races of the species that it is questionable whether the strains from all sources should be classed together or differentiated by separate names. The bacterium lives in the soil for at least a year or two. It also winters over in the living galls of perennial plants. One of the chief methods for dissemination is on infected nursery stock. When a soil once becomes infested, the amount of inoculum increases slowly if susceptible crops are grown until beets will no longer be profitable. The disease never appears quickly in epidemic form.

The bacterium enters the beet through wounds, and by abnormal stimulation causes a great increase in the number of cells. The vascular bundle is much distorted and often in cross-sections is almost lost among the great mass of cortical tissue. Much of the food that should be used for the formation of the normal root is diverted to the galls for the formation of new cells.

The bacterium will grow at any temperature between 32° and 98° F., although at both extremes the development is much slower than at 77° to 82° F. The thermal death point is 124° F.

Control.

The only control method necessary is the proper rotation of crops. If the soil once becomes infested, the bacterium can be starved out effectually by changing the crop to some immune host as corn, oats, or sorghum. Care should be taken

not to introduce the parasite with infected orchard nursery stock on the farm.

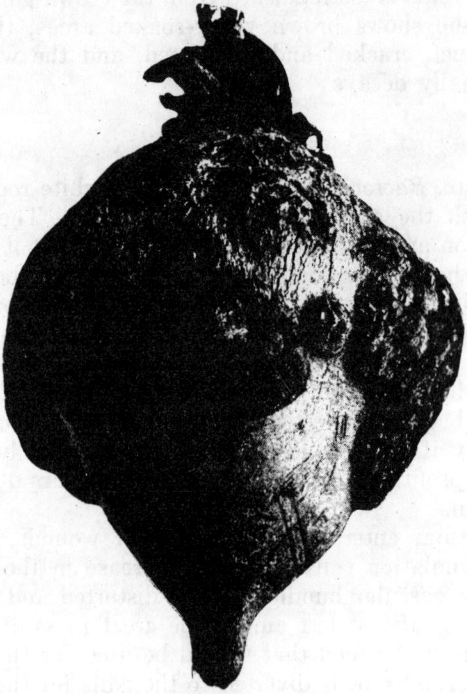

FIG. 24.—Common scab on beet.

REFERENCES

Levine, M. Studies of plant cancers III. Amer. Jour. Bot. 8: 507-525. 1921.

Smith, E. F., Nellie A. Brown, and C. O. Townsend. Crown-gall of plants, its cause and remedy. U. S. Dept. Agr. Bur. Plant Ind. Bull. 213: 1-215. 1911.

Smith, E. F., Nellie A. Brown and Lucia McCulloch. The structure and development of crown-gall. U. S. Dept. Agr. Bur. Plant Ind. Bull. 255: 1-60. 1912.

Townsend, C. O. Field studies of the crown-gall of the sugar beets. U. S. Dept. Agr. Bull. 203: 1-8. 1915.

Common Scab on Beets

Caused by *Actinomyces scabies* (Thaxter) Güssow

(For discussion see Common Scab of Potatoes, page 380.)

The common scab of potatoes is occasionally found on beets (Fig. 24) when the latter are grown in heavily infested soil. The symptoms on the two hosts are much alike except that the wart-like scab-spots on beets are somewhat more bulging than are those ordinarily found on the potato.

Reference

Lutman, B. F. and H. F. Johnson. Some observations on ordinary beet scab. Phytopath. 5: 30-34. 1915.

Rhizoctonia Root-Rot of Beet

Caused by *Rhizoctonia* sp.

(See Rhizoctoniose of Potato, page 367, and Bottom-Rot of Lettuce, page 249.)

The beet is affected at the crown of the root. The base of the leaf-petioles turn almost black, and the leaf finally falls over and dies. When the fungus has penetrated the crown of the root, the flesh becomes brown and cracks appear in the epidermis. The brown mycelium of the parasite grows over the surface of the affected portions.

Reference

Duggar, B. M. and F. C. Stewart. The sterile fungus Rhizoctonia as a cause of plant diseases in America. N. Y. (Cornell) Agr. Exp. Sta. Bull. 186: 51-76. 1901.

Root-Knot of Beets

Caused by *Heterodera radicicola* (Greef) Mül.

(See Root-Knot of Tomatoes, page 550.)

Damping-off of Beets

Caused by various fungi

(See Damping-off of Tomatoes, page 546.)

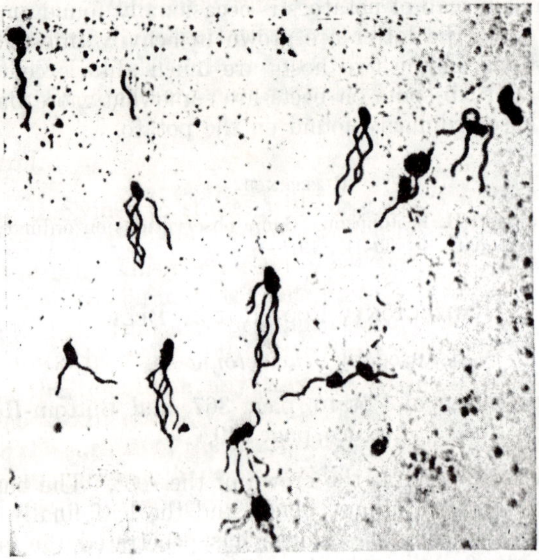

Fig. 25.—*Bacillus carotovorus* that causes soft-rot of vegetables.

Southern Sclerotium Rot of Beets

Caused by *Sclerotium rolfsii* Sacc.

(See Southern Sclerotium Blight of Sweet-Potato, page 510.)

Bacterial Soft-Rot of Carrot

Caused by *Bacillus carotovorus* Jones

The bacterial soft-rot of vegetables (Fig. 25) has been studied since 1891 when Halsted described the trouble on celery. A few years later it was recorded as injuring cabbage and other crucifers, and in 1901 was described in detail on carrots. Since then the organism, under various names, has been observed on the following additional hosts: artichoke, asparagus, brussels-sprouts, cauliflower, eggplant, horse-radish, kohlrabi, muskmelons, onions, parsnips, pepper, potato, radish, rhubarb, rutabagas, salsify, tomato, and turnips. The bacterial rot is of great importance not only in the field, but also in storage and during transportation. No actual figures of losses have been collected for all the hosts, yet it probably is safe to assume that few other phytopathogenic organisms cause a greater total loss.

Symptoms.

In a general way the symptoms on all the hosts are similar to those on carrots, and consist in a soft mushy or slimy disintegration of the affected tissues, a tendency for the epidermis to remain intact, a grayish or brownish discoloration of the rotted tissue, and a disagreeable odor of putrefaction. In the carrot the normal flesh being colored, the discoloration of the diseased portion is not so marked as is that on some other hosts; the cells, however, take on a water-soaked appearance, and after losing their middle lamellæ collapse into a slimy mass. If the rotting occurs on the root in the field and in the crown of the root where the trouble commonly has its origin, the outer older leaves turn yellow and wilt, followed by similar symptoms of the younger inner foliage. The effects on the leaves may not be apparent until the root is almost completely decayed. The rot in storage is similar to that in the field.

Cause.

Much confusion has arisen regarding the nomenclature of the parasite. The soft-rot bacteria from various sources differ just enough to make it questionable whether the strain from one host or one locality should be listed as a separate species or whether all the strains should be grouped under the one name *Bacillus carotovorus*. Among the other names which have been used are *Bacillus omnivorus* v. Hall, *B. olereaceæ* Harr., *Pseudomonas destructans* Potter, *Bacillus aroideæ* Townsend, *Bacillus* sp. Spiekermann, *B. melonis* Giddings, *B. apii* (Brizi) Migula, *B. apivorus* Worm., *B. cepivorus* and *B. brassicævorus*. Their variation is not because of parasitism or of a difference in the major characters by which bacteria are classified, but the species have all been based on such points as the reaction of each strain on dextrose, lactose, saccharose, starch, and glycerine. The resulting group numbers are 221.1113022, 221.1123022, 221.1213022, 221.2113022, 221.2123022, 221.2223022, and 221.2223032. There will probably always be differences of opinion in regard to the classification, yet, for convenience sake, all the proposed species are here classed with *Bacillus carotovorus*.

The bacterium is a relatively large, white, rod-shaped organism with large flagellæ, and usually very motile. It lives during the winter on stored plant parts, and no doubt also remains alive in roots and tubers left in the field. When the infected tissue is handled, the slimy mass is smeared on the healthy parts, and may thus be disseminated into the fields, or over long distances when the crop is shipped. Insects and cultivating tools carry the organism in the field. Evidently the bacteria enter the host tissue only through wounds. Progressing inwardly, they dissolve the middle lamellæ of the cells, and by the secretion of a toxic substance kill the protoplasm. The bacteria enter the cells only in the later stages of the disease.

When a high humidity is combined with a temperature of over 80° F., the bacillus is capable of causing its greatest injury. The optimum temperature for its growth is about 85° F., the maximum slightly over 100° and the minimum about 39° F. For this reason much of the loss due to the bacterial soft-rot occurs during the middle of the summer. For example, self-blanching celery in New Jersey can be grown successfully from April to July. After this period, however, there may be almost a total loss of the crop in infested soil. Like many bacteria, the parasite is killed quickly in bright sunlight.

Control of bacterial soft-rot.

Each year the storage-house should be cleaned thoroughly, every board, the walls, floors, and ceiling washed or sprayed with formaldehyde (1 pint in 10 gallons of water) or if the fumes are too unpleasant copper sulfate (1 pound in 5 gallons of water) may be substituted. The roots, as soon as they are dug, should be dried in the sun, any bruised or diseased ones sorted out, and only the perfectly healthy roots placed in storage. Carrots when piled together often decay very fast, while if they are stored in a slight amount of straw, forest leaves, or dry sand, so that the roots are held apart, the rot does not proceed from one root to the other.

The temperature and humidity of the storage-house are very important. There should be sufficient ventilation so that the humidity of the air can be kept reasonably low. During foggy or rainy weather the doors should be kept closed. At the same time the temperature should be as nearly 32° F. as possible. In the early autumn when the nights are cool and the days are warm, it may be desirable to open the doors and windows at night and close them during the day.

It is not always possible to control the rot in the field, nevertheless the trouble may be eliminated partly by long rotations with crops that are immune, as corn, cereals, and grasses. In certain cases it may be practicable to destroy the insects that aid in disseminating the bacterium, or which injure the host so that the parasite is able to enter the exposed tissue. If proper precautions are taken in cultivating and harvesting, also, the amount of injury and consequent infection may be lessened.

REFERENCES

Harding, H. A., W. J. Morse and L. R. Jones. The bacterial soft-rot of certain vegetables. N. Y. (Geneva) Agr. Exp. Sta. Tech. Bull.

11: 251-368. 1909. *Same in* Vt. Agr. Exp. Sta. Bull. 147: 243-360. 1909.

Harrison, F. C. A bacterial disease of cauliflower and allied plants. Ont. Agr. Exp. Sta. Bull. 137: 1-32. 1904.

Johnson, T. and J. Adams. Bacterial rot in turnips and other Brassicas in Ireland. Econ. Proc. Roy. Dublin Soc. 2: 1-7. 1910.

Jones, L. R. A soft-rot of carrot and other vegetables caused by *Bacillus carotovorus.* Vt. Agr. Exp. Sta. Ann. Rept. 13: 299-332. 1901. *Similar in* Centralbl. f. Bakt. u. Par. II 7: 12-21, 61-68. 1901.

Jones, L. R. The cytolytic enzym produced by *Bacillus carotovorus* and certain other soft-rot bacteria. Centralbl. f. Bakt. u. Par. II 14: 257-272. 1905.

Potter, M. C. Ueber eine bakterien-krankheit der rüben (*Brassica napus*). Centralbl. f. Bakt. u. Par. II 7: 282-288, 353-362. 1901.

Russel, H. L. A bacterial rot of cabbage and allied plants. Wis. Agr. Exp. Sta. Bull. 65: 1-39. 1898.

Spiekermann, A. Beitrag zur kenntnis der bakteriellen wundfaulnis der kulturpflanzen. Landw. Jahrb. 31: 155-178. 1902.

Phoma Root-Rot of Carrot

Caused by *Phoma sanguinolenta* Rostrup

For nearly a hundred years a Phoma root-rot has been present on carrots stored for seed in Denmark. Although very destructive there, sometimes causing the loss of an entire crop, it has seldom been reported from other European countries or from America.

The disease does not become destructive to carrots until the latter part of the summer, at which time it causes a sunken gray or brown canker near the upper part of the root. It is common to see a collapsed zone about the leaf rosette. Over the decayed area, which does not extend deeply into the flesh, are numerous black pycnidia, whose upper halves are exposed above the surface of the host tissue. In the presence of moisture, long tendrils of flesh-colored or blood-red spores ooze from the ostioles of the pycnidia, and serve as an excellent diagnostic symptom.

The disease does very little damage the first year, but when affected roots are placed in storage the rot not only progresses on such roots but the fungus is disseminated to healthy roots which in turn become diseased. Whole crops of carrots may be destroyed in this manner. The roots that are transplanted in the spring for seed may be only slightly affected, nevertheless the mycelium passes up into the stem often to the very tip, being revealed here and there, especially at the nodes, by black streaks. Pycnidia with their red tendrils may also appear on the parts of the plant above ground. The development of the fungus on the root and stem is sufficient to cause the whole plant to wilt just about blossoming time and without producing seed.

If all carrots are examined carefully after harvest, and only healthy ones placed in storage, much of the winter loss can be avoided. In a similar manner, a thorough inspection of the roots before they are transplanted will reveal many of the almost imperceptibly infected ones, thereby reducing the amount of the disease in the field. It has been discovered, also, that if the carrots meant for seed are left in place all winter, then permitted to fruit in the spring without transplanting, the Phoma rot can be almost entirely eliminated. The disease is worse on loose sandy soils than on clay; therefore, if in addition to the above mentioned precautions, a suitable soil for growing carrots is chosen, the disease need not be feared.

Reference

Rostrup, E. *Phoma sanguinolenta,* ein den samenertrag der möhre (*Daucus carota*) vernichtender pilz. Ztschr. f. Pflanzenkr. **4:** 195-196. 1894.

Alternaria Black-Rot of Carrot

Caused by *Alternaria radicina* M., D. and E.

When carrots are held over winter in storage, the roots are frequently injured by a soft depressed black-rot, which may begin at the crown and progress downward or spread from some wound on the side of the root. Much loss from this disease has been reported.

Pure cultures of the fungus causing the decay of carrot roots show it to be an Alternaria and as it differs in cultural characters, in size of spores, and in method of fruiting from the parasite usually found on carrot foliage, the organism is named *Alternaria radicina*. It has large, brown, catenulate, muriform spores. The spores, after germination, produce mycelium that enters an exposed place on the root, where gradually a decayed spot is formed. After the tissue has been penetrated for some distance, an aerial black mycelial weft is developed over the rotted tissue, and on which the spores are borne. Infection probably takes place, both in the field and in the storage-house. In either case, the injured roots are thrown away and serve as a source of inoculum for the succeeding crop.

No methods for controlling the rot have been devised. It is probable that if care is taken to sort out all bruised or diseased roots before they go into storage, to practice rotation of crops, and then follow the directions for storing, given for the control of bacterial soft-rot, the black-rot can, in a great measure, be eliminated.

Reference

Meier, F. C., Charles Drechsler, and E. D. Eddy. Black-rot of carrot caused by *Alternaria radicina* n. sp. Phytopath. **12**: 49, 157-167. 1922.

Macrosporium Leaf-Blight of Carrot

Caused by *Macrosporium carotæ* E. and E.

In many sections of the country where carrots are grown, and especially on Long Island, there is an appreciable loss resulting from the leaf-blight. By reducing the green leaf surface, the disease does not permit the full development of the root, and by being unsightly, reduces the market value of fresh carrots sold in bunches. The affected leaves and petioles first turn yellow then change to brown, so that during severe infection the whole top may be killed.

The fungus associated with leaf-blight has brown muriform spores, but differs from the root-rot organism in that the spores have exceedingly long, thin, permanent pedicels and are not borne in a catenulate manner. Consequently the name *Macrosporium carotæ* is applied. Presumably the fungus lives over winter in the discarded tops, and is able in the spring, after developing mycelial threads, to infect the new crop.

Some of the Long Island growers have practiced spraying their carrots with bordeaux mixture, and evidently obtained satisfactory results. So far as is known, no careful demonstrations have been conducted to determine the most profitable number of applications to make, and the strength of the spray best suited for carrot foliage. The turning under of the diseased tops by deep plowing and rotation of crops, no doubt, are aids in reducing the amount of leaf-blight.

References

Ellis, J. B. and A. B. Langlois. New species of Louisiana fungi. Jour. Mycol. **6:** 35-37. 1890.

Meier, F. C. Charles Drechsler and E. D. Eddy. Black-rot of carrot caused by *Alternaria radicina* n. sp. Phytopath. **12:** 49, 157-167. 1922.

Cercospora Leaf-Spot of Carrot

Caused by *Cercospora apii carotæ* Pass.

The Cercospora leaf-spot of carrots has been reported from Asia, Europe, and America. It is seldom of much importance, although in Indiana it was destructive when the overhead sprinkling system was used in the fields.

The spots on the leaves and petioles are gray to brown, being differentiated from other leaf-spots by the presence of long, thin, septate, Cercospora spores.

Ordinarily the disease does not require control measures, but if the loss incurred justifies the use of a fungicide, bordeaux mixture will perhaps be effective against the fungus.

References

Murata, Jutaro. Leaf blight of carrot and its control (Japanese). Jour. Plant Prot. 8: 185-188, 233-236. 1921.

Plant Disease Survey Office. Disease of carrots. Plant Dis. Survey Suppl. 22: 403. 1902.

Rhizoctonia Crown-Rot of Carrot

Caused by *Rhizoctonia* sp.

(See Rhizoctoniose of Potatoes, page 367, and Bottom-Rot of Lettuce, page 249.)

The carrots are affected at the crown by Rhizoctonia. The leaves die when the base of the petioles rot off and are covered with hyphæ of the fungus.

Violet Root-Rot of Carrot

Caused by *Rhizoctonia crocorum* DC.

(See Violet Root-Rot of Potato, page 439.)

Sclerotinia Rot of Carrot (Fig. 26)

Caused by *Sclerotinia libertiana* Fuckel

(See Lettuce-Drop, page 243.)

Downy-Mildew of Carrot

Caused by *Plasmopara nivea* (Ung.) Schroet.

(See Downy-Mildew of Parsnip, page 318.)

Sclerotinia Rot of Chicory

Caused by *Sclerotinia libertiana* Fuckel

(See Lettuce-Drop, page 243.)

Downy-Mildew of Chicory

Caused by *Bremia lactucæ* Reg.

(See Downy-Mildew of Lettuce, page 259.)

Fig. 26.—Sclerotinia rot of carrot.

Bacterial Leaf-Spot of Chicory

Caused by *Bacterium* sp.

In New York the chicory or French endive leaves occasionally are injured severely by a bacterial leaf-spot (Fig. 27). Looking at the leaf in reflected light, the small circular to angular spots appear water-soaked and dark. In direct light the same lesions are transparent as if the inner tissue had been eaten out by insects. In fact, the spots at first resembled insect injury so closely that an entomologist was asked to diagnose the trouble. When sections were made of the leaf, however, the infected tissue was filled with bacteria. Later another type of symptom was observed. Instead of the hundreds of small spots, one-sixteenth to one-eighth inch in diameter, a single lesion included a continuous area over a large portion of the leaf surface. Such water-soaked or browned areas do not have their origin at the edge of the leaf, but seem to spread from the leaf-blade adjoining the petiole and midrib, gradually approaching the margin of the leaf as the disease progresses. The leaf finally dies.

Isolations were made and a bacterium with polar flagella was obtained. Inoculations and re-isolations were also made successfully. It is not known how the parasite is disseminated, but apparently it remains alive in the soil for at least a year.

No control measures are known.

Gray Mold-Rot of Chicory

Caused by *Botrytis* sp.

During the autumn of 1923 when the amount of rainfall was above normal and the temperature was high until early in December, there was much rotting of chicory roots in New York due to Botrytis. The roots are harvested about the time cold weather is supposed to appear, the tops are cut off an inch above the crown of the roots, and the roots stored in outdoor trenches. They are covered with straw or leaves and protected from rain by roofing paper.

The flattened stubs of the petioles fit together almost as closely as scales in a bulb, and furnish an ideal place for the entrance of Botrytis, particularly in warm rainy weather. The outer petioles become infected first, the fungus gradually penetrating each layer of stems in succession. The decay of the inner petioles begins at their bases after the outer layer of stems is invaded. A soft light brown rot is formed, so that the petioles can be pulled out easily. The rot continues to

Fig. 27.—Chicory leaf destroyed by a bacterial blight.

invade the white stem tissue until the whole remaining top is included. When the base of the stem has become thoroughly infected, the crown of the root is attacked, the tissue softening and browning. The severity of the attack determines the extent to which the Botrytis penetrates the root. It is common to find the flattened sclerotia formed between the leaf-stalks, and to have the gray fruiting layer of the fungus cover the outside of the petioles and root.

There is no way of controlling the rot in outdoor trenches, for the growth of the parasite is dependent on moisture and temperature, which vary considerably in trenches. It has been found, however, that roots can be stored successfully in cold storage-houses where the humidity of the air and the heat can be governed. The crop usually is of enough value to justify the expense of artificially cooled rooms; therefore, such storage is recommended.

CHAPTER IV

DISEASES OF CELERY

MORE than twenty thousand acres of celery are grown in the United States; the states having the largest acreages are California, Michigan, New York, Pennsylvania, Ohio, Florida, New Jersey, and Massachusetts. A large percentage of this crop is destroyed by one or more field diseases and by storage rots. These losses could be much reduced if the growers put into practice all the recommendations that have proved effective.

LATE-BLIGHT OF CELERY

Caused by *Septoria apii* (Br. and Cav.) Rostrup

Although late-blight of celery may have been observed as early as 1840, the first authentic report was not made until 1890 when the causal fungus was discovered in Italy. The following year it was observed in a number of places in the eastern United States. Six years later it had spread across the country and was beginning to do serious damage in California. Notwithstanding the fact that it occurred so early in America and continental Europe, strangely enough it was not reported as present in England until 1906. Since that time it has been observed in all countries where celery is grown extensively, and is usually associated with much crop destruction. In one county in California there was a loss of over a half million dollars in a single season because of the disease. Similar damages to the crop could be reported from other intensive celery-growing areas.

The late-blight organism attacks only celery and the closely related vegetable celeriac. Formerly it was considered to be the same as the fungus on parsley but this has been disproved. Both wild and cultivated celery are affected. There is some difference in susceptibility of the varieties, but even the most

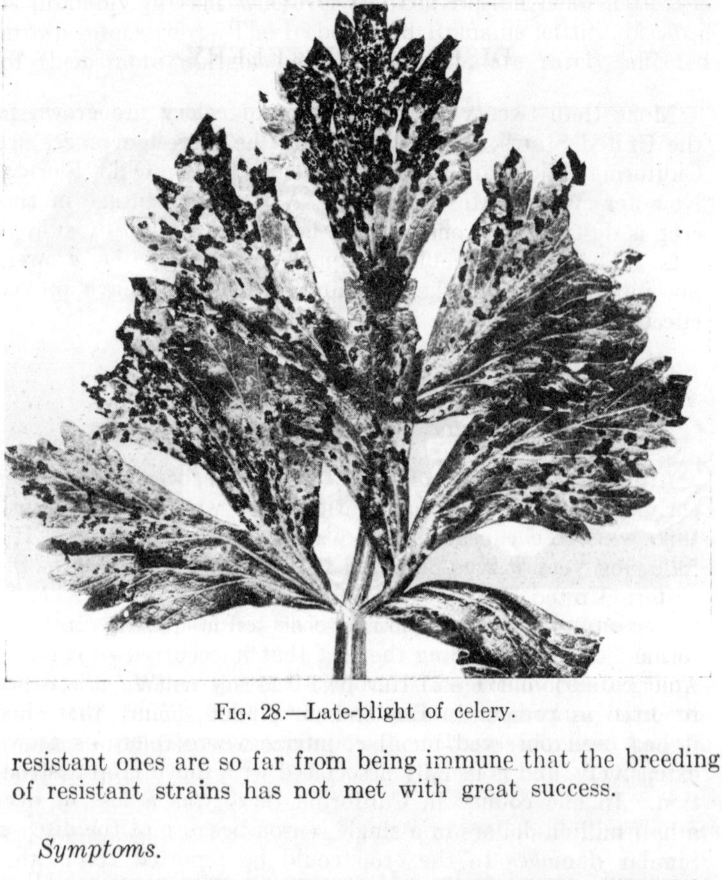

FIG. 28.—Late-blight of celery.

resistant ones are so far from being immune that the breeding of resistant strains has not met with great success.

Symptoms.

All parts of the celery plant above ground are affected by the late-blight. The lesions usually begin on the lower outer leaves and gradually spread upward and inward until not only the foliage but also the petioles and stems are badly injured or killed. The affected spots on the leaves (Fig. 28) begin as lighter yellow areas, the tissue of which later changes to brown and dies. The spots are inclined to be circular in outline, have a rather distinct margin, and are not large unless several of the lesions become confluent. When the spots become numerous or large enough, the leaves and petioles wither. The lesions on the stems are similar to those on the foliage except that on the former they are elongated rather than circular. On both the stem and leaf the chief diagnostic character of the lesion is the presence of numerous small black pycnidia protruding from the brown host tissue, and differentiating late-blight from the early and bacterial blights. No definite necrotic areas are produced on the seed, but it is not uncommon for pycnidia to appear on the seed-coat and on attached bits of pedicels.

When the diseased celery is stored after harvest, other organisms following the blight fungus cause a decay of the stalks. The most common secondary decay is pink-rot caused by Sclerotinia (see page 104.)

Cause.

The causal fungus, *Septoria apii,* has only the one spore stage, that of long, cylindrical, colorless, septate spores borne in comparatively large black pycnidia. During a rain or in heavy dew the spores are discharged from the mouth of the fruit-body in large numbers. They are disseminated by means of running or splashing water, wind, insects, men, tools, and horses. In California one of the principal agents for conveying the spores is the irrigation water flowing in the furrows. Favored by the presence of moisture and warmth, the spores germinate and send their germ-tubes into the celery tissue. At first the mycelium grows between the cells, but soon is able to break down the cell-walls, so that in an affected area of the leaf or stem hardly more than the host epidermis remains.

In the meantime the parasite proceeds with the formation of fruit-bodies and sclerotium-like masses of hyphæ. The whole process of incubation and infection requires from one to three weeks, after which the fungus is ready to be disseminated in an ever-widening circle, and if left undisturbed during the average season will find its way to nearly every plant within any continuous celery area.

The fungus lives over winter either in the form of pycnidia or as hyphal masses in the diseased leaf and stem refuse. In many muck fields where rotation rarely is practiced and where the affected leaves and petioles, which are cut away during harvesting, are left on the soil, there is every opportunity for the parasite to propagate until it becomes a serious pest. In a similar manner it may be disseminated from the compost or manure heap not only to the field but also to the seed-bed.

Aside from wintering over in the discarded parts of the plant, the mycelium is found among the cells of the seed where pycnidia may often be observed breaking through the seed-coat. The hyphæ evidently do not penetrate deeply enough to enter the embryo. Although the spores in the old leaves die in eight to eleven months, those on the seed remain alive for more than two years, and serve as a means of contaminating the seed-bed. Consequently the seedlings are often inoculated soon after they break through the ground and are thoroughly infected when it is time for them to be set into the field.

The Septoria grows slightly better at temperatures ranging from 60° to 70° F. than it does at those ranging from 70° to 80°F. It is most destructive in low poorly drained fields. Sodium nitrate and manure seem to increase the amount of the disease, while calcium sulfate and lime decrease the spotting. The younger the leaf the more readily infection takes place, although the older leaf never becomes wholly immune.

Control of late-blight.

From the foregoing discussion it is evident that particular attention should be given to the seed-bed. The diseased plants should either not be trimmed in the field, or if the trimmings are scattered over the ground they should be removed later and destroyed. Likewise, when the undesirable leaves are cut away in the crating shed, the refuse should not be placed on the compost heap unless the process of composting can be made complete at which time the fungus is killed.

Since the pathogene is carried on the seed, it is natural to suggest seed treatment. This control measure would be commendable were it not for the fact that in most celery districts the fungus is already widespread in plant refuse, so that treating would produce no great reduction in seedling infection; besides, celery seed, especially in warm weather, is very susceptible to injury while being disinfected. The treatments that have been suggested are the soaking for ten to thirty minutes in corrosive sublimate, 1-1000 (1 ounce in 7.5 gallons of water) or by placing the seed, tied loosely in cheese-cloth for thirty minutes in water heated to 118° F. and then rinsing in cold water. In both cases the seed is planted while it is still wet. Before making such treatments, however, a small amount of the seed should be used as a test in an attempt to determine the possible deleterious effects this control measure might have on the viability of the celery seed. Investigational work that has been done suggests that if the celery seed is held until it is three or four years old, the fungus within the tissue as well as that on the surface dies, so that treating is not necessary for seed that has been aged. This may be another way of avoiding seed treatment injury and the trouble it entails.

The most important recommendation is that of spraying with 5-5-50 bordeaux mixture or dusting with 20-80 or 15-85 copper-lime dust. The first two applications should be made when the plants are in the seed-bed, beginning when the seedlings are about an inch tall and again a few days before they are transplanted. Continuing the spraying or dusting when the plants in the field are six to eight inches tall, thorough applications are made each week until a week or ten days before harvest time. Usually five applications are sufficient. When the plants are large, about a hundred gallons of the bordeaux mixture are required to cover an acre at each application, with a corresponding decrease while the celery is small. When dust is used, the first two field applications each require about twenty-five pounds of the fungicide, and the last three about thirty-five pounds each. The dust should be applied when the air currents are moving slowly and preferably when the celery leaves are wet with dew. No matter whether spray or dust is the choice, the fungicide should be applied so carefully that all parts of the plant are covered, even if this necessitates driving twice over the same row in opposite directions. If the blight gets a start in the field, it is well to apply the fungicide twice at three-day intervals in order to check further infection by the parasite. The time to spray or dust is before and not after a continued rainy period, for it is during the wet weather that the Septoria is disseminated and grows into the host. The poison on the leaf during such a period will prevent germination of the spore.

References

Duggar, B. M. Late-blight of celery. N. Y. (Cornell) Agr. Exp. Sta. Bull. 132: 206-220. 1897.

Dye, H. W. and A. G. Newhall. Spraying and dusting for the bacterial and late blights of celery in western New York. Phytopath. 12: 48. 1922.

Klebahn, H. Die krankheiten des Selleries. Ztschr. f. Pflanzenkr. 20: 1-40. 1910.

Krout, W. S. Treatment of celery seed for the control of Septoria blight. Jour. Agr. Research 21: 369-372. 1921.

Pethybridge, G. H. The possible source of the origin of the leaf-spot disease of cultivated celery. Jour. Roy. Hort. Soc. 40: 476-480. 1914.

Rogers, Stanley S. The late-blight of celery. Calif. Agr. Exp. Sta. Bull. 208: 83-115. 1911.

Sherbakoff, C. D. Seed disinfection. Fla. Agr. Exp. Sta. Ann. Rept. 1916: 86R-87-R. 1917.

Thomas, H. E. The relation of the health of the host and other factors to infection of *Apium graveolens* by *Septoria apii*. Bull. Torrey Bot. Club. 48: 1-29. 1921.

Early-Blight of Celery

Caused by *Cercospora apii* Fres.

More than twenty years after its discovery in Europe, the early-blight of celery was observed in Missouri in 1884. Its exact distribution is not known, since earlier workers often confused it with other blights of this crop. Evidently it is found in varying abundance wherever celery is grown, being most severe in the states with the latitude of New Jersey and farther south, although it does considerable damage as far north as Boston and has been reported as present even in Canada.

There is a wide variation in the susceptibility of the different varieties of celery, among which White Plume and Golden Self Blanching are listed as severely infected. Aside from celery the disease is present in celeriac. Formerly the wild parsnip was suspected of acting as host for the fungus, but this has since been disproved.

Symptoms.

The name early-blight is misleading, for the disease rarely occurs as early as does either the blight caused by Septoria or the one produced by bacteria. Ordinarily lesions do not appear until the plants are more than six weeks old. The first noticeable symptom is the yellowing in minute spots of both sides of the foliage (Fig. 29). These spots enlarge rapidly and as they do so the affected tissue changes from yellow to an ashen-gray color, and a dry-papery texture. The lesions often are irregularly circular in outline, but may be angular or lunate, and in most cases have a poorly defined border. In the presence of humidity the injured leaf surface is covered with an ashen-gray mold that can be seen only by close observation, but which serves to differentiate this disease from the bacterial blight with the halo-bordered spot, and Septoria blight with its pycnidium-studded lesions. When the spots on a leaf become large or numerous enough, it wilts and dies. Similar lesions are produced on the stems and leaf-petioles.

Cause.

The early-blight is caused by one of the imperfect fungi known as *Cercospora apii.* It is characterized by having no

FIG. 29.—Early-blight of celery.

inclosed fruit-body, and by long, colorless, multi-septate spores, slightly enlarged at the base and tapering almost to a point at the tip. They are splashed by rain, carried with the manure, dragged by cultivators, or blown to neighboring plants where the germ-tube is able to penetrate the uninjured tissue. The mycelium grows between the cells and in the intercellular spaces forms sclerotial growths, which are able to remain alive for a year or at least as long as the host tissue is not completely disintegrated. The fungus fruits on both sides of the leaves

FIG. 30.—Bacterial blight of celery.

and on the stem. The spores are short-lived, so that they do not live over from one season to another. So far as is known the fungus does not penetrate the seed nor is able to live on the dry surface of the seed-coat, thus eliminating the possibility of its being seed-borne. The fungus reproduces most prolifically and the disease is most destructive when the weather is hot and wet after a period of drought.

Control.

The recommendations given for the control of late-blight of celery apply in all their details to the early-blight (see page 92.)

REFERENCES

Atkinson, G. F. Notes on the Cercospora of celery-blight. N. Y. (Cornell) Agr. Exp. Sta. Bull. 49: 314-316. 1892.

Duggar, B. M. Two destructive celery blights. N. Y. (Cornell) Agr. Exp. Sta. Bull. 132: 201-220. 1897.

Klotz, L. J. A study of the early-blight fungus, *Cercospora apii* Fres. Mich. Agr. Exp. Sta. Tech. Bull. 63: 1-43. 1923.

Krout, W. S. Report on diseases of celery. N. J. Agr. Exp. Sta. Ann. Rept. 37: 584-603. 1917.

Scribner, F. L. Celery leaf-blight. U. S. Dept. Agr. Ann. Rept. 1886: 117-120. 1887.

Sturgis, Wm. C. On the prevention of leaf-blight and leaf-spot of celery. Conn. Agr. Exp. Sta. Ann. Rept. 1897: 167-171. 1898.

Townsend, C. O. Notes on celery-blight. Md. Agr. Exp. Sta. Bull. 74: 167-182. 1901.

BACTERIAL BLIGHT OF CELERY

Caused by *Bacterium apii* Jagger

Several of the earlier investigators have mentioned a bacterial disease of celery, but their descriptions are either so meager or so mixed with references to the bacterial soft-rot that it is impossible to state definitely whether or not this particular disease was observed before Jagger of New York gave it his attention in 1909. It has since been reported as general and destructive in the above named state, and as present in Michigan, Indiana, and Minnesota. It may be even more widespread, but its similarity to early-blight and late-blight has probably caused it to be mistaken for these diseases. Apparently it attacks all varieties of celery, but is not known to affect any other host.

Symptoms.

The spots (Fig. 30) can be distinguished with certainty from the late-blight only by the absence of pycnidia, and from early-blight by not having the ashen-gray fruiting mold. Furthermore, the organism is confined almost wholly to the leaves. The spots first are noticeable by their turning yellow then changing to brown. The affected area may be circular in outline, but often is irregular in shape, and seldom is over 5 millimeters in diameter. A striking characteristic of many of the spots is a yellowish border or halo, although this symptom may be absent often enough to make the diagnosis uncertain when based on this point alone. When enough lesions appear on a single leaf it dies. The disease occasionally is severe enough to give a very pronounced blighted appearance to an entire field of celery.

Cause.

This blight is caused by *Bacterium apii* having the group number 211.2322033. The organism lives over winter in the diseased refuse left in the field or on the compost heap. In the spring it is disseminated by means of insects, tools, rains, and in any other manner by which a microscopic body may be moved from one place to another. In the presence of moisture the bacteria are able to enter the stomata, and in a few days cause visible symptoms. A warm rainy period is ideal for the inception of an epidemic. Plants may be attacked at any time from the seed-bed stage to maturity.

Control.

The recommendations given for the control of late-blight of celery are equally applicable to the bacterial blight. The cultivating should be done after the spraying so that plants are not infected from the spores in the turned-up soil.

REFERENCES

Dye, H. W. and A. G. Newhall. Spraying and dusting for the bacterial and late blights of celery in western New York. Phytopath. 12: 48. 1922.

Jagger, Ivan C. Bacterial leaf-spot disease of celery. Jour. Agr. Research 21: 185-188. 1921.

PHOMA ROOT-ROT OF CELERY

Caused by *Phoma apiicola* Klebahn

A leaf-spot of celery was described in New Jersey in 1891, and a root-rot of the same host was observed in Ohio in 1907, both of which were supposed to have been caused by an organism similar to if not the same as the one associated with a root-rot found in Michigan in 1914 and reported a few years earlier as present also in Germany, Holland, and France. Root-rots of celery have been mentioned in literature from other countries and from several sections of the United States, but because no exhaustive study was made of the pathogene in each case, it is not known whether the descriptions refer to the decay caused by Phoma or to some disease bearing a resemblance to the Phoma root-rot.

The disease is not of great importance but occasionally may become serious, particularly on Golden Self Blanching celery which is quite susceptible. Other varieties are not wholly immune. The disease has also been found on celeriac, and the fungus has been inoculated successfully into parsley, parsnip, carrot, and caraway.

Symptoms.

Under very moist conditions the parasite has been made to infect the leaves and floral parts, causing spots with pycnidia that might be mistaken for the Septoria late-blight were the shape of the spores not examined under the microscope. This stage of the disease, however, is not commonly found. The more characteristic symptoms of the parts of the plant above ground are the wilting and drooping of the leaves, beginning with the outer ones, and the general unhealthy appearance of the stems and foliage (Fig. 31).

The direct attack of the fungus is nearly always at the base of the stem or the crown of the root. The affected tissue turns bluish-green when first invaded, then gradually changes to dark brown or black. The epidermis at first remains intact, but as the spots enlarge it cracks, exposing the discolored tissue below and giving a corky or scabby appearance to the lesion. When the outer part alone is affected, the inner tissue of the root or stalk continues to grow until fissures are produced in the darkened surface which admit secondary organisms that cause soft-rots. The rot seldom extends down the roots far into the soil nor far up the stem. The major injury results at or just below the surface of the ground. The disease may produce only a black zone about the crown of the root without any further injury to the plant, or the entire crown may rot off. In the latter case the host may attempt to force a new top from below the affected portion.

FIG. 31.—Phoma root-rot of celery.

Cause.

Halsted mentioned *Phyllosticta apii* as the cause of a leaf-spot of celery, but his description was too meager for Klebahn to make sure that the root parasite was the same fungus; therefore, the latter furnished a technical description and named the fungus on the root *Phoma apiicola.*

As the pycnidia of the parasite form mostly underground, and the one-celled oval spores are discharged in long gelatinous tendrils, the wind has no effect on the dissemination of the pathogene. The spread of the disease in the field depends almost altogether on washing rains. It sometimes happens that seedlings pulled for transplanting are all placed together in a pan of water, and under such circumstances if one plant is affected the entire lot becomes inoculated. The spores are able to germinate and with their germ-tubes penetrate the host tissue within forty-eight hours. The mycelium grows rapidly both between and through the cells, causing the death of the invaded tissue. Any of this tissue left in the field may serve as a source of inoculum for the following season. The fungus does not often grow in or on the seed, but does so occasionally, and probably in this manner was brought from Europe to America.

The Phoma thrives best in the cool temperatures of spring and fall, while its growth is retarded during the hot summer months. The optimum temperature is between 60° and 70° F. and the maximum is slightly over 80° F. It has been found growing in temperatures that were almost low enough to freeze water.

Control of Phoma root-rot.

No wholly satisfactory methods of control have been devised, but the following recommendations are enumerated: two- or three-year rotations of crops, obtaining clean seed if that is possible, burning all available diseased refuse, guarding the seed-bed against infestation, removal of diseased seedlings, care in handling healthy seedlings so that they do not become inoculated, use of well drained soil both in the seed-bed and in the field, and if seedlings are grown in the green-house keeping the soil as dry as good growth will permit, and the temperature relatively high.

References

Bennet, C. W. A Phoma root-rot of celery. Mich. Agr. Exp. Sta. Tech. Bull. 53: 1-40. 1921.

Klebahn, H. Krankheiten des Selleries. Ztschr. f. Pflanzenkr. 20: 1-40. 1910.

Quanjer, H. M. and N. Slagter. De Roest- of Schurftziekte van de Selderieknol en enkele opmerkingen over andere Selderieziekten. Tijdschr. o. Plantenziekten 20: 13-27. 1914.

Yellows of Celery

Caused by *Fusarium* sp.

The yellows of celery was first observed in Michigan in 1914, and since then has been found also in Indiana, Ohio, Pennsylvania, New York, Massachusetts, Connecticut, and New Jersey. As yet it has not become of general importance in the above named states, but percentages of loss which have been recorded show that the disease at times may become epidemic in nature. In Indiana certain fields were observed where the reduction in yield was 10 per cent; in Pennsylvania, 5 to 25 per cent; and in New York one man lost a hundred thousand plants in one season because of yellows. Fortunately all the green varieties are resistant, only the Golden Self Blanching types being very susceptible.

Symptoms.

The term yellows describes admirably the symptoms of the disease. The plants are off-color, stunted, the tissues are brittle and bitter in taste. When young plants are affected they may die soon; older plants, however, live as long as normal individuals even though the diseased ones are much dwarfed. A longitudinal section of the stem shows the vascular tissue yellow or red beginning with the crown of the root and in severe cases extending as far as the leaf-veins. The roots, also, are affected.

Cause.

A Fusarium has been isolated from the diseased tissue and shown to be the causal organism. It lives in the soil, gradually becoming more abundant as celery is grown year after year in the same fields. Like many other Fusaria, it requires a relatively high soil temperature to cause severe injury on the host.

Control.

It is seldom possible to use fungicides in fighting a soil organism under field conditions. Resistant varieties have proved the only solution of the problem. With this fact in mind, the pathologists in Michigan selected plants of the Golden Self Blanching celery which appeared to be resistant to yellows. For several years the selections have been tested until at present a strain is available that has given 90 per cent of a crop in heavily infested soil. The seed has not yet been placed on the market, but no doubt soon will be.

References

Coons, G. H. and Ray Nelson. Celery yellows. Phytopath. **11**: 54-55. 1921.

Coons, G. H. and Ray Nelson. First progress report on "yellows" resistant Golden Self Blanching celery. Phytopath. **12**: 48. 1922.

Coons, G. H. and Ray Nelson. Yellows-resistant celery, 3rd progress report. Phytopath. **13**: 56. 1923.

Pink-Rot of Celery

Caused by *Sclerotinia libertiana* Lib.

(See Lettuce-Drop for discussion of the fungus, page 243.)

The pink-rot (Fig. 32) or watery soft-rot of celery frequently causes great loss in the field, in the storage-house,

FIG. 32.—Pink-rot destroying the top of a stored celery plant.

and even during transportation. The rot may begin when the seedlings are small, the disease then being known as damping-off. When the plants are older, a watery soft-rot attacks the base of the stem. When the plants are nearly mature or are being blanched preparatory for harvesting, incipient infections occur about the small growth cracks or insect stings on the stems and petioles. In the presence of moisture in the field, the plants may decay before they are harvested. Often the disease is not destructive until the celery is placed in the storage-room. The affected tissue first is slightly pink in color, then gradually changes to gray or brown. In severe cases every stalk in the crate becomes soft and watery. Whole carloads of the crop may thus be destroyed before it can be placed on the market.

Thorough spraying of the plants as suggested for the control of late-blight of celery and careful handling when harvesting will protect the crop against insect and mechanical injury, and therefore will make the plant more resistant to the attacks of the Sclerotinia. Celery is commonly grown on muck in rotation with lettuce, carrots, and other root-crops, all of which are very susceptible to the parasite and help to increase the amount of infestation in the soil. If onions, sweet-corn, cereals, grasses, or other immune crops can be grown occasionally, the inoculum will be much decreased and the next crop of celery correspondingly healthy.

As the fungus grows at low temperatures, it will be necessary to keep the storage-room fairly dry and at a temperature slightly below freezing. Celery stored for the winter may be decayed if the temperature is 33° F., but a similar lot subjected to 30.5° or 31° F. may remain in good condition.

References

Poole, R. F. Some recent investigations on the control of *Sclerotinia libertiana* in the green-house on the muck farms of Bergen county, New Jersey. Phytopath. **12**: 16-20. 1922.

Reddick, D. Decay of celery in storage. Phytopath. **4**: 45. 1914.

Winters, R. Y. Report of assistant in Botany. Fla. Agr. Exp. Sta. Ann. Rept. **1909**: LXXIX-LXXXV. 1910.

Black-Heart of Celery

Cause physiological

Under certain unfavorable conditions in the field, celery is susceptible to a malady known among growers as black-heart. The first symptom is a slight tip-burn on one of the inner younger leaves. This quickly spreads to the greater part, or possibly all, of the immature foliage, including the heart leaves. The browning or blacking progresses until the entire leaflet is dead, then continues down the petiole, shriveling and killing it. In the worst cases the entire heart is killed and blackened, and even many of the older leaves and petioles are badly damaged.

The disease is similar in its nature to tip-burn or sunscald, occurs only during the hot weather of midsummer, and probably is worse where plenty of moisture is not available. It has been proved, also, that the types of fertilizers used have a direct bearing on the amount of injury. Nitrate of soda and kainit augment the trouble; dried blood, fish-scrap, and bone-meal decrease it.

No definite control measures are available. Care in cultivation, choosing a suitable type of soil, and the application of prohibiting fertilizers are precautions that may well be practiced. Hearst suggests that nicotine added to the bordeaux mixture sprays kills the tarnished plant-bugs, which he believes are responsible for the beginning of black-heart.

References

Beach, S. A. Center blight. N. Y. (Geneva) Agr. Exp. Sta. Bull. 51: 134-135. 1893.

Halsted, B. D. A bacterial disease of celery. N. J. Agr. Exp. Sta. Bull 2: 10-12. 1892.

Hearst, W. H. Black-heart of celery. Ont. Dept. Agr. Ann. Rept. 1917: pt. 1. 61. 1918.

Kinney, L. F. Black-heart, blights and blast. R. I. Agr. Exp. Sta. Bull. 44: 19-22. 1897.

Winters, R. Y. Influence of fertilizer on black-heart. Fla. Agr. Exp. Sta. Ann. Rept. 1908: XCIX-CIII. 1909.

Mosaic of Celery

Cause undetermined

It seems that nearly every species of plant is afflicted by mosaic, and celery is no exception to the rule. The exact distribution of this disease is not known, since many observers are not yet acquainted with the symptoms; furthermore, the appearance of the trouble is similar to that caused by drought injury. No doubt it is present to a slight extent in most of the larger celery districts, at least where aphids that feed on this host are prevalent.

The affected plant stands stiffly upright and for this reason is rather conspicuous even though there is no marked discoloration of the foliage. The air of erectness is due to the fact that the petioles do not spread outward, but grow upward almost tightly against the stem. This brings the leaves closely together and causes the diseased top to have a bushy effect. The leaves may become filiform in shape like that described for tomato mosaic, or may shrivel and show blister-like pustules that do not burst. In addition to these symptoms the tissues of the vascular bundles become so hard and brittle that by a wrench of the hand the whole plant may readily be snapped off near its base.

The virus of the disease may be transmitted from diseased to healthy plants by aphids, and possibly in any other way by which affected sap can be transferred from plant to plant.

The disease has not yet become serious enough to demand control measures. It is of interest in this connection to know that some varieties have proved more nearly immune than others, and might be made the parent stock for breeding resistant strains if the mosaic ever becomes serious enough to require attention.

Reference

Poole, R. F. Celery mosaic. Phytopath. 12: 151-154. 1922.

Rust of Celery

Caused by *Puccinia bullata* (Pers.) Winter

A rust on umbelliferous plants is rather common both in America and Europe, and as celery belongs to this group it may be attacked by the same fungus. The disease, however, is not of importance on the latter host except in rare cases. The trouble may be recognized by the yellowish rust pustules, which resemble somewhat the well known rust pustules on cereal plants. The fungus has been given many specific names among which are *Puccinia bullata, Puccinia apii* Corda, and *Puccinia castagneii* Thüm. Since it causes no economic loss control measures have never been formulated.

Reference

Halsted, B. D. Some fungous diseases of celery. N. J. Agr. Exp. Sta. Ann. Rept. 12: 250-259. 1891.

Damping-off of Celery

Caused by various fungi

(See Damping-off of Tomatoes, page 546.)

Root-Knot of Celery

Caused by *Heterodera radicicola* (Greef) Müll.

(See Root-Knot of Tomato, page 550.)

Bacterial Soft-Rot of Celery

Caused by *Bacillus carotovorus* Jones

(See Bacterial Soft-Rot of Carrots, page 76.)

Gray Mold-Rot of Celery

Caused by *Botrytis* sp.

(See Gray Mold-Rot of Lettuce, page 253.)

Rhizoctonia Root-Rot of Celery

Caused by *Rhizoctonia* sp.

(See Bottom-Rot of Lettuce, page 249.)

CHAPTER V

DISEASES OF SWEET-CORN

THERE are many diseases of corn, some of which are very destructive. Nevertheless a large share of the pathology work connected with this crop has to do with field corn and is not concerned with the garden varieties. For this reason only the three most prominent troubles of sweet-corn are discussed here. They are the only ones which the gardener will find at all bothersome.

CORN SMUT

Caused by *Ustilago zeæ* (Beckm.) Unger

The smut of corn probably was present when the white man first came to America. It was not described, however, until 1754. This was due partly to the fact that the crop was considered unimportant, as is shown in the case of an early author who apologizes for discussing such a lowly valueless plant. It is now present in nearly all countries where corn is grown, being of great economic importance in North America. Although the disease affects all varieties of corn, it is most injurious to sweet-corn. The average loss is estimated at 1 to 3 per cent, and individual fields of sweet-corn have been observed where more than half of the seedlings were affected as shown in Fig. 33, or where 40 per cent of the ears were disfigured.

Symptoms.

The corn plant may be infected at any time in the early stages of its development, but gradually grows less susceptible after the formation of the ear. Any part of the plant above ground may be invaded, although it is more common to see smut boils on the ears (Fig. 34), tassels, and nodes than on the leaves, internodes, and aerial roots. The boils are similar in appearance on all the host parts, but are larger on the ear and nodes where more food is available for the fungus. They are also very prominent on the succulent seedling. The boil is composed of a white smooth covering inclosing a great mass, sometimes four or five inches in diameter, of black greasy or powdery spores. After the maturity of the spores, the covering becomes dry and brittle, breaks open and permits the black powdery contents to escape. On the tassel the galls are made up of much swollen and distorted stamens, often giving the male flowers the appearance of a group of distorted fingers. The boils on the stem have no definite shape. In their original form they are inclined to be globose until the unequal pressure

FIG. 33.—A seedling plant of sweet-corn being killed by smut.

of surrounding galls changes them into all manner of shapes. On the leaf-blades the boils are small and firm and frequently do not break open. The ears ordinarily are most conspicuously affected. The silks and kernels are invaded, the cob never being injured. Ordinarily not all the kernels are malformed, yet those that do become diseased are so swollen that the husks burst open exposing the great white galls.

FIG. 34.—Smut boils replacing kernels of corn.

Cause.

Notwithstanding the fact that the early botanists always found fungous spores in the boils, they attributed the cause to excessive sap of the corn plant, too much moisture, or too large amounts of manure. Later the big boils were considered as puff-balls. It was nearly a hundred years after the disease was described before the causal fungus, *Ustilago zeæ,* was studied and named. It is closely related to the smut organisms on cereals, and its life history is much the same. However, there is one fundamental difference; the corn parasite is never carried in the seed. The boils are composed of innumerable brown roughened spores inclosed in a white membrane resulting from the stimulation of the host tissues to abnormal growth. When the membrane is broken, the spores sift to the ground or are blown long distances. The galls on ear and fodder also find their way into the manure where the spores have a means for dissemination and after reaching the field have an ideal medium in which to germinate.

The large brown spores do not cause infection directly, but on germination produce a septate basidium on which are borne a large number of basidiospores. The latter may further increase in number by budding as do yeast cells. In this manner one gall may furnish enough inoculum, if rightly distributed, to infest an entire field. In a section where corn is grown generally, the basidiospores are so abundant that every plant is sprinkled with them. Many lodge on the tassels, among the silks, and especially between the leaf-sheath and stem. When water is present, the spores germinate, the germ-tube entering the host tissue. It is doubted by some workers whether the parasite can pass through uninjured tissue, except through the corn silk into the kernel. It frequently has been observed after a heavy wind in which the plants have been damaged or after a hail storm that the amount of infection was increased greatly. The stimulation of the parenchyma and epidermal cells begins at once after the entrance of the parasite, which proceeds to form an enormous mass of knotted mycelium. The spores are formed within these knotted and swollen branches of the hyphæ, the walls of which later become gelatinous and finally disintegrate, leaving only the mature reproductive bodies.

Control of corn smut.

Obviously seed treatment is of no value. Eradication in the field is the only direct remedy. Furthermore, it is a community matter, for if one grower attempts to destroy the smut boils in his field and his neighbor continues to harbor the inoculum, the time spent in eradicating the fungus is wasted. If the whole community acts in concert, the smut may be controlled. It will mean an inspection of all the fields at least three times during the season. This will cost approximately a dollar an acre, or more if the disease is severe. A man with a large basket and corn-knife can inspect two rows at a time. When smut is found, the diseased part of the plant is carefully cut off and placed in the basket. The cutting, to be effective, is made before the boils burst, and the diseased refuse is destroyed at once. This may be done by throwing the plants into the fire, or in boiling water, or burying them deeply in the soil. All the smut cannot be eliminated in one year, still it is remarkable how greatly it can be reduced by the work of one season.

In addition to eradication, it is well to take other precautions as care against infesting manure, avoiding the use of contaminated manure in fields where corn is to be planted, and practicing long rotations of crops. Fortunately, no other host is susceptible.

References

Arthur, J. C. and W. Stuart. Corn-smut. Ind. Agr. Exp. Sta. Ann. Rept. **12**: 84-135. 1900.

Clinton, G. P. Smuts of Illinois agricultural plants. Ill. Agr. Exp. Sta. Bull. 57: 321-335. 1900.

Güssow, T. Smut diseases of cultivated plants: their cause and control. Canada Cent. Exp. Farm (Ottawa) Bull. 73: 37-40. 1913.

Hartman, E. C. Spore germinations in the cereal smuts. Minn. Agr. Exp. Sta. Tech. Bull. 133: 34-38. 1913.

Hitchcock, A. S. and J. B. S. Norton. Corn smut. Kan. Agr. Exp. Sta. Bull. 62: 169-212. 1896.

Knowles, E. C. A study of the abnormal structures induced by *Ustilago zea-mays*. Jour. Myc. 5: 14-18. 1889.

MacMillan, H. G. An epidemic of corn smut following hail. Phytopath. **8**: 584, 585. 1918.

Piemeisel, F. G. Factors affecting the parasitism of *Ustilago zeæ*. Phytopath. **7**: 294-307. 1917.

Root-, Stalk-, and Ear-Rot of Corn

Caused by various fungi

The ear-rots of corn have been observed for many years, yet only comparatively recently has this trouble been associated in any way with seedling wilt, root-rot, and stalk-injury. The trouble may be found on all varieties of field-corn, sweet-corn, flint-corn, and pop-corn, the first two being the most susceptible. The loss results from poor stands caused by diseased seed, from stunting in growth of many plants, from barren stalks, from prostrate stalks, from poorly filled ears, from irregularly matured seed, and from rotting of the ears. The disease probably is present wherever corn is grown, being of greatest economic importance in the corn-belt, and particularly where wheat and corn are grown in continuous rotations.

Symptoms.

All parts of the plant may be infected, and during any time of its life cycle. When diseased seeds are planted, they may not grow, or if they do sprout produce seedlings that are weak, yellow in color, often with a reddish rot on the roots and a dark-colored decay at the base of the stem. In severe cases the young plant dies. When the corn stem attains a height of two or three feet, the difference between the healthy and diseased individuals frequently is very striking. The healthy stalks are taller than the neighboring stunted diseased ones, which are yellow and spindling as if the soil lacked nutrient material. In poor soil, however, the weak corn is usually in definite areas in the field, while the root-rotted plants are scattered promiscuously. They may also be recognized by the dying of the tips of the leaves, and their wilting on hot days. The inside of the base of the stalk is discolored and more or less rotted, although the outside may appear healthy.

The injured plants sometimes grow so slowly that they do not come into silk and tassel until after the pollen has been

FIG. 35.—Poorly pollinated sweet-corn on stalks that were retarded in growth because of the root-rot.

shed from the healthy stalks; therefore, the ears are not always pollinated fully and thus do not bear well-filled ears (Fig. 35). The roots are weakened so that the added weight of the ear causes the stalk to lean or even fall prostrate in a brisk wind. The stalks and ear-shanks are so brittle that they may be broken with only slight effort. The shanks are often not strong enough to bear the weight of the ear, and on breaking permit the latter to hang straight down instead of lopping over in a graceful curve or standing nearly upright as is true of those on a healthy plant. If such a shank is broken off near the cob, the fibers separate unevenly causing a shredded end in contrast to the smooth surface of the shank stub when a normal ear is broken out of its husks.

FIG. 36.—*Fusarium moniliforme* causing rot of sweet-corn kernels.

The diseased ears may be partly or wholly rotted (Figs. 36,

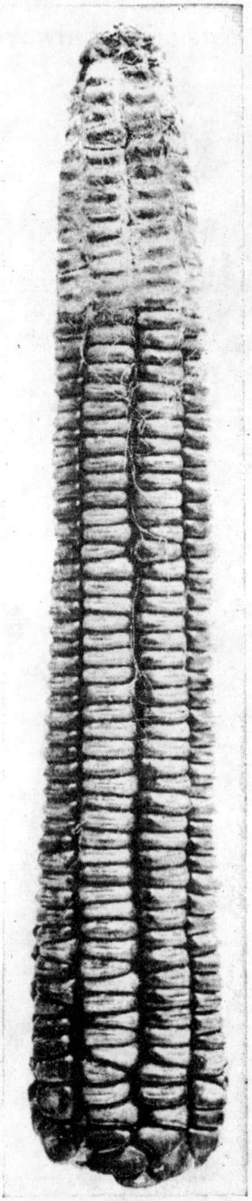

FIG. 37.—Ear-rot of corn.

37), and disfigured by white or pink mold, or so immature that they remain soggy when placed in storage. Such immature kernels, when dried, wrinkle more and have less of a luster than do the healthy ones. Some diseased stalks bear ears that appear normal. Occasionally such infected corn can be detected by cutting across the shank near the base of the cob. The cut surface may show the discoloration of the tissue. In other cases the only sure method of making a diagnosis is in germinating the seed in a specially constructed germinator.

Cause.

The fungi associated with the corn trouble are so closely connected with the environmental conditions that it is difficult to determine whether they or the disposing factors are most important. The fact remains that when the weather and soil conditions are favorable for infection, the weak parasites living in the soil may take advantage by entering the host tissue. Hoffer has explained the relationship of environment and parasite by showing that the corn plant, when grown in unfavorable soil, will collect in its vascular tissue, particularly at the lower nodes, abnormal amounts of metals, such as iron (Fig. 38) and aluminum, which do not permit free circulation of sap, thereby weakening the roots until they are susceptible to the attacks of various organisms.

Among the fungi that cause one or more phases of the root-, stalk-, and ear-rot are *Gibberella saubinetii* (Mont.) Sacc., *Fusarium moniliforme* Sheldon, *Diplodia zeæ* (Schw.) Lev., and *Cephalosporium sacchari*. *Gibberella saubinetii* also causes

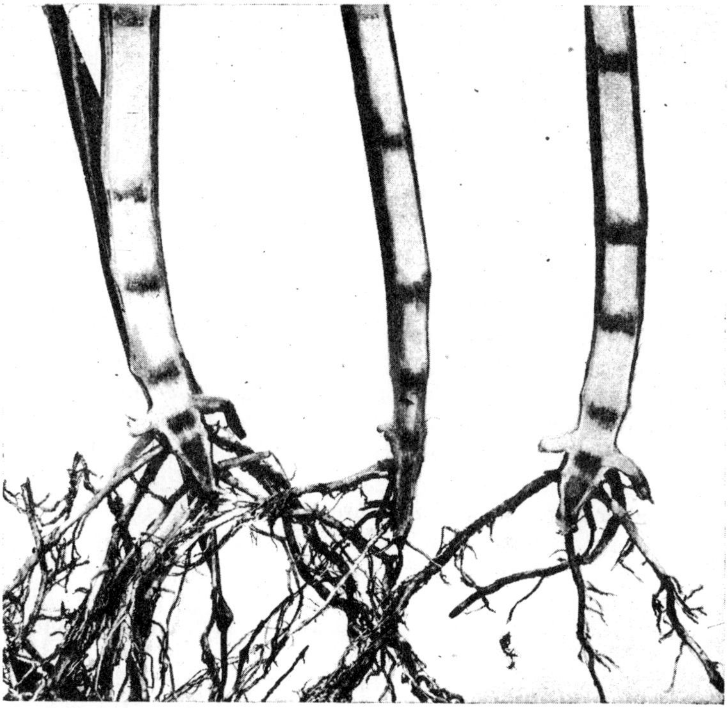

FIG. 38.—Corn root- and stalk-rot. Iron accumulation in the nodal plate tissues and decay of roots.

wheat-scab. The four fungi are carried on or in the seed and in the soil. They may be disseminated also by wind, water, and insects. In the late autumn and early spring the fungi grow in profusion on corn-stalks left in the field, it hardly being possible to find a stalk which does not show about the nodes the black fruit-bodies of either Diplodia or Gibberella. When a diseased seed is planted, the seedling may be infected directly from the mother kernel, or the fungus grows into the soil and penetrates the weakened roots. The longer wheat or corn is grown in the same soil, the more heavily the soil will be infested, although it is possible for virgin soil to contain one or more of the root-rot organisms.

The optimum temperature for all the different corn parasites has not been determined. The Gibberella grows best at 76° F. Sunlight is injurious to its spores when exposures are made during all the winter months. The four parasites are favored by the presence of moisture.

Control.

Good cultural methods are the first requisite in the control of the root-, stalk-, and ear-rot. If the soil is supplied with sufficient humus, lime, and fertilizer, and well drained, the plant will grow rapidly enough to avoid much of the trouble. Long rotations of crops, being careful not to plant wheat and corn in succession, will keep the amount of the inoculum at a minimum.

The principal control measure consists in obtaining healthy seed. This is much more complicated than the mere statement might lead one to believe. The selection is begun in the fall before the first heavy frost. The seed is chosen from sturdy upright stalks with firm shanks, well covered, healthy appearing, ripened ears. It is advisable to harvest two or three times as much seed as is required, for a large part of it may be discarded in later tests. After the selection, the ears are placed in a dry well ventilated storage. This is of particular importance inasmuch as the parasites may continue to infect and injure corn as long as it contains any appreciable amount of moisture. In the spring the ears are sorted again, using only those that are of the correct type and show a healthy condition after the winter storage.

The final test is made in a germinator, which may be built

by boring holes into the bottoms of two wooden boxes, one of which is three or four inches smaller in each dimension than the other. and which is at least eighteen inches in depth. Wire cross rods are placed at three-inch intervals each way in the upper part of the smaller box. A layer of sawdust is put in the bottom of the larger box and the other box set on this, after which the space between the two is filled with sawdust. When the germinator is in use, the sawdust is kept thoroughly wet.

The modified rag-doll used in such a germinator is made by stretching on a table ordinary glazed wrapping-paper about sixteen inches wide and at least sixty inches long. On this is placed a muslin cloth that has been boiled in water, and which measures twelve by fifty-four inches. Eight kernels are removed from different parts of each ear and put in a row with the germ sides down. Their location is given a number that will correspond with that on the ear. The tip of each kernel is so placed that it will point toward the lower end of the rag-doll when the latter is set in the box. After the muslin is thoroughly wet, preferably with water that has been boiled and cooled, the paper, cloth, and corn are carefully rolled up, the top marked with a pencil, rubber bands placed around each end, a tag attached bearing ear numbers and date, and the whole stood upright in the germinator. The temperature is kept at 75° to 85° F., and the dolls wet twice a day with water that has previously been boiled and cooled.

The dolls are ready for examination after seven to ten days. The paper and cloth are unrolled carefully, and note taken not only of those kernels that failed to germinate, but also those that germinated weakly, or of those whose sprouts are rotted. The ears with corresponding numbers are discarded. Only that seed which has germinated vigorously and sprouts of which are free from rot are retained for planting in the seed-plot or the field.

Not every one growing sweet-corn can spare so much effort in producing healthy seed. One can, however, be careful in procuring seed from the most reliable sources, and from the grower who has made an effort to eliminate the root-, stalk-, and ear-rot.

References

Burrill, T. J. and J. T. Barrett. Ear rots of corn. Ill. Agr. Exp. Sta. Bull. 133: 64-109. 1909.

Duddleston, B. H. The modified rag-doll and germinator box. Ind. Agr. Exp. Sta. Bull. 236: 1-12. 1920.

Hoffer, G. N. Disease-free sweet-corn. Ind. Agr. Exp. Sta. Bull. 233: 1-12.

Hoffer, G. N. and R. H. Carr. Accumulation of aluminum and iron compounds in corn plants and its probable relation to root-rots. Jour. Agr. Research. 23: 801-824. 1923.

Hoffer, G. N. and J. R. Holbert. Selection of disease-free seed corn. Ind. Agr. Exp. Sta. Bull. 224: 1-13. 1918.

Holbert, J. R. and G. N. Hoffer. Control of the root, stalk, and ear-rot diseases of corn. U. S. Dept. Agr. Farmers' Bull. 1176: 1-24. 1920.

Hyslop, J. A. *Triphleps insiduosus* as a probable transmitter of corn ear-rot (Diplodia sp. Fusarium sp.). Jour. Econ. Ent. 4: 435-438. 1916.

Manns, T. F. and J. F. Adams. Corn root-rot diseases. Del. Agr. Exp. St. Bull. 128: 1-24. 1921.

Pammel, L. H., C. M. King and J. L. Seal. Studies on a Fusarium disease of corn and sorghum. Iowa Agr. Exp. Sta. Res. Bull. 33: 115-136. 1916.

Sheldon, J. L. A corn mold. Nebr. Agr. Exp. Sta. Ann. Rept. 1903: 23-32. 1904.

Taubenhaus, J. J. A study of the black and the yellow molds of ear corn. Texas Agr. Exp. Sta. Bull. 270: 1-38. 1920.

Valleau, W. D. Seed corn infection with *Fusarium moniliforme* and its relation to root and stalk rots. Ky. Agr. Exp. Sta. Res. Bull. 226: 27-50. 1920.

Stewart's Bacterial Wilt of Sweet-Corn

Caused by *Aplanobacter stewarti* (EFS) McC.

The bacterial disease of corn has been found only in the United States. It was first observed on Long Island in 1894. Since then it has been reported from twenty states, most of which are in the latitude of Long Island or farther south.

Where it does occur it usually causes a large percentage of loss; its control, however, is so simple that the total amount of injury in the whole country is very slight. It is most destructive on sweet-corn, but may also attack field-corn and pop-corn. The earlier varieties of sweet-corn, like First of All and Golden Bantam, are very susceptible, while the later maturing ones, such as Stowell's Evergreen and Country Gentleman, show much less disease. Among the varieties of field-corn, the early flint sorts receive the greatest injury.

Symptoms.

The disease may appear at any time in the growth of the plant, usually not being conspicuous until the seedlings are one or two feet tall. The affected plants may then be stunted or so severely injured that they die. In larger plants, the tassel forms prematurely, dies, and in drying turns white. The death of the tassel is accompanied or followed by the wilting and drying of the leaves as if injured by frost. The whole plant may finally wilt and succumb or linger during most of the summer. In rare cases the plant seemingly recovers.

When the stem is cut open and left exposed to the air for a few minutes, a chrome-yellow exudate collects at the open ends of the vascular ducts. A similar exudate is often observed on the inner side of the husks next to the kernels and occasionally on the inner side of the leaf-sheath. The vascular bundles turn yellow or brown. Whitish diseased spots bordered by dark zones appear on the kernels (Fig. 39). No soft-rot is produced in any part of the plant. Apparently the roots are not invaded.

Cause.

The parasite was first named *Bacterium stewarti;* later investigations found the organism to be non-motile, therefore, it is now classified as *Aplanobacter stewarti.* It is a yellow, rod-shaped, non-flagellate, non-liquefying bacterium. The attempts that have been made to procure infection from contaminated soil have resulted negatively. The parasite is carried in and on the seed, and from there is able to enter the water-pores of the very small seedling. In the field it is carried from plant to plant by insects. The slightly diseased ears may be gathered for seed, and in this manner the disease is propagated year after year.

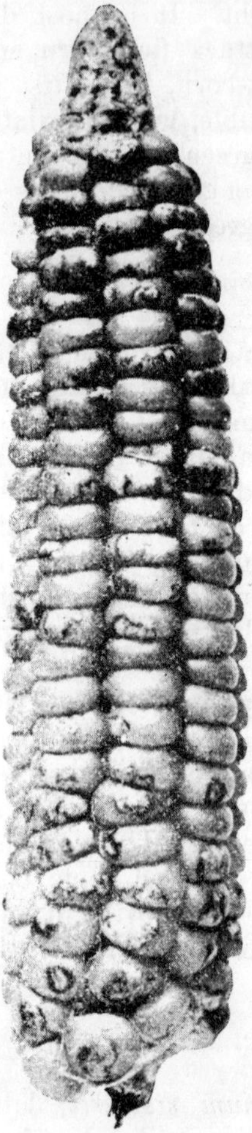

Fig. 39. — Stewart's sweet-corn disease causing spotting of kernels.

Whenever the rainfall is heavy during or following the planting period, the crop has more chances for infection than when the weather is dry. High temperatures are necessary for the development of the disease. Infected seed sown at the same time in Maine and in Maryland resulted in infection of the corn in the latter state, but none in the former. Light sandy soils are less favorable than rich loam in aggravating the trouble. In other words, any environment that hastens germination in the soil aids the bacterium in migrating from the seed to the water-pores of the seedlings.

Control.

Since the organism does not live over in the soil, the only requirement for disease control is the use of healthy seed. This may be procured from a locality far enough north to avoid the bacterial trouble. If southern grown seed is sown, the ears should be selected carefully and if there is any danger of the trouble being present the seed should be treated.

The treating may be accomplished by dry heating the kernels for one hour at 140° to 158° F. or soaking in a corrosive sublimate solution (1 ounce in 7.5 gallons of water) for fifteen minutes, then rinsing thoroughly and drying or planting at once.

References

McCulloch, Lucia. A morphological and cultural note on the organism causing Stewart's disease of sweet corn. Phytopath. **8**: 440, 441. 1918.

Rand, F. V. and L. C. Cash. Stewart's disease of corn. Jour. Agr. Research **21**: 263-264. 1921.

Rand, F. V. and L. C. Cash. Further evidence of insect dissemination of bacterial wilt of corn. Science **59**: 67-69. 1923.

Smith, E. F. Stewart's disease of sweet-corn (maize). *In* Bacteria in Relation to Plant Diseases **3**: 89-147 1914.

Stewart, F. C. A bacterial disease of sweet-corn. N. Y. (Geneva) Agr. Exp. Sta. Bull. 130: 423-439. 1897.

CHAPTER VI

DISEASES OF CRUCIFERS

SINCE most growers are acquainted with the term crucifers and know that it includes all the vegetables of the mustard family, it was thought best to treat the diseases of this group in one place rather than to string them along alphabetically, having brussels sprouts, cabbage, and cauliflower near the beginning of the Manual while radishes and turnips were near the end. Most of the diseases of one crucifer host are found on the other related vegetables.

Cabbage and cauliflower are two important vegetables with an annual value in the United States of more than twenty million dollars. If there were no diseases to reduce the yields or to make the crops unsalable, there would be at least a 10 per cent increase in value or more than two million dollars, all of which is destroyed annually by the various troubles. Not all of this loss can be avoided, yet more than a million dollars might be saved if proper precautions were always taken against diseases.

BLACK-LEG OF CABBAGE

Caused by *Phoma lingam* (Tode) Desmazieres

Black-leg, also known as foot-rot and Phoma wilt, was first recorded in France in 1849. Since that time it has spread to nearly every part of the world where cabbage or cauliflower is grown. The first report of its presence on this side of the Atlantic was in Ohio in 1911. It is now common in all of Europe, in Australia, and in most parts of the United States.

The fungus seems to attack all varieties of cabbage as well as many other plants of the mustard family, such as cauliflower, brussels sprouts, rutabaga, kohlrabi, kale, rape, turnip, Chinese mustard, black mustard, charlock, radish, garden cress, pepper-grass, stock, sweet alyssum, tumble mustard, and hedge mustard.

Black-leg causes extremely large losses to cabbage and cauliflower growers (Fig. 40). It is not uncommon to find 50

FIG. 40.—Showing effect of planting cabbage from bed infested with black-leg. The cabbage at the right was taken from a clean bed.

to 90 per cent of a crop destroyed in a season favorable for the organism. It is especially prevalent in the trucking section of tidewater Virginia, on Long Island, and in Ohio.

Symptoms.

The plants may become infected in the seed-bed or any time later in the summer. Usually the first symptom is an

FIG. 41.—Black-leg spot on cabbage leaf showing pycnidia.

oval, depressed, light-brown canker near the base of the stem. The canker enlarges until the stem is girdled. This is the foot-rot stage in which the disintegrating tissue turns black and to which is applied the term black-leg. Circular light brown spots may also appear on the leaves (Fig. 41). Similar lesions, slenderly elliptical in shape, appear on the seed-stalks (Fig. 42) and pods of seed plants. Soon after the cankers on the stem and seed-stalk and the spots on the leaves begin to form, the lesions are marked with numerous black dots, the fruit-bodies of the parasite. The presence of these fruit-bodies is an important diagnostic symptom since no other cabbage disease produces lesions with similar distinctive markings. When the plants are badly diseased they may wilt or the edges of the leaves may occasionally turn a bluish-red in color. When the stem and roots decay late in the season the weight of the head causes the plant to lean or fall over. This has caused the disease to be called "drop" in some countries. Usually the affected plant soon dies, but if the soil is damp, adventitious roots may arise from the stem above the canker, permitting the plant to survive and even to form a head.

FIG. 42.—Black-leg rot with pycnidia on cabbage flower-pedicels.

Cause of black-leg.

The fungus was first described on dead cabbage stems in 1791. Later it was found causing a disease and named *Phoma lingam.* It passes the winter as mycelium in the seed and as spores and mycelium on the seed and with cabbage refuse in the soil. A small canker on the stem of the mother seed plant, unobserved at harvesting, enlarges in storage so that in the second season when the seed is developing, the mycelium grows upward through the plant or forms pycnidia with pycnospores for dissemination. When it grows up through the seed-pod stem or penetrates from without the seed-pod it enters the seed. When infected seed is planted, the cotyledons, pushed above the soil by the stem, serve the parasite as a place for fruiting. Infection also occurs at the base of the new stem from the mycelium harbored under the seed-coat. Dissemination and inoculation take place rapidly. A few infected seeds may be the source of an epidemic later. A common method by which this is brought about is the practice of pulling up a large number of young plants and placing them together in water. If spores are present every seedling may become contaminated under such conditions.

At any place in the lesions on the plant masses of hyphæ grow just below the epidermis until they form pycnidia. The pycnidia are flask-shaped with thick dark walls. Later the upper part elongates forming a short-necked opening or ostiole through which the spores are discharged in long, snake-like, rose-colored tendrils. Pycnospores continue to form as long as the host tissue supplies sufficient nutriment. The fungus remains alive in the soil as long as any particles of cabbage stalks or leaves are left intact. It is easily disseminated with soil or manure, containing cabbage refuse. Heavy rains, which cause washing of the soil or result in rivulets of water, carry the spores over large areas of a field or from one field to another. Cultivating tools and insects parasitic on cabbage also aid in the dissemination.

Another fungus occurring in Europe and known as *Phoma napobrassicæ* Rost. is now considered distinct from the black-leg organism.

The Phoma is very susceptible to environmental conditions. In the Puget Sound district where the rainfall from May to July is extremely light there is very little, if any, of the disease present. In the eastern states it may be rare for several seasons, then become epidemic. The severity of the disease is in direct proportion to the amount of rainfall in the early summer.

Control of black-leg.

Since the fungus lives as long as cabbage refuse remains in the field, it is necessary to practice at least three-year rotations. The diseased stalks and leaves should not be fed to cattle, and great care should be exercised in preventing the spread of such cabbage refuse to other parts of the farm. If one field is diseased and another healthy, different sets of tools should be used in cultivating the soil of each, for the fungus may be carried readily from one place to another. No seedlings should be transplanted from a seed-bed which shows even the slightest infestation. If possible the origin of the seed should be known and only that which is disease-free used. In the past such a practice has not been feasible because the seed-growers were careless in their selections. It is now possible to obtain clean cabbage seed from the Puget Sound section where the disease does not normally occur. A number of the eastern seed companies also are exerting every effort to grow cabbage seed free from Phoma, so that as time goes by it will become easier to procure clean strains of cabbage.

In the meanwhile the seed should be treated. The method of treatment generally advocated is the use of corrosive sublimate at the rate of one tablet to a pint of water, or an ounce of the powder in seven and one-half gallons of water. Only wooden, earthenware, or glass containers should be used. After the powder is well dissolved in a little hot water and enough cold water added to make up the required amount, the seeds tied loosely in cheese-cloth bags are placed in the solution for twenty to thirty minutes. Not over a quarter of a pound of seed should be placed in each bag, thus insuring a ready penetration of the disinfectant. After the treatment the seeds are rinsed thoroughly in clean water, then spread out to dry. It is unsafe to do the drying over a hot stove or in hot sunlight.

The corrosive sublimate will kill any part of the parasite, which adheres to the outside of the cabbage seed-coat, but will not kill the mycelium within the seed. Therefore, a more drastic measure is necessary if black-leg is to be eradicated entirely. A large tub or washboiler is partly filled with water heated to 122° F. The bags of seeds are placed in the water for thirty minutes, the water and bags being stirred constantly with a thermometer. As soon as the temperature begins to lower, a small amount of boiling water is added slowly so that the temperature is kept constant throughout the treatment. At the end of thirty minutes the bags are removed, dipped in cold water to cool, and the seed spread out to dry. This treatment will reduce the germination considerably, especially if the seed is more than one year old, but a large loss is justified in eliminating an organism which may destroy the entire crop. Inasmuch as some lots of cabbage seed are more susceptible to heat injury than others, it is better to test a sample of the seed before treating the entire amount.

References

Henderson, M. P. The black-leg disease of cabbage caused by *Phoma lingam* (Tode) Desmaz. Phytopath. 8: 379-431. 1918.

Manns, Thos. F. Black-leg or foot-rot of cabbage. Ohio Agr. Exp. Sta. Bull. 228: 276-297. 1911.

Ritzema Bos, J. "Krebstrünke" und "Fallsucht" bei den Kohlpflanzen, verursacht von *Phoma oleracea* Sacc. Ztschr. f. Pflanzenkr. 16: 257-276. 1906.

Ritzema Bos, J. and H. M. Quanjer. Het Langendijker Koolziektenvraagstuk. Tijdschr. Plantenziekten 16: 101-148. 1911.

Walker, J. C. Seed treatment and rainfall in relation to the control of cabbage black-leg. U. S. Dept. Agr. Bull. 1029: 1-27. 1922.

Walker, J. C. The hot water treatment of cabbage seed. Phytopath. 13: 251-253. 1923.

Black-Rot of Cabbage

Caused by *Bacterium campestre* (Pammel) EFS.

Black-rot of cabbage is an extremely destructive disease which was first observed about 1890. It evidently was present long before that time, for soon after it was mentioned by Garman in Kentucky it was found in many other states as well as in European countries. It has since been reported from nearly every country where cabbage is grown commercially. The disease not only affects cabbage, but is found on nearly all related species of the crucifers, such as turnip, cauliflower, kale, rape, radish, and other economic plants and weeds.

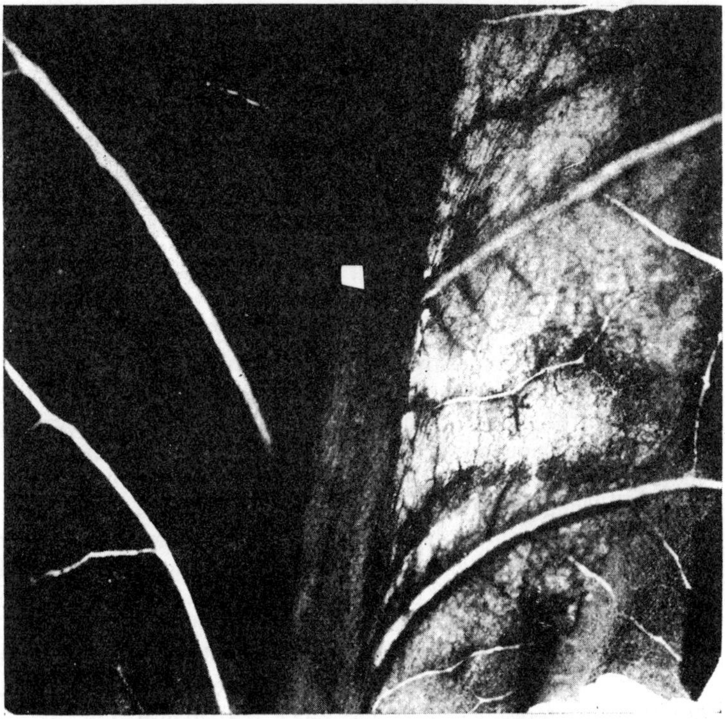

FIG. 43.—Darkened veins on cabbage leaf caused by black-rot.

Symptoms.

The host may be affected at any time during its growth from the youngest seedling until the time of harvesting. On the very small plant a blackening of the tissue appears along the margin of the sinus of the cotyledon, which later shrivels and drops off. Similar infections occur on the margins of the leaves. At first only a small wilted V-shaped area is noticeable. As the diseased area enlarges the veins show a black or dark brown discoloration (Fig. 43). The entire leaf may become yellow and drop off. Occasionally it is possible to find affected plants with a long bare stalk having only a small tuft of leaves at the top. If the petiole of the affected leaf or the diseased stem is cut open, the vascular bundles will appear as darkened areas. The black-rot alone does not cause any disagreeable odor, but when the disease is followed by soft-rot the smell is often offensive. With the advent of the soft-rot the head of the plant may drop off or become soft and watery. This stage is known as stump-rot.

Cause.

Black-rot is caused by the one-flagellate rod-shaped organism, *Bacterium campestre*. When grown in colonies, the bacterium is light yellow. It lives over winter on the seed-coat and in the refuse left in the field, particularly in the surviving plants of cabbage and brussels sprouts, as well as in the stored mother seed plant. In the spring when the young plants have broken through the ground, the bacteria on the broken seed-coat come in touch with the expanding cotyledons, and easily gain entrance through the stomata. They pass downward through the water-ducts until they reach the stem, from which they migrate to the roots below and to the leaves above. This infection takes place so early in the life of the seedling that for a long time the organism was supposed to gain entrance through the rootlets. This has since been disproved. Later infections take place along the margins of the leaves, the bacteria being splashed from the soil, or the leaves blown down so that they touch the soil and offer an opportunity for the parasite to gain an entrance through the relatively large water-pores. When the very young plant is attacked it frequently is killed rapidly, but infections on old plants may permit the latter to survive throughout the season even though no head is formed or is so marred that it cannot be marketed.

When the mother seed plants are affected, many of the bacteria remain in the harvested stalk, which is dried and threshed. In the threshing process the bacteria float with the dust and are caught on the oily coats of the seed where they remain over winter. During the growing season the organism may be disseminated by various insects, but the splashing and washing of rain are the most important methods of conveyance.

The parasite thrives best in high temperatures in the presence of plenty of moisture.

Control of black-rot.

The results in the control of black-rot have been so encouraging that it seems justifiable to predict that in the coming years this disease will not be of importance. Particularly is this true if the seed-producers will take every precaution to eliminate it from growing stock. Aside from the obtaining of clean seed it is necessary to rotate crops, eradicate cruciferous weeds, use well drained soil, handle the seed-bed in the manner outlined for the control of club-root, and never use seedlings from any seed-bed where even one diseased plant is found. In order to make sure that the seed is clean, it should always be treated with corrosive sublimate or with hot water as suggested for the control of black-leg of cabbage.

References

Drechsler, Charles. Cotyledon infection of cabbage seedlings by *Pseudomonas campestris*. Phytopath. 9: 275-282. 1919.

Harding, H. A. and F. C. Stewart. Vitality of *Pseudomonas campestris* on cabbage seed. Science n. s. 20: 55-56. 1904.

Harding, H. A., F. C. Stewart and M. J. Prucha. Vitality of the cabbage black-rot germ on cabbage. N. Y. (Geneva) Agr. Exp. Sta. Bull. 251: 177-194. 1904.

Pammel, L. H. Bacteriosis of rutabaga (*Bacillus campestris* n. sp.) Iowa Agr. Exp. Sta. Bull. 27: 130-134. 1895.

Russell, H. L. A bacterial rot of cabbage and allied plants. Wis. Agr. Exp. Sta. Bull. 65: 1-39. 1898.

Smith, Erwin F. *Bacterium campestris. In* Bacteria in Relation to Plant Diseases. 2: 300-334. 1911.

Stewart, F. C. and H. A. Harding. Combating black-rot of cabbage by the removal of affected leaves. N. Y. (Geneva) Agr. Exp. Sta. Bull. 232: 43-65. 1903.

Club-Root of Crucifers

Caused by *Plasmodiophora brassicæ* Wor.

It is not known definitely where club-root of crucifers first appeared for it was observed almost simultaneously in England and Scotland about 1780. It evidently spread rather slowly, not becoming a serious menace until 1820. During the next fifty years the agricultural journals contained numerous articles about the disease, the editors offering prizes for the best essays on the subject. The authors often reasoned very ingeniously regarding the cause of the trouble, so that their explanations are now interesting to read. It was not until the classical work of Woronin in 1878 that the cause was really known.

The disease is now present in every country where cabbage is grown, and because of its wide distribution is known by many local names. In England it has been named finger-and-toes, anbury or ambury, and in America the terms club-root and clump-root are generally used. In other countries the translation of cabbage hernia is applied.

Club-root attacks nearly all species of the crucifer family, such as cabbage, turnip, radish, mustard, rutabagas, kohlrabi, cauliflower, brussels sprouts, shepherd's purse, candytuft, sweet alyssum, and many other cultivated and wild cruciferous plants. There is much difference in the susceptibility of the various species, some of them being almost immune, while others have been found with a hundred per cent of infection. Varieties of radishes vary from almost perfect immunity to severe infection, while all cabbage varieties are quite susceptible. Turnips and rutabagas show an even greater range of susceptibility than do radishes.

FIG. 44.—Club-root of cabbage.

Symptoms.

The most striking symptom above ground is the wilting of the tops during the hot days, followed by a partial recovery at night. Frequently the affected plants are stunted and do not head; the older and outer leaves turn yellow and drop off. This condition is most pronounced in the latter part of the season, although symptoms of the disease may appear on the seedlings in the seed-beds. The affected plant often dies, or if it lives is weak and unproductive.

The true diagnostic symptom is the swelling of the roots (Fig. 44). There is no uniformity in the shape or size of the galls produced, except, as would be expected, that those on the older tap-root are much larger than those on the younger lateral roots. The shape and position of the swelling also depend on the host on which the disease is found. The affected parts are often five times the diameter of the healthy adjoining portions, and are easily broken whereas the healthy roots are tough and fibrous. At first the swellings are dirty-grey or pale yellow on the surface and cheesy-white within. Later they become soft, flabby, and dark colored, with numerous cracks in the surface. Saprophytic organisms gain entrance through these openings and cause decomposition of the root and production of a foul odor. New lateral roots may grow out from the root above the swellings, or even from the leaf-scars on the stems, and thus brace the injured plant, enabling it to survive the remainder of the season.

Cause.

Club-root of crucifers is one of the few diseases of economic plants caused by a slime-mold. The organism was named *Plasmodiophora brassicæ* by Woronin. In its life cycle innumerable small spherical spores are liberated into the soil or manure when the diseased roots are decomposed. Stimulated by suitable temperature and moisture, as well as by the presence of the host, the spores germinate, each giving birth to a swarm-spore, which soon loses its cilium and penetrates the host tissue. The entrance may be forced through the wall of a root-hair or even directly through the epidermis. The small amœba continues to grow and divide as it penetrates farther into the root until it finally reaches the fast growing cambial cells. From the cambium it passes around and up and down the root until the mature malformation is formed. As soon as the amœba or naked mass of protoplasm has developed to a certain stage it divides into small particles, each of which becomes inclosed in a wall and forms the spore. Like other soil organisms, the slime-mold is disseminated with manure or with soil. Such agents as over-flowing water, agricultural implements, feet of men or animals, and possibly wind serve to scatter the spores in an ever enlarging area. When the organism once becomes established, it remains in the soil indefinitely.

Many years ago the fact was established that club-root was not prevalent in the soil which was influenced by the lime cliffs of England and Scotland. This led to experimentation which proved that the parasite does not thrive in soil which is highly alkaline, but is most active in poorly drained acid soil. The spores germinate and infection takes place readily only when the soil is warm and wet.

Control of club-root.

Unlike most of the other cabbage parasites, the organism is not carried on or in the seed. The place where the greatest precaution is necessary is in the seed-bed. The bed should always be located on a plat elevated enough so that it is free from any inoculum that may have been washed about by flooding waters, and on land where diseased cruciferous crops have never been grown. If it is not certain that the manure to be used on the plat is free from contamination, commercial fertilizers are much safer to apply. Lime need not be applied to the seed-bed, for it is taken for granted that the bed is free from disease, and if it is not the seedlings should never be transplanted. If even one clubbed seedling is found in the bed, the entire planting should be discarded, for there are sure to be many more plants infected which show no evidence until they are set into the field.

The fields used for cabbage-growing should be kept as free from cruciferous weeds as possible, and every precaution practiced to prevent inoculation of the soil. When infestation once takes place, it is better if the field can be used for other crops for many years, even though some careful growers feel that a six-year rotation with no cruciferous weeds present will eradicate the slime-mold. If cabbage must be grown on infested soil, a very liberal application of lime applied three to six months before planting is recommended. The amount of lime required will vary for different fields, depending on the acidity present. From two to five tons should be worked thoroughly into the soil. Such a large amount of lime often is undesirable, especially when potatoes are in the rotation, and should not be used if uninfested fields are available for the cruciferous crops.

Experiments with corrosive sublimate (1 ounce dissolved in 10 gallons of water) applied around the bases of the young plants in the seed-bed have given encouraging results in the control of club-root. The applications are begun as soon as the plants are through the ground and repeated four or five times at weekly intervals. When the plants are very small, one gallon will treat forty feet of the row, but later only about half of that space can be covered with the same amount of solution. Cauliflower is injured by such a concentrated solution, but may safely be treated with one ounce of the poison dissolved in fifteen gallons of water.

References

Chupp, Charles. Studies on club-root of cruciferous plants. N. Y. (Cornell) Agr. Exp. Sta. Bull. 387: 421-452. 1917.

Cunningham, G. C. Studies of club-root II. Disease resistance of crucifers: methods of combating club-root. Vt. Agr. Exp. Sta. Bull. 185: 67-96. 1914.

Eycleshymer, A. C. Club-root in the United States. Jour. Myc. 7: 79-88. 1892.

Gloyer, W. O. and H. Glasgow. Cabbage seed-bed diseases and Delphinium root-rots: their relation to certain methods of cabbage maggot control. N. Y. (Geneva) Agr. Exp. Sta. Bull. 513: 1924.

Halsted, B. D. Club-root of cabbage and its allies. N. J. Agr. Exp. Sta. Bull. 98: 1-16. 1893.

Kunkel, L. O. Tissue invasion by *Plasmodiophora brassicæ*. Jour. Agr. Research **14**: 543-572. 1918.

Lutman, B. F. Studies on club-root. The relation of *Plasmodiophora brassicæ* to its host and structure and growth of its plasmodium. Vt. Agr. Exp. Sta. Bull. 175: 3-27. 1913.

Nawaschin, S. Beobachtungen über den feineren bau und Umwandlungen von Plasmodiophora. Flora **86**: 404-427. 1899.

Woronin, M. *Plasmodiophora brassicæ*. Jahrb. f. wiss. Bot. **11**: 548-574. 1878.

Cabbage Yellows

Caused by *Fusarium conglutinans* Woll.

Although cabbage yellows was found in New York as early as 1895, no serious attention was given to it until fourteen years later when it became so destructive in other states that it caused alarm in many of the important cabbage sections. Limited by the soil temperature, the disease has caused by far the larger losses in a zone extending from southern Wisconsin to Kansas on the west and eastward to Pennsylvania and Virginia, with slighter losses north and south of this area. It has not been reported from any other continent.

When the soil is infested severely, most of the planting may be destroyed, and in many localities the cabbage industry would almost have ceased had it not been for the resistant varieties which have been selected. The various cabbage varieties show considerable difference in susceptibility to this disease, yet no variety is wholly immune. Aside from being found on cabbage, the causal fungus has been isolated from cauliflower and kale, but is not of economic importance on the two last hosts.

Symptoms.

Cabbage yellows might easily be mistaken for black-rot, for the symptoms are the same in many respects. The two diseases, however, may readily be distinguished from each other when examined carefully. When the host is infected in the seedling stage, the cotyledons wilt and the whole plant dies. Later infections may permit a plant to live for some time, or even through the entire season, but its sickly-yellow dwarfed appearance makes it easily distinguished from its healthy neighbors (Figs. 45, 46). Usually the lower leaves drop one by one, beginning at the lower part of the stem. If a head has formed already, it will be left bare on the top of a naked stalk. Often where such defoliation occurs, the elongated bare stalk will bear merely a few small leaves at its tip in place of the solid head. Occasionally the disease may affect only one side of the host with the result that the healthy side, continuing to grow, produces a deformed lopsided plant.

FIG. 45.—Yellows on cabbage seedling. One side is more dwarfed than the other.

A cross-section of the cabbage stem affected with yellows

shows discolored vascular bundles similar to those found in the stems of crucifers injured by black-rot, although in the latter case the diseased tissue is usually more nearly jet-black.

FIG. 46.—Cabbage yellows. The lower leaves die and fall. The attack is often worse on one side, warping or curling the stems.

After the host dies, pinkish masses of fungous spores may collect on the surface of the stem, particularly if there is much moisture in the air or in the soil.

Cause.

Some time after the fungus had been isolated by different workers and its pathogenicity proved, it was named *Fusarium conglutinans*, a genus known by its sickle-shaped septate spores and whose species cause wilt of many other plants. It is a typical soil organism, which apparently can live over winter and even for a number of years in the soil without being associated with any of the host parts. It has been suspected of being disseminated over long distances by means of the cabbage seed, although it has been proved that, if true, the dissemination by this method is rare. After the pathogene once becomes established in a locality, it spreads rapidly by means of particles of soil conveyed from one place to another. Rivulets after rain, floods, tools, and other agents may carry the inoculum to neighboring uninfested soil. When the cabbage is planted in such "sick" soil, the parasitic mycelium enters the root-hairs, and by penetrating the cell-walls succeeds in forcing its way through the tissue until it reaches the vascular bundles where it easily can follow the water-ducts to all parts of the plant. After the death of the tissue, the fungus fruits both within and on the outside of the stem. The myriads of spores find their way back into the soil to menace any succeeding crop of cabbage that may be grown there.

The temperature requirements of yellows have been determined quite accurately by Tisdale, who explains the lack of severe infestation in the North and in the extreme South. The fungus does not develop at temperatures lower than 63° F., and reaches its maximum rate of growth only when the temperatures reach 80° to 90° F. At 95° its growth again is inhibited. It, therefore, has a fairly narrow margin in which it can be destructive. During a cool summer or in a latitude where the soil temperature is lower than 63° F. or higher than 95° F., there is little danger that the pathogene can do much harm. In the central states, however, it finds the ideal conditions for infecting the host, so that it must be held in check if cabbage is long to be continued as a crop where temperatures are favorable.

Control of cabbage yellows.

Such control measures as seed treatment, isolated fields, crop rotation, destruction of infected cabbage refuse, discarding infested manure, and similar stock recommendations have been advised in the past largely on general principles. Unfortunately the recommendations give little or no relief, and have to be laid aside for something more satisfactory in the form of resistant strains.

For some years Jones and his co-workers observed that even in fields where the disease was most destructive, a few heads invariably escaped injury; consequently in 1910 he made selection of Hollander (Danish Ball Head) cabbage plants, and after growing seed sowed some of the stock in comparative tests with commercial strains. The results were so gratifying that not only were the selections with this variety continued, but also promising strains were obtained of the varieties Brunswick, All Seasons, Glory of Enkhuizen, and All Head. Additional strains were developed at other state stations until at present the solution of the yellows problem seems to be near at hand. The grower who has trouble with this disease may apply to his state pathologist for information regarding sources of seed of resistant strains, and no doubt in those states in which the disease is destructive, he will find a supply available at once, or at any rate in the process of being made available. It will be some time before a suitable organization for the distribution of resistant seed will be developed, as there is so much danger of reversion of the type or admixture with susceptible varieties. It is hoped that a reliable seed firm having the services of some experienced pathologist will assume the responsibility of supplying the market with the most resistant strains of seed that can be procured and will put forth every effort to keep them pure.

References

Gilman, Joseph C. Cabbage yellows and the relation of temperature to its occurrence. Ann. Mo. Bot. Gard. 3: 25-82. 1916.

Jones, L. R. and Joseph C. Gilman. The control of cabbage yellows through disease resistance. Wis. Agr. Exp. Sta. Res. Bull. 38: 1-70. 1915.

Jones, L. R., J. C. Walker and W. B. Tisdale. Fusarium resistant cabbage. Wis. Agr. Exp. Sta. Res. Bull. 48: 1-34. 1920.

Manns, T. F. Two recent important cabbage diseases in Ohio. Ohio Agr. Exp. Sta. Bull. 228: 255-297. 1911.

Tisdale, W. B. Influence of soil temperature and soil moisture upon the Fusarium disease in cabbage seedlings. Jour. Agr. Research 24: 55-86. 1923.

Alternaria Leaf-Spot and Blight of Crucifers

Caused by *Alternaria brassicæ* (Berk.) Sacc.

The leaf-spot of crucifers has been known for many years, as it is distributed in nearly every part of the world. In the United States it is very common on cabbage and cauliflower and has been reported also on turnip, collards, horse-radish, and Chinese cabbage. The leaf-spot is not of great economic importance in the field, but may cause severe injury when it gains access to the cabbage in storage. The most important stage, however, is the pod-spot of seed cabbage, and the discoloration of blanched cauliflower. On Long Island where I. H. Vogel has done considerable work, he estimates that in an epidemic year there may be a loss of twenty-five thousand pounds of cabbage seed, and 10 per cent of the cauliflower crop.

Symptoms.

The leaf-spot may occur on crucifers at any stage in the development of the foliage, but is more common on the lower older leaves. It begins as a small, circular, yellowed area, which as it enlarges in concentric circles takes on a black sooty color (Fig. 47). The spots may be no larger than a pin-head and very numerous or may be few in number and of all sizes up to two inches in diameter. In storage the spots may unite, forming a black moldy growth over the entire leaf.

FIG. 47.—Alternaria leaf-spot of cabbage, showing slight concentric markings.

According to Vogel, the spots on the seed-pod (Fig. 48), and beginning at its tip, are purplish in color when first formed, but later turn brown. They depend for their size on the weather conditions; when there is abundant rainfall the whole pod may be included in the lesion. Under such a severe attack the seeds do not develop in the young pod, or they shrivel in size in the older one.

The injury to cauliflower (Fig. 49) is described by Vogel as a browning of the head, beginning at the margin of the individual flower or of a cluster of flowers while they are still joined in a compact head. The discoloration may finally cover a part or all of the head, and although only superficial it spoils the sale of the crop.

Cause.

The fungus, *Alternaria brassicæ,* (earlier described as *Sporiodesmium* or *Polydesmus exitiosum* Kühn) is characterized by

FIG. 48.—Alternaria rot on seed-pods and pedicels of cabbage.

large, brown, muriform, catenulate spores, which are borne in great numbers on the diseased host tissue. They are blown, splashed, or carried by tools and animals to all parts of the field, and when conditions are ideal cause infection on the normal green tissue of the host. More often, however, they find the old leaves less resistant and become established in the aged tissue before the fungus seems able to enter the newer

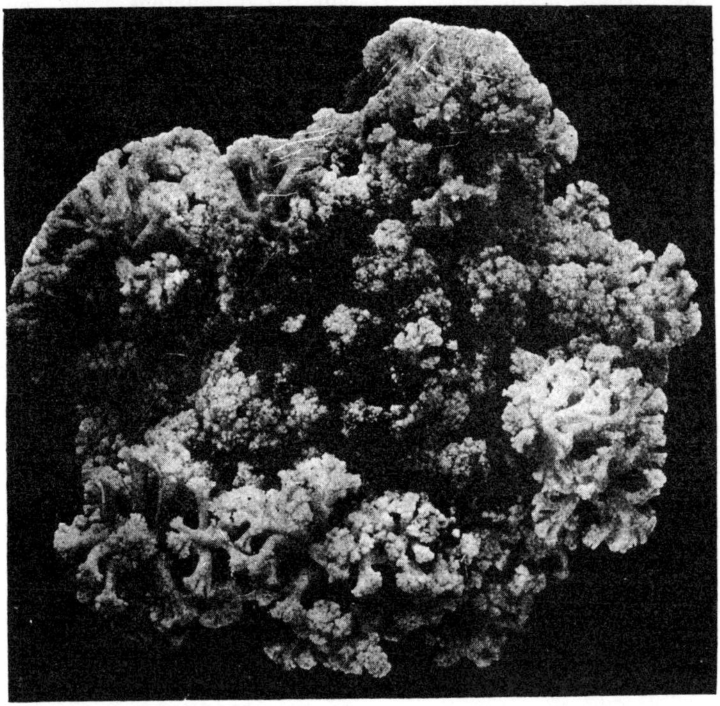

FIG. 49.—Alternaria blackening the blanched head of cauliflower.

foliage. When they infect the pods, the spores are mixed with the seed during threshing, and apparently the mycelium grows through the pods into the seeds. In this position, or in the old refuse left in the field, the fungus passes the winter in readiness to attack the new crop.

A very closely related cabbage parasite, *Alternaria oleracea*, has been described by Milbrath as occurring in California.

A much smaller leaf-spot caused by *Macrosporium herculeum* E. and M. has been reported on turnip, horse-radish, and cabbage. When first discovered, it was causing injury to flat turnips. The fungus is considered different from the Alternaria.

Control.

The storage-house should be disinfected with formaldehyde or blue-vitriol solution and all the old refuse destroyed. The heads meant for storage should be free from the leaf-spot and as little bruised as possible. In addition, if the temperature is kept near the freezing point and the humidity is reduced to a minimum by proper ventilation, the stored cabbage will suffer little or no injury.

As the organism is carried with the seed, seed treatment with hot water as suggested for the control of black-leg will kill the mycelium beneath the seed-coat. Mercuric chloride is not effective. It has not been proved, however, that clean seed will reduce the amount of disease in the field. No doubt there are occasions when disinfection is beneficial, since more than half of some lots of seed are affected.

Little can be done to eliminate spotting of cauliflower. Long rotations, with an avoidance of other crucifers, and care in tying the heads for blanching may aid in reducing loss during wet weather late in the fall.

References

Harter, L. L. and L. R. Jones. Cabbage diseases. U. S. Dept. Agr. Farmers' Bull. 925: 1-30. 1918.

Kühn, Julius. Das befallen des rapses und rübsens. *In* Krankheiten der Kulturgewachsen, pp. 151-177. 1858.

Milbrath, D. G. Alternaria from California. Bot. Gaz. **74**: 320-324. 1922.

Vogel, I. H. Unpublished thesis.

FIG. 50.—Wire-stem of cabbage.

WIRE-STEM OF CRUCIFERS

Caused by *Corticium vagum* B. and C.

(See also Rhizoctoniose of Potatoes, page 367.)

The wire-stem of cabbage (Fig. 50), cauliflower, brussels sprouts, and kale is really a stage of damping-off appearing on older plants. When the host tissues have become too woody to be affected by a soft-rot, the cortical cells at the base of the stem collapse, leaving the fibrous central cylinder either bare or partly encased by the dry leathery epidermis. The withering reduces the size of the stem so that there is a marked constriction at its base and frequently extending for more than an inch above the surface of the soil. The plant, still supplied by the conducting tissues, continues to live, the stem often becoming distorted when growth takes place above the lesion. No soft-rot is apparent in any stage of true wire-stem except during a rainy period. Frequently the above named crucifers together with radishes are not girdled by the attacks of the fungus but exhibit depressed brown or black cankers on the stem or tap-root somewhat similar to those found on potato stems (Fig. 51).

FIG. 51.—Rhizoctonia on radish.

The disease is most destructive to the very small plants that are still susceptible to damping-off. The trouble, however, may cause much loss on older plants, since the stems may be killed, or even if they do bear heads the latter are dwarfed and probably late in coming to maturity.

The fungus causing wire-stem is *Corticium vagum.* Long lists of hosts for this organism have been compiled. Its vegetative stage, known as *Rhizoctonia solani,* apparently is composed of numerous strains, some of which are pathogenic to a certain group of hosts that are not affected by other strains. For example, the common strain found on cabbage in New York will not infect potatoes. In general, all strains have the same morphological characters, although there may be differences in the rate of growth in culture, in their reaction toward acid or alkaline media, and in the number of sclerotia produced. All strains live over winter in the soil, and by production of mycelium in the spring are able to infect the plant. The spore stage is distinguished by the delicate, white, fungous weft found in wet weather near the base of the stem. The weft is composed of hyphæ and short club-shaped basidia on the sterigmata of which the spores are borne. From all indications, the spore stage is of little use in the life history of the fungus. The minute black sclerotia and the short barrel-shaped cells of the mycelial threads are able to survive in the soil without the presence of the host, and to cause infection when crucifers are present.

The disease has been found most abundant when the soil-moisture and temperature are relatively high. The minimum and maximum temperatures for the growth of the organism are approximately 47° and 89° F. respectively, and the optimum lies between 60° and 85° F.

When the disease occurs in the seed-bed, soil sterilization with steam or formaldehyde will kill the fungus (page 600). When sterilization cannot be practiced, or there is danger of early infection in the field, cabbage, cauliflower, brussels sprouts and radish may be treated successfully with corrosive sublimate as advised for the control of club-root (page 140), and as is generally recommended for the eradication of cabbage maggot. Five applications at weekly intervals, beginning as soon as the plants are through the ground, have proved effective. At the first application one gallon of the poison solution (1 ounce of the powder in 10 gallons of water) is sufficient for treating about forty feet of the seedling row. When the plants grow older, the amount must be increased until finally a gallon is applied to twenty feet of the row. A more dilute solution is required for cauliflower.

Other precautions that are worth observing are rotation of

crucifers with corn, cereals, grasses, or other immune crops, and keeping the cabbage plants in a rapidly growing condition by means of proper cultural practices.

References

Duggar, B. M. and F. C. Stewart. The sterile fungus Rhizoctonia as a cause of plant diseases in America. N. Y. (Cornell) Agr. Exp Sta. Bull. 186: 1-30. 1901.

Gloyer, W. O. and H. G. Glasgow. Cabbage seed-bed diseases and Delphinium root-rots; their relation to certain methods of cabbage maggot control. N. Y. (Geneva) Agr. Exp. Sta. Bull. 513: 1924.

Gratz, L. O. Wire-stem of cabbage. Unpublished thesis. 1923.

Stewart, F. C. Notes on New York plant diseases I. N. Y. (Geneva) Agr. Exp. Sta. Bull. 328: 386. 1910.

Ring-Spot of Cauliflower

Caused by *Mycosphærella brassicicola* (Duby) Lindau

The organism causing ring-spot was described nearly a hundred years ago. It was then found in France. It is now present in most of Europe, including England with adjacent islands, in Australia, New Zealand, and in 1915 was found on cauliflower shipped from California to the Boston market. It affects cabbage, brussels sprouts, cauliflower, broccoli, rape, and some other cruciferous plants. It formerly was not important, but recent surveys in England mention considerable injury. Similar damage has occurred in New Zealand; and in shipments to Boston dealers report a loss of three hundred to four hundred dollars on each injured carload. In 1922 there was 15 per cent loss on two thousand acres in San Mateo County, California.

Symptoms.

The leaves and seed-pods of the crucifers are attacked. The spots on the leaves are present on both the laminæ and the veins. The outer lower leaves are more seriously affected than the inner ones. The brown dead areas are bordered by a green zone that retains its color even after the remainder of the leaf turns yellow. The spots vary in size from those just visible to some nearly a half inch in diameter. Still larger lesions are made when two contiguous spots unite. Scattered over the dead areas, and more numerous about the margin than in the center, are many small black fruit-bodies so deeply embedded in the tissue that they are barely perceptible to the unaided eye. In some cases these fruit-bodies are arranged in concentric rings.

Cause.

The causal organism has been burdened with a number of Latin names, but those which are commonly applied now are *Mycosphærella brassicicola* and *Phyllosticta brassicicola* for the ascerigerous and pycnidial stages respectively.

In the presence of moisture the spores ooze in pink tendrils from the pycnidia and are ejected from the perithecium, being scattered by flooding or splashing rains and possibly by insects to healthy plants. Infection takes place readily on any part of the leaf, the spots usually showing within ten days. The fruit-bodies are formed in great numbers. In Europe the prevailing stage seems to be the pycnidial, while in New Zealand and in the United States the perithecia are more numerous. The fungus evidently lives over winter in diseased material left in the field or placed in storage. It has not been determined whether it will live in the soil after the host has decayed.

Control.

No successful control measures have been demonstrated. Burning diseased refuse and spraying with bordeaux mixture 4-4-50 have been advised, but their efficacy has not been proved.

REFERENCES

Grove, W. B. A cabbage disease. Jour. Roy. Hort. Soc. 40: 76-77. 1914.

Kirk, T. W. Cabbage leaf-spot. N. Zeal. Agr. Dept. Rept. 14: 348-349. 1916.
Osmum, A. V. and P. J. Anderson. Ring-spot of cauliflower. Phytopath. 5: 260-265. 1915.

Bacterial Leaf-Spot of Cauliflower and Cabbage

Caused by *Bacterium maculicolum* McCulloch

The bacterial leaf-spot was first recognized in 1909 when specimens of cauliflower leaves were sent from Virginia to the Department of Agriculture at Washington. The disease has since been found on cauliflower and cabbage in New York, Florida, California, Algeria, Denmark, and probably Alabama, Mississippi, Louisiana, Texas, and Australia. Ordinarily it is of minor importance.

Fig. 52.—Bacterial leaf-spot of cauliflower.

The leaves are covered with numerous small brown or purplish spots ranging from mere points to lesions one-eighth inch in diameter (Fig. 52). The dead areas may coalesce and produce irregular blotches somewhat larger in size. The spots appear first and most abundantly on the lower side of the leaf, where each originates at a stomatal opening. The infection may occur either on the vein or on the part of the leaf between the vein, and when on the former often retards growth and puckers the infected foliage. When badly diseased, the leaves turn yellow and drop off. The spotting may also be found on the white part of the cauliflower.

The parasite, *Bacterium maculicolum*, is a white, motile, 1-5-polar flagellate rod-shaped organism with the group number 211.3332023. Very little of its life history is known. In the presence of moisture it enters the stomata of middle-aged leaves, the very young and the old foliage being almost immune. Visible spots are formed within three to six days after inoculation. The method of its dissemination and overwintering has not been determined. It grows at low temperatures but will not develop at a point above 84° F.

No control measures have been suggested.

References

Ferdinandsen, Carl. *Bacterium maculicolum* McCulloch paa europæish grund. Nord Jordbrugsforskn. No. 5-8, pp. 467-474. 1923.

McCulloch, Lucia. A spot-disease of cauliflower. U. S. Dept. Agr. Bur. Plant Ind. Bull. 225: 1-15. 1911.

Plant Disease Survey Office. The plant disease Bulletin. Suppl. 10: 239. 1920.

Smith, E. F. McCulloch's cauliflower spot. *In* An introduction to bacterial diseases of plants, pp. 300-313. 1920.

Bacterial Soft-Rot of Crucifers

Caused by *Bacillus carotovorus* Jones

(For a detailed discussion, see Bacterial Soft-Rot of Carrot)

Many, if not all, of the cruciferous plants are susceptible to the bacterial soft-rot, as is shown by its being reported on cabbage, cauliflower, brussels sprouts, kohlrabi, turnip, rutabaga, radish, rape, and horse-radish. Among forty varieties of turnips that were tested for comparative susceptibility, nearly all showed some rotting, the Jersey Navel being the only one listed as immune. The disease is extremely destructive in some instances and every year causes considerable injury to cabbage seed plants. It often follows black-rot, causing on cabbage the well-known stump-rot disease.

Symptoms.

The disease has much the same appearance on each host. Infection usually takes place through a wound and at any place on the plant, the affected tissue becoming soft and slimy without much discoloration but often accompanied by an offensive odor. On cabbage and cauliflower the rot frequently begins just below the head and finally rots the stem, so that the head falls off or may easily be pushed over leaving only a stump of the host. When such a decay takes place on the mother seed plant, the floral stalk wilts and withers. The rot may also have its inception at some insect puncture on the upper part of the stem. The root crops are affected in a similar manner, the disease known as white rot of turnips representing a common form of the decay.

Cause.

The organism, *Bacillus carotovorus*, is common on so many hosts, both in storage and in the field, that it lacks no opportunity of becoming established on the new crop in the summer. Carried by insects or in any other manner in which microscopic bodies may be disseminated, the bacterium gains an entrance into injured tissue where it soon dissolves the middle lamellæ, leaving only a slimy mass of free cells.

Control.

No direct control methods are known. It is good insurance to practice such well known precautions as long rotations which include cereals or other immune crops, care in not feeding cattle with diseased plants, making sure that the inoculum is not carried in the manure, thorough disinfection of the storage-house with formaldehyde or a blue vitriol solution, selection of healthy plant parts for storing, keeping the

FIG. 53.—Sclerotia of Sclerotinia on cabbage seed-pods.

FIG. 54.—Sclerotinia rot on cabbage seed-stem.

temperature of the store-room as near freezing as vegetables will bear without deterioration, and the humidity of the air as low as possible.

Sclerotinia Rot of Crucifers

Caused by *Sclerotinia libertiana* Fuckel

(See also Drop of Lettuce, page 243.)

The Sclerotinia rot may affect a number of the crucifers; it has been reported particularly on marrow cabbage, thousand-headed kale, cauliflower, and on the common field cabbage. On the last named host it frequently is destructive as a head-rot of late harvested plants, or as a stem and pod-rot of seed cabbage (Figs. 53, 54). In all cases the disease is distinguished by the presence of white, fluffy, aerial mycelium, and large black sclerotia either on the surface or embedded in the tissue of the host. Almost a half pint of such sclerotia may sometimes be grown on a single cabbage head affected by this watery soft-rot.

In controlling the disease the first requisite is rotation with cereals or other immune crops, thereby reducing the amount of inoculum in the soil. When cabbage is grown in small gardens or in intensive trucking sections, it should not be alternated in rows with lettuce or other very susceptible hosts. The field cabbage should be harvested early if there is danger of infection, as only the cabbage which is mature seems to be infected easily. The cabbage that is set out for seed should be as free from wounds as possible, and placed far enough apart in the row so that the stems dry quickly after a dew or rain.

References

Hedgecock, G. G. A disease of cauliflower and cabbage caused by Sclerotinia. Mo. Bot. Gard. Ann. Rept. 16: 149-152. 1905.

Lawrence, W H. Plant diseases induced by *Sclerotinia perplexa* n. sp. Wash. Agr. Exp. Sta. Bull. 107: 1-22. 1912.

White Rust of Crucifers

Caused by *Albugo candida* (Pers.) Ktz.

The pustules caused by white rust are so prominent and distinct on cruciferous leaves that the attention of interested observers was drawn to them more than a hundred years ago. The disease is world-wide in its distribution, and probably affects every known cruciferous host. It is less commonly found on cabbage, brussels sprouts, and cauliflower than on shepherds-purse, radish, horse-radish, mustard, and turnip. It is of slight economic importance except in localities where radish seed is grown. The floral parts of the radish are attacked and a large percentage of the seed crop may be destroyed.

The pustules are formed on the leaves, smaller stems, and floral parts. They are formed under the host epidermis and soon rupture exposing a white chalky dust in small circular cavities. Frequently the only visible symptoms are the presence of the pustules, but occasionally the affected portion is swollen and much distorted.

The parasite, though morphologically identical on the different cruciferous plants, apparently is selective in its host requirements. The form on shepherds-purse is distinct from that on common vegetables, and the race on mustard does not infect cabbage, nor will the one on radish attack turnip. Evidently all the forms are highly specialized, so that weeds are probably not sources of inoculum for any of the common vegetable crops.

When liberated in the form of a white dust, the spores, borne in chains within the pustules, are scattered bv rain and insects to neighboring plants. They germinate by giving birth to five to eight swarm-spores, which in turn germinate by the protrusion of a germ-tube. Infection takes place through the stomata. The younger the plant the more readily the parasite can gain admission. The mycelium grows between the cells and obtains its food by forcing haustoria through the cell-walls into the living protoplasm. When the tissue has become thoroughly invaded, not only are fruiting pustules produced on the surface of the plant, but oospores develop in large numbers throughout the tissue. The fungus lives over winter by means of these thick-walled fruit-bodies and by perennial mycelium in plants remaining alive from one season until the next. The fungus occasionally grows in the same tissue with Peronospora.

The white rust thrives in wet weather but is able to develop with only a slight rainfall. The water obtained from dew is sufficient for germination and infection. The spores germinate better when slightly chilled, although when placed in an ice-box for ten to twenty-four hours germination is inhibited.

Control measures usually are unnecessary. When radish seed is grown, it is well to burn all the old affected tops. If the mother seed plants of the radish or any cruciferous seedlings in the seed-bed are in danger of infection, it may be a profitable precaution to spray at weekly intervals with bordeaux mixture 4-4-50, beginning before there has been an opportunity for the plants to become inoculated.

References

Eberhardt, A. Contribution à l'étude de *Cystopus candidus* Lev. Centralbl. Bakt. u. Par. abt. 2. **12**: 235-245, 426-439, 614-631, 714-725. 1904.

Harter, L. L. Diseases of cabbage and related crops and their control. U. S. Dept. Agr. Farmers' Bull. 488: 1-32. 1912.

Melhus, I. E. Experiments on spore germination and infection in certain species of Oomycetes. Wis. Agr. Exp. Sta. Res. Bull. 15: 25-91. 1911.

Melhus, I. E. Perennial mycelium in species of Peronosporaceæ related to *Phytophthora infestans*. Jour. Agr. Research **5**: 59-70. 1915.

Downy-Mildew of Crucifers

Caused by *Peronospora parasitica* (Pers.) DeBary

The conspicuous mildewed appearance of crucifer foliage has been observed for many years, and in nearly every country where pathologists have made surveys. The mildew is not a serious disease, except occasionally in the seed-bed and on pods of mother seed plants. It attacks cabbage, cauliflower,

FIG. 55.—Downy-mildew on cabbage leaf causing large tan spots on the upper surface.

turnip, radish, and possibly every other cruciferous vegetable and weed.

Symptoms.

The mildew may be found on nearly every part of the plant above ground, but ordinarily is present only on the foliage or the seed-bearing parts. The first signs of the disease on the leaf are small, almost imperceptible, light green lesions, which are most prominent on the upper side, but which soon show on the lower epidermis. As the spot enlarges it becomes more yellow, and during this transition period the fungus fruits over the spots and on both sides of the leaf with a downy white mold (Fig. 55). The mold appears only when there is dew or rain to moisten the surface of the foliage. The fungus does not fruit on the discolored dead tissue, but continues its growth in ever widening circles made by the yellow border of the lesion. The dead tissue is somewhat shrunken, and on the upper side of the leaf is light tan or ashen-gray in color, while on the lower side the tissue is gray with a faint tinge of purple, especially near the margin of the spot. Black sooty mold may soon invade the dead tissue and cause it to appear black. The spots when fully grown may be of any size from a quarter inch in diameter to several inches. The larger and more irregular shapes are usually bounded on the sides by the leaf-veins. Occasionally the mildew may be located along the margin of the leaf and then cause lesions which appear like tip-burn. On young seedlings the whole leaf may be blighted.

The spots on the other parts of the plant are similar to those on the foliage, except that they may cause swollen distortions or other malformations of the affected tissue. The fleshy roots of turnips and radishes may have an internal irregular region of discoloration, extending from the crown of the root downward. The flesh may be brown or black, or show a form of net-necrosis. In either case the tissue remains firm and turgid.

Cause.

No group of fungi has aroused more interest or has been studied with more care than the family to which the downy-mildews belong. According to Gaumann, one of the facts that has been revealed in this study of *Peronospora parasitica* is that the fungus is divided into many strains, all of which differ in their size of spores and in their ability to attack a given species or genus of crucifers. For example, one strain attacks only shepherds-purse; another only *Sisymbrium officinale;* and still another attacks both *Brassica oleracea* (cabbage, cauliflower, brussels sprouts, kale) and *Brassica Rapa* (turnip). This is of importance since it has been considered that the vegetables often receive inoculum from weeds that might be growing nearby.

The fungus lives over winter in roots and in old diseased parts. In the former the mycelium is perennial, growing up with the new shoot in the spring. Thick-walled oospores, formed in the old infected plants and among the seeds, are able to withstand the winter months. When infection takes place, the mycelium grows between the cells, but sends through the cell-walls haustoria which sap the strength from the host. Long branched conidiophores with conidia borne on their tips comprise the white mold that covers the yellowed tissue. The conidia germinate with a long germ-tube, which causes infection. Often the Peronospora and the white rust parasite are growing in the same lesion.

The downy-mildew is favored by the presence of moisture. The spores germinate most rapidly at 46° to 55° F. and are greatly inhibited at 77° F. The development of the fungus after spore germination requires higher temperatures, so that when the nights are cool and the days are relatively warm infection may take place readily.

Control.

Downy-mildew is not often serious enough to require attention. Since it is more likely to be injurious in seed-beds of cauliflower and cabbage, and on mother seed plants, the seedbed should be changed from year to year, the seedlings planted far enough apart for the dew to dry quickly, and both the seedlings and the mother seed plants that show infection should not be transplanted. If there have been frequent losses due to the disease, spraying with bordeaux mixture 4-4-50 or dusting with 20-80 copper-lime dust may be profitable. Applications can be made on the seed-bed easily and cheaply.

References

Anderson, H. W. *Peronospora parasitica* on *Arabis lævigata*. Phytopath. 4: 338. 1914.

Gardner, M. W. Peronospora in turnip roots. Phytopath. 10: 321-322. 1920.

Gaumann, E. *Peronospora parasitica*. Centralbl. Bakt., etc., 2 abt. 45: 575-577. 1916.

Melhus, I. E. Perennial mycelium in species of Peronosporaceæ related to *Phytophthora infestans*. Jour. Agr. Research 5: 59-70. 1915.

Schrenk, H. von. On the occurrence of *Peronospora parasitica* on cauliflower. Mo. Bot. Gard. Ann. Rept. 16: 121-124. 1905.

Powdery-Mildew of Crucifers

Caused by *Erysiphe polygoni* DC.

Races of the fungus causing powdery-mildew of crucifers are found on an endless variety of plants. The organism is common on clover and garden pea, as well as on nearly every cruciferous plant as cabbage, turnip, rape, kohlrabi, broccoli, and charlock. This does not signify that when cabbage is grown beside infected peas or even turnips the cabbage will become diseased. The numerous races of the pathogene are highly specialized, so that they seldom pass from one host genus to another, or from one species to another. Some varieties of crucifers are less susceptible than others; for example, some turnips are almost immune while others are severely affected. Even at its worst, the disease is not of economic importance.

The plant tissue is not changed in its appearance by the fungus. The attacked portion is covered with a white, powdery, often almost imperceptible mold, which growers liken to talcum or chalk dust. Rarely black fruit-bodies, just large enough to be seen, are scattered among the white wefts.

The pathogene lives from autumn until spring in the form of perithecia or as mycelium in volunteer host plants. The white powdery growth is composed of great masses of short conidiophores on which the spores are borne in chains. They are blown about by the wind, and in the presence of slight moisture and high temperature cause a superficial infection.

No control measures are recommended. If the disease should become serious, it no doubt could be controlled by dusting the plants with sulfur, as the powdery-mildews are very susceptible to the fungicidal effect of this chemical.

References

Harter, L. L. and L. R. Jones. Cabbage diseases. U. S. Dept. Agr. Farmers' Bull. 925: 1-30. 1918.

Searle, G. O. Some observations on *Erysiphe polygoni*. Trans. Brit. Mycol. Soc. 6: 274-293. 1920.

Malnutrition of Cabbage and Cauliflower

Malnutrition, or whip-tail as it is known on cauliflower, manifests itself in a change of the normal green of the leaves to a light green or yellow between the veins and around the margins. The lower outer leaves are first and most severely affected. Later the upper inner leaves also show the abnormality. The injured leaves have a "quilted" appearance, being thicker than are the healthy ones and so brittle as to be easily crushed between the fingers. The seriousness of the injury to the plant may vary from the production of a full-sized head to the entire absence of one. The lateral roots may be reduced in number and in length and have dead areas at their tips. The surface of the stem at the point where it emerges from the soil may also show lesions.

The disease is caused by an unbalanced condition of the food supply. It may be brought about by an excessively acid soil, by the lack of humus, or by extravagance in the use of mineral fertilizers. These conditions are readily obtained by cropping the land continuously, and each year removing all the plant growth. If instead of turning under cover-crops, especially legumes, the attempt is made to furnish all the plant-food by increased applications of certain types of mineral fertilizers, a barren acid soil may soon be expected.

The correcting of the above named faults will control malnutrition. The program for improvement of the crop includes the proper use of lime, turning under leguminous cover-crops, care in selecting the fertilizers, and the exercise of judgment in its application.

References

Clayton, E. E. Investigations of cauliflower diseases on Long Island. N. Y. (Geneva) Agr. Exp. Sta. Bull. 506: 1-15. 1924.

Harter, L. L. and L. R. Jones. Cabbage diseases. Malnutrition, a physiological disease. U. S. Dept. Agr. Farmers' Bull. 925: 23-25. 1918.

Œdema of Cabbage and Cauliflower

Cause non-parasitic

Small excrescences appearing like broken pustules have been observed on the lower side of leaves of many plants, especially cabbage, cauliflower, tomato and potato. This malformation is referred to as œdema.

On the lower side of a cabbage leaf there appear small wart-like prominences in varying degrees of abundance (Fig. 56). Occasionally the upper side of the leaf will show a corresponding depression. The warty outgrowths may be located closely together, so that they unite, forming ridges or irregularly-shaped elevated areas. At first the epidermis covers the swelling, but finally the pressure becomes strong enough to burst it, permitting the abnormally enlarged mesophyll cells to push outward forming white frosty-like excrescences. The exposed tissue finally changes to yellow or brown in color, and becomes corky in appearance.

The œdema is caused by any factor which stimulates groups of mesophyll cells to abnormal growth. The hypertrophy has been caused experimentally in, at least, three ways. First, when the roots are made to take up more water than is given off by the leaves, the pressure within the plant is sufficient to produce œdema. Such a condition is rarely, if ever, found

FIG. 56.—Œdema on cabbage leaf.

under field conditions, but may be observed occasionally in greenhouses in the winter when plants are growing in a warm soil while the tops are surrounded by cool air saturated with water, and are lighted only by the feeble rays of a winter sun during the short days.

The second method is the use of copper sprays. When the

lower sides of cauliflower leaves are sprayed with ammoniacal copper carbonate, dilute enough not to kill the tissue but so concentrated that the cells react to the poison, typical œdema is obtained.

On sandy soil the lower side of cabbage leaves often show intumescences. It has been demonstrated that when the sand particles are blown against the foliage with great force a bruise results and the plant, in response to this injury, produces the abnormally elongated mesophyll cells that burst the epidermis.

The œdema is not serious under field conditions, and therefore needs no attention there. It may be destructive in the greenhouse if proper precautions are not taken. The soil should not be too warm nor too wet, and the excess moisture should be removed from the air by proper heating and ventilation. The glass house should be so constructed that no necessary rays of the sun are obstructed, and the plants placed far enough apart to avoid any undue shading.

References

Atkinson, G. F. Œdema of the tomato. N. Y. (Cornell) Agr. Exp. Sta. Bull. **53**: 77-108. 1893.

Schrenk, Hermann von. Intumescences formed as a result of chemical stimulation. Mo. Bot. Gard. Ann. Rept. **16**: 125-148. 1905.

Wolf, F. A. Intumescences, with a note on mechanical injury as a cause of their development. Jour. Agr. Research **13**: 253-259. 1918.

Mosaic of Crucifers

Cause undetermined

The characteristic mosaic mottling has been observed on Chinese cabbage, mustard, and turnip. The leaves have light green areas that gradually shade into the normal appearing parts. In addition to the mottling, the foliage is ruffled or distorted in various ways. When slightly infected the plant may be normal in size, but often is much dwarfed, and the number of its blossoms is reduced. It has been shown that the causal agent is able to reduce the nitrates to ammonia and nitrites. In the chemical change the host loses much of the nitrogen which is needed in its development. This loss may account for the dwarfing of the affected plants.

Evidently the virus is not carried in the seed, for seeds collected from diseased plants give healthy progeny. The mosaic can be transmitted by rubbing affected leaves against those that are healthy, or by letting aphids feed on the mottled foliage, then transferring them to normal plants. The manner of its surviving the winter and the method best adapted for controlling the disease have not been discovered.

References

Gardner, M. W. and J. B. Kendrick. Turnip mosaic. Jour. Agr. Research **22**: 123-124. 1921.

Jodidi, S. L., S. C. Moulton, and K. S. Markley. A mosaic disease of cabbage as revealed by its nitrogen constituents. Science n. s. **52**: 588. 1920.

Schultz, E. S. A transmissible mosaic disease of chinese cabbage, mustard, and turnip. Jour. Agr. Research **22**: 173-178. 1921.

Anthracnose of Crucifers

Caused by *Colletotrichum higginscanum* Sacc.

Anthracnose has been observed on crucifers for many years, but has not been found destructive. In cross inoculations the disease was produced on turnip, radish, cabbage, and collards, the first two being most susceptible.

The lesions are very small circular gray spots on the leaf, and more elongated ones on the stem. Occasionally the spots coalesce producing a larger more irregularly-shaped area. Under moist conditions it sometimes may be possible to observe a minute mass of pink spores over the dead tissue.

The parasite in America has been named *Colletotrichum higginscanum.* Evidently it is different from the fungus found in England and known as *Glœosporium concentricum* (Grev.) Berk. and Br. The latter has white masses of spores, which are exuded from the acervuli on the under side of the leaf. The two pathogenes evidently cause similar types of lesions. The small unicellular spores borne in acervuli are washed or splashed to healthy plants where they affect not only the foliage, but also the seed-pods. No indication has been found, however, that the fungus is carried in the seed.

No control measures have been necessary.

References

Cotton, A. D. Fungus diseases of crops, 1920-1921. Min. of Agr. and Fish. Misc. Publ. 38: 49. 1922.

Higgins, B. B. A Colletotrichum leaf-spot of turnips. Jour. Agr. Research 10: 157-162. 1917.

Bacterial Root-Rot of Horse-Radish

Causal organism not fully determined

A root-rot of horse-radish has been destructive in New Jersey for some time. Its distribution in other states and in foreign countries has not been determined.

The first symptoms are a water-soaked appearance and a yellowish discoloration of the core of the root. The infected tissue finally becomes soft, and in old roots may disappear entirely leaving a hollow center. The outer cylinder of tissue is rarely, if ever, invaded. The diseased roots when planted are mutilated by rodents, or rotted by saprophytic fungi and bacteria. In hot dry days the roots may wilt and die without producing a new growth.

The decay spreads most rapidly in storage, particularly just before planting time when the high temperature and moisture favor its development. The larger stored roots are more susceptible than are the smaller ones. Very few roots that are healthy when placed in the field become infected after active growth has begun.

If all the roots used for planting are clipped at both ends, and those discarded which show any internal discoloration, the amount of disease can be much reduced. The clipping should be followed by soaking the apparently healthy roots in corrosive sublimate, one ounce in seven and one-half gallons of water, for fifteen minutes, and drying them before planting.

References

Poole, R. F. Horse-radish root-rot investigations. Report of the Department of Plant Pathology. N. J. Agr. Exp. Sta. Ann. Rept. **1920**: 610. 1921; Ann. Rept. **1921**: 464, 465. 1922.

Poole, R. F. Bacterial root-rot of horse-radish in New Jersey. Phytopath. **12**: 49. 1922.

Black-Root of Radish

Caused by *Nematosporangium aphanidermatum* (Edson) Fitz.

The fungus causing black-root also produces a damping-off of radishes and beets. When injured by black-root, the radish becomes brown or black either in part or over its whole surface. When only small areas of the root are affected, the injured tissue does not grow while the healthy portions continue their development, thus producing ill-shapen roots. A favorite place for the beginning of infection is at the point where the secondary roots unite with the tap-root. At first the invaded cells remain turgid and retain their normal flavor and odor, but as infection progresses secondary organisms, gaining an entrance, cause the root to decompose. The pathogene may invade and blacken also the cortex of the stem, petiole, and leaf midrib. It is slightly more severe on white radishes than on the colored varieties.

The causal fungus was formerly considered as being *Aphanomyces lævis* DeBary. The result of later work suggests that the injury is due to another organism known by the two long names, *Rheosporangium aphanidermatus* Edson, and more correctly, *Nematosporangium aphanidermatum*. It is composed of delicate mycelial threads between the host cells. The fruit-bodies are formed in the presence of abundant moisture, and consist of a primitive sporangium, from which arises the true zoosporangium, and oogonia with oospores. The zoosporangium bears numerous zoospores which together with the oospores cause new infections.

No practicable control measures have been devised. In

small beds soil sterilization can be practiced if the value of the crop warrants the expense of applying steam or formaldehyde (see pages 602 and 610).

References

Barrett, J. T. A serious root disease of radish. Phytopath. 2: 96. 1912.

Edson, H. A. *Rheosporangium aphanidermatus,* a new genus and species of fungus parasitic on sugar beets and radishes. Jour. Agr. Research 4: 279-292. 1915.

Fitzpatrick, H. M. Generic concepts in the Pythiaceæ and Blastocladiaceæ. Mycologia 15: 166-173. 1923.

Kendrick, J. B. Infection court in radish black-root. Phytopath. 14: 66. 1924.

Plant Disease Survey. Disease of crucifers. U. S. Dept. Agr. Plant Dis. Bull. Suppl. 22: 366. 1922.

Leaf-Spots of Horse-Radish

Caused by Ramularia, Septoria, and Cercospora

The Ramularia leaf-spot of horse-radish has been reported from Germany, Great Britain, France, Austria, Italy, Finland, and many places in North America. As the host is not of great economic importance, the disease has received very little attention.

The first symptoms usually appear early in the summer in the form of few to extremely numerous light green or yellowish spots on the foliage (Fig. 57). The invaded areas quickly turn white and papery, so that the affected leaves may be recognized from a long distance. The spots are irregular in shape and seldom over a third of an inch in diameter unless they coalesce, which they frequently do. Later in the season the dead areas by shrinking tear away from the healthy tissue and drop to the ground, leaving a shot-hole effect on the foliage. The spots often drop out in large irregular circles, thus engulfing patches of healthy tissue, which when cast off result in ragged holes sometimes almost an inch in diameter. When such lesions are on the edge of the leaf, it is reduced to a much frayed condition. Whole leaves may be so badly affected that they wither and die, although usually a leaf with three-fourths of its surface included in the diseased areas will still remain alive. In the dead papery tissue are innumerable small stromata or sclerotia-like bodies, which may be seen on the outer surface and which are misleading in that, at the first glance, they have the appearance of pycnidia.

FIG. 57.—Ramularia leaf-spot on horse-radish.

The fungus, *Ramularia armoraciæ* Fuckel, evidently lives over winter as sclerotia in affected host tissue. Conidiophores with spores are grown on the sclerotia in the spring, and a little later infection of the young leaf is produced. The new conidiophores break through the epidermis of both leaf surfaces or push through the stomata of the host in clumps. They are short and knobby and bear the usual Ramularia type of spore. The longer the clumps endure the larger is the stromatic tissue formed at their base, until a sclerotium visible to the unaided eye is formed.

An Ascochyta leaf-spot with similar symptoms is caused by *Ascochyta armoraciæ* Fckl., or as the fungus was previously known, by *Septoria armoraciæ* Sacc. The only difference is that pycnidia replace the sclerotia, and the spores are two-celled. The two types of spotting have been confused so much that it is impossible to say how widely the Ascochyta is distributed, whether it is known at all in America, or whether it causes as much injury as does the Ramularia.

A third leaf-spot reported from West Virginia, New Jersey, Illinois, and Missouri, and similar in appearance to the two former types, is caused by *Cercospora armoraciæ* Sacc., which also spots the foliage of *Brassica chinensis*. It is not so severe, however, on the latter host as is *Cercospora brassicicola* Henn. *Brassica chinensis*, known as Pechay, is a common vegetable grown in the Philippines.

No control measures for the three leaf-spots have been demonstrated, but it is probable that if the loss were severe they could be held in check by applications of bordeaux mixture 4-4-50, beginning when the leaves first enlarge and continuing at weekly intervals until danger of infection is past.

References

Halsted, B. D. Fungous diseases of various crops. N. J. Agr. Exp. Sta. Ann. Rept. **11**: 360. 1891.

Reinking, Otto. Philippine plant diseases. Phytopath. **9**: 114-140. 1919.

White-Spot of Turnip

Caused by *Cercosperella albomaculans* (E. and E.) Sacc.

This common leaf-spot of turnips, specimens of which have been collected in Connecticut, New York, South Carolina,

Louisiana, Mississippi, and California, and which probably is present in many other states, was described in 1894. Small circular ashen-colored spots, from one-eighth to one-fourth inch in diameter, show on both sides of the leaves. When the spotting becomes severe, the leaf turns yellow prematurely and dies. Apparently the fungus lives over winter in the old diseased plant refuse. It is seldom, if ever, serious enough to demand control measures.

References

Clinton, G. P. Report of the Botanist for 1906. Conn. Agr. Exp. Sta. Ann. Rept. **1906**: 330. 1907.

Ellis, J. B. and B. M. Everhart. New species of fungi from various localities. Proc. Acad. Nat. Sci. Philadelphia. Nov. 30, **1894**: 378. 1894.

Ascochyta Root-Rot of Radish

Caused by *Ascochyta* sp.

In Massachusetts a rather severe injury on seedling and young radish plants was observed in the greenhouse. An undescribed Ascochyta was constantly associated with the lesions. It was shown by experiment that the parasite is carried in the soil and not with the seed.

Soil sterilization evidently will prove effective if the disease should ever become harmful enough to demand attention (see page 600).

Reference

Stone, G. E. A disease of the radish. Mass. Agr. Exp. Sta. Ann. Rept. **21**: 50-51. 1909.

Olpidium in Cabbage

Caused by *Olpidium brassicæ* (Wor.) Dang.

When examining thin sections of cabbage roots under the microscope, it is not uncommon to observe in the epidermal cells and possibly the adjoining tissue, fungous fruit-bodies almost as large as the cell-cavity and often with a neck that extends through the tissue to the surface of the root. Zoospores are produced in the sporangium and gain their freedom by passing through the opening in the elongated neck.

The same fungus, *Olpidium brassicæ* has been observed in roots of tobacco and tomato. In no case is there any visible injury to the root, or any external symptoms by which the presence of the organism may be detected.

References

Bensaude, Mathilde. A species of Olpidium parasitic in the roots of tomato, tobacco, and cabbage. Phytopath. 13: 451-454. 1923.

Chupp, Charles. Studies on clubroot of cruciferous plants. N. Y. (Cornell) Agr. Exp. Sta. Bull. 387: 419-452. 1917.

Damping-off of Crucifers

Caused by various fungi

(See Damping-off of Tomatoes, page 546.)

Gray Mold-Rot of Crucifers

Caused by *Botrytis* sp.

(See Gray Mold-Rot of Lettuce, page 253.)

Reference

Potter, M. C. Rotting of turnips and swedes. Jour. Agr. Bd. Agr. (British) 3: 120-131. 1896.

Black Root-Rot of Horse-Radish

Caused by *Thielavia basicola* (B. and Br.) Zopf

(See Black Root-Rot of Bean, page 37.)

Root-Knot of Crucifers

Caused by *Heterodera radicicola* (Greef) Müll

(See Root-Knot of Tomato, page 550.)

CHAPTER VII

DISEASES OF CUCURBITS

THE diseases that affect cucumber are found also on many of the other cucurbits, so that it is considered wise to discuss the maladies of this group of vegetables in one chapter, rather than to scatter the information throughout the Manual. The term cucurbits here includes the cucumber, muskmelon, watermelon, squash, pumpkin, gourd, gherkins, citron, and several other less important hosts.

More than one hundred and fifty thousand acres of watermelons, more than seventy-five thousand acres of muskmelons, and more than fifty thousand acres of cucumbers are grown in the United States. Nearly half of the watermelons are produced in the four states: Georgia, Texas, Florida, and Missouri. Most of the muskmelons are grown in California, Arkansas, Maryland, New Jersey, Indiana, and Colorado. The states having the largest acreage of cucumbers are Michigan, New York, Wisconsin, and Florida.

BACTERIAL WILT OF CUCURBITS

Caused by *Bacillus tracheiphilus* EFS.

The bacterial wilt was first observed near Washington in 1893. Since then it has been reported from thirty-seven states, and from Canada, Germany, Russia, Transvaal, and Japan. It is interesting to note that the disease is not mentioned in the plant disease survey of England. In the United States the greatest losses are sustained in the latitude of Long Island, Pennsylvania, Kentucky, Indiana, and Kansas. The disease evidently decreases in virulence as the distance increases north

FIG. 58.—Field of cucumbers showing bacterial wilt on some of the vines.

and south of this line, although northern states have recorded heavy losses in epidemic years. In individual fields the entire crop may be destroyed, and the average reduction in yield for a state may be more than 20 per cent.

The bacterium affects plants belonging only to the family Cucurbitaceæ. It has been found on cucumbers, muskmelons, pumpkins, squashes, gherkins, *Cucumis odoratissimus, Benincasa cerifera, Cucurbita fœtidissima, C. californica, Sicyos angulatus,* and *Echinocystis lobata,* the last of which are wild hosts. Watermelons are almost wholly immune. Among the cucumbers, the American varieties are much less susceptible than are those from Europe. For example, the Chicago Pickling, in tests, gave 31 per cent infection, while the Short Green Early, a German variety, gave 85 per cent. A number of other varieties gave similar percentages.

Fig. 59.—Cucumber fruit on plant injured by bacterial wilt.

Symptoms.

As the disease is not accompanied by any soft-rot, the term wilt expresses exactly all the symptoms that may be observed on the outside of the vine. The infection begins on one or a few leaves, which droop at once. The disease spreads from the foliage downward into the petioles and stems until the entire plant has wilted and died (Fig. 58). Even the fruit is included in the wilting and shriveling (Fig. 59). Sometimes less susceptible plants like the squashes, not succumbing so quickly, will show a dwarfing in growth, occasionally accompanied by an excessive blossoming and branching. Mechanical injury to the base of the plant, a foot-rot, or stem-borers may cause symptoms resembling those of wilt, but an examination of the affected plant will easily reveal the true cause of any trouble other than wilt.

An excellent method of diagnosis is testing the sap for viscosity. The sap in a healthy stem is watery and not at all stringy. On the other hand, the sap of a diseased plant, when permitted to collect on the cut surface of a stem, will adhere to the fingers when touched and string out in delicate threads when the hand is drawn away. In severe infection the sap may also be milky-white instead of the normal colorless liquid, and will ooze out in white drops if one end of a cut stem is placed in water.

Cause of bacterial wilt.

The parasite is *Bacillus tracheiphilus,* a white rod-shaped motile organism with the group number 222.2322023. There are at least two strains of this organism, one of which is common on cucumber but which will not affect squash, and a second which may be inoculated successfully into both hosts. The bacterium is of special interest inasmuch as it must depend on the striped and 12-spotted cucumber beetles for aid in its life cycle. The beetles and the pathogene are so intimately related that if the former are eradicated the latter will disappear. Evidently the bacterium lives during the winter in the digestive tract of a few of the hibernating beetles, and in the spring finds its way through the feces of the carriers to the young plant. The parasite is never disseminated with the soil or the seed.

Infection does not take place through the stomata, but only through wounds that are deep enough to reach the vascular tissue. When infested feces are dropped on a leaf injured by the feeding beetles, infection takes place if the moisture from dew or rain is present. The bacteria swim through the droplets of water into the xylem ducts, through which they can migrate to all parts of the plant. When the beetles chew on invaded leaves, their mouth-parts become smeared with the bacteria that are carried to the next three or four plants on which the beetles feed or leave their droppings. Only a small percentage of the beetles are carriers, and infection takes place only when there is a film of water sufficient for the bacterium to reach the inner leaf tissue. Visible infection takes place in less than a week and the whole plant is invaded in twelve to fifteen days.

The wilt is not much influenced directly by temperature or rainfall, although it has been observed to increase in warm wet weather and to decrease during a drought. The real reason for this variation is the difference in the succulence of the plant, the sappy growth being more easily invaded. The organism, however, is very sensitive to heat. Its minimum temperature requirement is 46° F., its optimum 77° to 86° F., its maximum 95° F., and its thermal death point 110° F.

Control of bacterial wilt.

While the plants are young control depends almost altogether on the eradication of the striped and 12-spotted cucumber beetles, but as the vines grow older the application of bordeaux spray or copper-lime dust aids directly in lessening the number of infections. A few plants in the garden may be protected by inclosing them in cheese-cloth tents. Specially prepared sticks or wooden barrel hoops cut in halves are placed at right angles to each other with their sharpened ends inserted in the ground. The cheese-cloth is stretched over the hoops and its lower edges weighted with stones or soil so that no insect can crawl under the covering.

Large plantings of cucurbits must be protected by insecticides and fungicides. The very young seedlings are slightly stunted by hydrated lime; therefore, the entomologists have suggested several possible combinations of dusts that may be used for the first two applications: 1. Calcium arsenate one pound, gypsum twenty pounds; 2. paris-green one-half pound, lead arsenate one pound, and gypsum, ground lime-stone, hydrated lime, or air-slacked lime fifteen pounds; 3. copper-lime-arsenate of lead dust, 20-72-8; 4. black-leaf-40 five pounds, lead arsenate fifteen pounds, and gypsum, ground lime-stone, or air-slaked lime eighty pounds. The black-leaf-40 or nicotine dust should be applied on a bright, warm, wind-still day. If it is placed on the surface of the leaves without any precautions, many of the beetles will get away or hide under the soil until the nicotine fumes have been dissipated. In order to avoid this possibility, a large tin funnel is attached to the nozzle of the hand duster, and when in operation is set with some force over the plant and the necessary amount of dust liberated. The chamber keeps the beetles inclosed and at the same time retains the fumes for a few seconds in such small space that its insecticidal value is increased many times. The funnel need not be kept over the plant much longer than is necessary to manipulate the duster handle, after which it may be moved to the next plant in the row.

If a horse-drawn duster is employed, the nozzles can be inserted into trailing caverns made by bending into the shape of an inclosed arch strips of tin or galvanized iron four feet long and wide enough to make the arch of the necessary height. The front lower part of the cavern is cut into sled-runner shape so that it may be drawn over the ground without catching on stones, crop-refuse, or other impediments. The caverns are hitched by wire or rigid iron bars to the axle of the duster in such a manner that each will straddle a row. The slower the forward motion of the duster, the longer will the nicotine have to react under confinement.

The first application of dust is made as soon as the plants are through the ground, and succeeding ones at intervals of four or five days. New nicotine dust is bought each time or is freshly mixed at home, as it soon loses strength. When the plants begin to vine, applications are made at weekly intervals and usually without nicotine. The liquid bordeaux 4-4-50 with one and one-half pounds of powdered arsenate of lead, or copper-lime-lead arsenate dust 20-70-10, may be used on cucumbers and in many cases are more desirable than the nicotine dust even for the first applications. Muskmelons are more susceptible to injury so that the bordeaux mixture should be weakened to 3-4-50, and the amount of insecticide reduced as much as possible.

When intending to spray or dust any cucurbits, it is well to have in mind the width of the machine and plan the width of the rows accordingly. Ordinarily it is advisable to have the plants close together in the row and the rows far apart, so that the wheels will not injure the vines. While the plants are still small, one or two nozzles may be used for each row; but after vining begins, three or four nozzles are necessary. An ordinary four-row potato sprayer or duster can have its boom so arranged that one complete row and the inner half of each adjoining row are covered at one time. The nozzles are directed forward and all the pressure generated that the machine will bear, so that the leaves will be blown over and their under sides covered. Driving twice over the same row in opposite directions has at times been profitable.

Under certain conditions it may be advantageous not to have a spray boom but to employ two to four leads of rubber hose attached to the sprayer and as long as can be handled conveniently. Roadways every six to eight rows can be left open through the field, and one man handle the end of each hose, directing the spray toward all parts of the plant. The outer and longer hose may be fastened to an overhead projecting pole, so that it will not drag on the vines.

In addition to the application of a fungicide and an insecticide, as soon as a plant begins to show symptoms of wilting it should be removed from the field. It may also prove beneficial to have a catch-crop such as squashes sown thickly about the edge of the field a few days before the regular planting. The beetles are attracted to these. One or two later sowings may be made in order to have succulent growth for the feeding beetles. The catch-crop can be destroyed when it is no longer needed.

References

Doolittle, S. P. Overwintering of the bacterial wilt of cucurbits. Phytopath. **11**: 299, 300. 1921.

Doolittle, S. P. Comparative susceptibility of European and American varieties of cucumbers to bacterial wilt. Phytopath. **12**: 143-146. 1922.

Rand, F. V. Dissemination of bacterial wilt of cucurbits. Jour. Agr. Research **5**: 257-260. 1915.

Rand, F. V. and Lillian C. Cash. Some insect relations of *Bacillus tracheiphilus* Erw. Sm. Phytopath. **10**: 133-140. 1920.

Rand, F. V. and Ella M. A. Enlows. Transmission and control of bacterial wilt of cucurbits. Jour. Agr. Research **6**: 417-434. 1916.

Rand, F. V. and Ella M. A. Enlows. Bacterial wilt of cucurbits. U. S. Dept. Agr. Bull. 828: 1-43. 1920.

Smith, E. F. *Bacillus tracheiphilus* n. sp., a cause of wilting in various cucurbits. Centralbl. Bakt. u. Par. Allg. **1**: 364-374. 1895.

Smith, E. F. Wilt of cucurbits. *In* Bacteria in Relation to Plant Diseases. **2**: 209-299. 1911.

Downy-Mildew of Cucurbits

Caused by *Peronoplasmopara cubensis* (B. and C.) Clint.

In 1868 specimens of a cucurbit affected with downy-mildew were sent from Cuba to England for examination. Nothing more was heard of the disease until twenty years later when material was sent to Harvard University from Japan. The following year Halsted observed the disease in New Jersey, after which reports were made in quick succession of the damage it was doing in many places. It has now been reported from every state east of the Mississippi River and from at least nine states west of it, as well as from Europe, including England, from Java, Dutch East Africa, and from Brazil. It apparently is most destructive in the states along the Atlantic shore, especially those north of Virginia not including Maine.

The downy-mildew does not seem to be so virulent as when it was first discovered in the United States. In 1904 nearly all the muskmelon crop of Connecticut was lost. Three-fourths of the cucumbers on Long Island were destroyed in 1896, and in the succeeding year there was great loss in Ohio. It is true that individual states have reported 5, 10, and 25 per cent losses in recent years, but probably because of better control measures it is no longer considered a serious menace.

The downy-mildew occurs only on members of the Cucurbitaceæ, and mostly on those which are cultivated, although it has been observed on the wild cucumber and a few other weed hosts. It affects severely cucumbers and muskmelons, less severely pumpkins and squash, and rarely attacks watermelons. All the varieties of each host are about equally susceptible.

Symptoms.

Irregularly shaped yellowish spots appear on the upper sides of the foliage, usually of that nearest to the center of the hill. If the leaf is examined on the opposite side when dew or rain is present, the brown lesion will be covered, or at least bordered, by a purple fungous growth. The spots increase rapidly in size until the whole leaf withers and dies. In a badly diseased field it is not uncommon to find all the foliage dead near the center of the hill with only the young leaves remaining at the tips of the vines. If fruit is being formed during the attack, it will remain dwarfed and have a poor flavor. Cucumbers so affected are known as nubbins.

Cause.

As the fungus resembles morphologically both Peronospora and Plasmopara, it has been classed in both genera. It was later given the name Pseudoperonospora, but the author failed to give a description of the new genus, so that his nomenclature is not generally recognized. It is now known as *Perono-plasmopara cubensis*. It belongs to the very destructive order of Peronosporales, so that its life history is much the same as that of the Phytophthora on potato or the one on lima-beans.

The purple fungal growth on the lower side of the leaf is made up of long dichotomously branched conidiophores, on the tips of which are borne singly large lemon-shaped conidia, each with a conspicuous papilla. The conidia germinate by liberating zoospores, which swim about for a brief time then become stationary. They germinate by means of a germ-tube and penetrate any part of the leaf surface. As the mycelium grows between the host cells, it obtains its food through button-like haustoria extended into the cell. Within a few days it is established and sends out conidiophores through the stomata either singly or in small bunches.

Oospores have been reported by a Russian worker, but his statements have not been verified by other investigators. It is not known in what manner the fungus lives through the winter, although several possibilities have been suggested: First, the fungus fruits all the year in Florida and other southern states, and may migrate north in easy stages during the spring months. There seems to be a degree of probability in this assumption, since the disease does not become prevalent in the North until July or August. Secondly, it may survive in greenhouses in the North and accidently spread to outdoor hosts when the latter begin to grow. It does not live in the soil.

Most authors suggest that Peronoplasmopara requires a hot wet season. No doubt they are correct about the wetness, for zoospores must always have plenty of water. Apparently, however, the parasite is favored by a rather low minimum temperature, becoming epidemic when rainy periods are accompanied or followed by cool nights and comparatively warm days.

Control.

Spraying with bordeaux mixture or dusting with copper-lime dust as suggested for the control of bacterial wilt of cucurbits will eliminate downy-mildew. If only the latter is bothersome, applications with nicotine are unnecessary. The weekly applications should be made with high pressure and with at least four nozzles to the row.

References

Clinton, G. P. Downy-mildew, or blight, *Peronoplasmopara cubensis* (B. and C.) Clint., of muskmelons and cucumbers. Conn. Agr. Exp. Sta. Ann. Rept. 1904: 329-362. 1905.

Geise, F. W., H. H. Zimmerley and Herbert Spencer. Spraying and dusting vegetable crops in 1922. Va. Truck Exp. Sta. Bull. 41: 237-256. 1922.

Humphrey, J. E. The cucumber mildew—*Plasmopara cubensis* (B. and C.) Mass. Agr. Exp. Sta. Ann. Rept. 8: 210-212. 1891.

Jarvis, C. D. Spraying cucumbers and melons. Conn. (Storrs) Agr. Exp. Sta. Bull. 72: 85-123. 1912.

Rostowzew, S. J. Beiträge zur kenntniss der Peronosporeen. Flora 92: 405-430. 1903.

Stewart, F. C. The downy-mildew of the cucumber: what it is a how to prevent it. N. Y. (Geneva) Agr. Exp. Sta. Bull. 119: 1 183. 1897.

Mosaic of Cucurbits

Cause undetermined

The peculiar wartiness of cucumber fruits affected with mosaic is so noticeable that the trouble was recognized more than twenty years ago. It is now prevalent in all parts of the United States and Canada where cucumbers are grown extensively, and has recently been reported from England. The disease is known by various names, all of which are descriptive of the symptoms, such as mosaic, white-pickle, white-wart, nubbin disease, and chlorosis.

Next to bacterial wilt it is probably the most serious trouble affecting cucurbits. It kills the vines in greenhouses, it stunts them in the field, reduces the number of fruit, gives canteloupes a poor flavor, and causes the cucumber to develop into misshapen nubbins that do not cure properly in the pickling vat. It may become so serious in a field that the

FIG. 60.—Mosaic on fruit of cucumber.

entire crop is ruined, and in certain of the older pickling sections it has become such a menace that growers are forced to discontinue cucumber-growing.

Nearly every member of the Cucurbitaceæ is susceptible to mosaic. Cucumber and canteloupe are most severely affected, pumpkin and squash less so, citron rarely, and watermelon only slightly if at all. The wild species, *Echinocystis lobata* and *Sicyos angulatus*, are often diseased and serve as points for dissemination of the inoculum. A number of hosts not included in the Cucurbitaceæ are also listed, among the most important of which is the milkweed. Other suggested hosts are pepper, pokeweed, tobacco, pigweed, ground-cherry and potato. It does not mean necessarily that the potato mosaic, tobacco mosaic, and cucumber mosaic are the same. In fact, it will probably be found that they are different and cause different symptoms on the same host.

Symptoms.

The plants may be infected at any stage in their growth. If the seedlings are diseased, the cotyledons turn yellow, the leaves show mottling of yellow and green, and the plant is much dwarfed, bears no fruit and usually dies early. More often the plants are attacked after they have grown out of the seedling stage and are developing most rapidly. The peculiar yellow and green mottling on the younger leaves as described for the potato mosaic is the first symptom. The upper surface of the leaf may become wrinkled and its edges curl downward. The distortion and mosaic pattern of the leaf are accompanied

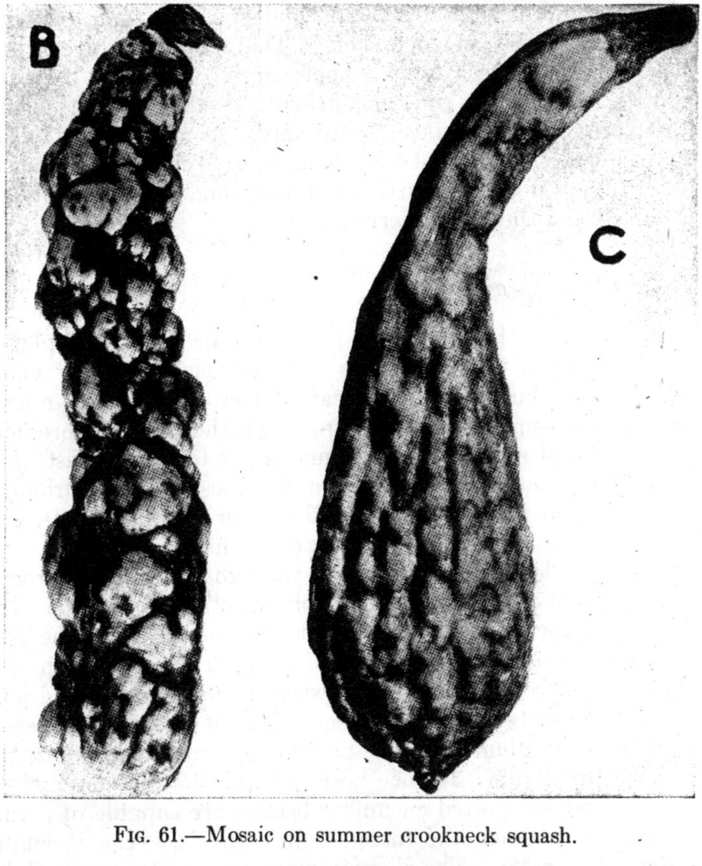

FIG. 61.—Mosaic on summer crookneck squash.

by dwarfing of the plant and reduction in the set of the fruit. If the cucumber fruit is affected it shows alternate irregular yellow and green spots, the latter being raised on wart-like projections (Fig. 60). As the plant grows older, the mottling is not so evident, but the roughening of leaves and fruit is more marked. Often the younger internodes of the vines are shortened so that the young tip leaves are bunched or in a rosette.

The symptoms on pumpkins and squashes (Fig. 61) are similar to those on cucumbers. The leaves are even more roughened, having a quilted appearance. The fruit begins to mottle at the stem-end, and gradually takes on a typically discolored warty surface, or it may become russeted. The seeds of such fruit are usually small and shrunken. The musk-melon vines may show little or no mottling, but the plants are dwarfed and the fruit inferior in quality.

Cause of mosaic.

No cause has been assigned to the mosaic diseases of plants. The trouble is very infectious as has been proved by many experiments, but whether the causal factor is an ultra-microscopic organism in the plant sap, or whether it is the presence or absence of some chemical remains yet to be demonstrated. Evidence points to the fact that the mosaic on cucurbits is distinct from that on many of the other common hosts, and that possibly there are two types on cucurbits alone.

The mosaic is not transmitted through the soil or through the seeds of the common cucurbit vegetables. It has been shown, however, that it can live during the winter in the seed of the wild cucumber, *Echinocystis lobata,* and in the perennial roots of the milkweed, pokeweed, ground-cherry, and possibly other hosts. During the early part of the summer insects carry the inoculum from the previously named sources to the cultivated plants. It has been proved that melon aphids, striped and 12-spotted cucumber beetles are capable of acting as carriers. When the insects injure the leaf, the inoculum enters the tissue. The disease may also be transmitted by rubbing together a healthy and a diseased leaf until there is bruising, or by any other method in which plant sap may be transferred from one vine to another. The juice loses its power of infection when heated to 70° F. or when left standing forty-eight hours or more after expressing from the plant. It may be diluted one part in ten thousand parts of water and still be slightly infectious.

The mosaic is not influenced much by rainfall nor temperature. It is able to develop in any temperature in which the cucurbit will grow, although in cool weather the incubation period is longer than when it is warm. Temperature, also, has an effect on the type of symptoms which appear on the diseased host.

Control of mosaic.

Definite and satisfactory control measures have not been demonstrated. Spraying or dusting for the eradication of insects as suggested for the control of bacterial wilt, and the removal for a considerable distance from the field of all wild cucumbers, bur cucumbers, milkweed, poke-berry, and ground-cherry, are measures which, if faithfully adopted, will control the disease. The eradication of the weeds is almost impossible in many sections; nevertheless, it should be attempted in districts in which cucumbers or muskmelons are grown intensively. Diseased plants in the greenhouse should be pulled up by the roots, but left in place on the trellis until dead and dry so that the neighboring intertwining vines will not be inoculated through injury when the affected plant is removed.

References

Doolittle, S. P. The mosaic disease of cucurbits. U. S. Dept. Agr. Bull. 879: 1-69. 1920.

Doolittle, S. P. and W. W. Gilbert. Seed transmission of cucurbit mosaic by the wild cucumber. Phytopath. 9: 326-327. 1919.

Doolittle, S. P. and M. N. Walker. Cross-inoculation studies with cucurbit mosaic. Science 57: 477. 1923.

Doolittle, S. P. and M. N. Walker. Experiments on the control of cucurbit mosaic. Phytopath. 14: 56. 1924.

Jagger, Ivan C. Two transmissible mosaic diseases of cucumber. Phytopath. 7: 61. 1917.

Anthracnose of Cucurbits

Caused by *Colletotrichum lagenarium* (Pass) E. and H.

No disease of cucurbits is more destructive than anthracnose. It has been known for more than fifty years, and is now present in every country where cucurbits are grown. In many markets it is difficult to find watermelons or muskmelons that do not have at least a few spots. The loss of an entire field is not rare, and individual states report to the Plant Disease Survey Office average reductions as great as 50 per cent in the yield or value of the crop.

Although the watermelon is most severely infected, the disease may appear in epidemic form on muskmelons and cucumbers. The fungus also is able to infect gourds, citrons, several cucurbits of no economic importance, and possibly squash and pumpkin fruits. There is considerable difference in varietal susceptibility of certain hosts. For example, one case was observed in which the White Pearl watermelon, a round white variety growing among some long dark green ones, was killed about the beginning of the picking season, while the latter were only slightly affected, even at the end of the summer.

Symptoms.

All parts of the plant above ground may be infected. The spots on the foliage begin as small yellowish or water-soaked areas, which enlarge rapidly and turn brown in most cucurbits, but black on the watermelon. The dry dead tissue may break and shatter or the whole leaf die. In the case of muskmelons the petioles are attacked so often that defoliation of the vines frequently results. Elongated lesions similar to those on the leaf are present on the stems, and these lesions together with the destruction of the foliage may kill the whole vine. When the fruit-pedicels are attacked the young fruit may darken, shrivel and die. It is not uncommon to see such small black fruit in an invaded watermelon patch.

The most noticeable symptoms are formed on the fruit, where circular, black, sunken cankers appear. Such depressions vary in size, depending on the host on which they are found, their age, and their tendency toward becoming con-

FIG. 62.—Anthracnose spots on watermelon fruit.

fluent. On the watermelon the spots may measure one-fourth inch to two inches in diameter, and from nearly level with the surface to almost a third of an inch in depth (Fig. 62). When moisture is present the black center of the lesion is covered with a gelatinous mass of salmon-colored spores. The cankers lined by this characteristic color can never be mistaken for any other disease. The canker does not penetrate deeply enough to discolor much of the edible flesh, but a muskmelon or watermelon with a large number of lesions is usually not delicious in taste, and may in extreme cases be bitter. When the vines are killed after the melon is partly mature, the fruit may ripen but the flesh is tough and insipid. Cankered fruit is likely to be destroyed by soft-rot organisms that gain entrance through the broken rind.

Cause of anthracnose.

Like many other fungi which have been known for a long time, the parasite has been given so many names that it is difficult to choose among them. Probably the most commonly accepted name now is *Colletotrichum lagenarium,* although the English writers still seem to prefer the appellation, *Colletotrichum oligochœtum* Cav. The setæ not always being present in the fruit-body, there likewise is some ground for the term Glœosporium.

The same variety of fungus is found on all the infected cucurbit hosts, so that it will spread from cucumbers to melons or from melons to gourds. The fungus remains alive in the old diseased tissue in the soil during at least one winter, as well as in the living seed. When infected fruit is passed through a grinder then permitted to remain for a day or more in the fermenting juices, in order to liberate the seed for future planting, there is every opportunity for the spores to become attached to the sticky surface of the seeds. Furthermore, cultures of the fungus have been obtained from surface water in an infested field, and from water in shallow wells near which diseased plants grew.

The young plant may be inoculated successfully in any stage of its growth. The spores are washed by water, splashed by rain, carried on tools by the workmen to the host. In the presence of moisture the spore germinates, and after producing an appressorium pushes its germ-tube through the cell-wall into the tissue below. The process of germination and penetration requires about three days. The mycelium develops rapidly. When the tissue has collapsed, numerous acervuli are developed in the dead area, and produce pink colored spores so abundantly that the mass in the bottom of the canker is the most striking symptom of the disease.

Infection may take place just before the melons or cucumbers are picked, producing the unsightly lesions while the crop is being shipped to market. It is not known definitely that spores are disseminated after the fruit is in the car, or that lesions result from any inoculations taking place during transportation.

The fungus occurs in epidemic form only when there is more than the average rainfall. In addition it requires a relatively high temperature, growing best at 75° F. It, however, will develop slightly at temperatures as low as 50° and as high as 86° F.

Control of anthracnose.

Notwithstanding the seriousness of the disease when permitted to run its course, satisfactory control measures have been discovered. As the pathogene lives for at least one season in the diseased plant refuse left on the soil, a two-year rotation of crops is essential. To this is added seed treatment. The seeds are dipped in corrosive sublimate (1 ounce in 7.5 gallons of water) for five minutes, then rinsed thoroughly in running water and dried. In order to avoid infection from other sources, spraying with bordeaux mixture as outlined for the control of bacterial wilt and stem-end rot of watermelon is recommended.

REFERENCES

Bewley, W. F. Anthracnose of cucumber under glass II. Jour. Min. Agr. (Gt. Brit.) **29**: 558-562. 1922.

Cardinell, H. A. One season's dusting against cucumber beetles and anthracnose of watermelons in Missouri. Proc. Amer. Soc. Hort. Sci. **18**: 123-130. 1922.

Gardner, M. W. Anthracnose of cucurbits. U. S. Dept. Agr. Bull. 727: 1-68. 1918.

Link, G. K. K. and F. C. Meier. Anthracnose of muskmelons. U. S. Dept. Agr. Circ. 217: 1-4. 1922.

Meier, F. C. Control of watermelon anthracnose by spraying. U. S. Dept. Agr. Circ. 90: 1-11. 1920.

Sheldon, J. L. Diseases of melons and cucumbers during 1903 and 1904. W. Va. Agr. Exp. Sta. Bull. 94: 121-138. 1904.

Angular Leaf-Spot of Cucurbits

Caused by *Bacterium lachrymans* Sm. and Bryan

The disease was probably first observed in 1904 when Smith isolated an organism from an affected cucumber, but was unable to obtain reinfections. Since then it has been observed in nearly every state east of the Mississippi River, and in Iowa, Minnesota, North Dakota, Colorado and California, and in Canada and Europe. It is most destructive in Indiana, Michigan and Wisconsin, but even in these states the disease is not important since definite control measures have been suggested.

The malady attacks cucumbers, West Indian gherkins, Mandera gourd, hedgehog gourd, calabash gourd, and *Bryonopsis laciniosa,* a tropical cucurbit. It failed to infect muskmelon, watermelon, citron, squash, pumpkin, and common gourd.

Symptoms.

The disease appears on the leaves, stems, and fruit. The spots on the foliage are irregular in shape, angular, and have a water-soaked appearance. In the presence of moisture bacteria ooze from the spot in tear-like droplets, which dry down into a white residue. The water-soaked area later turns grey and dies. The drying and shrinking of the dead tissue may tear it away from the healthy portion, leaving large irregular holes in the affected leaves. The spots on the fruit are much smaller and nearly circular. When the diseased portion dies, the tissue becomes white in color and may crack open. The lesions are quite superficial, but the injured epidermis permits the entrance of soft-rot organisms that may cause the decay of the entire fruit.

Cause.

The angular leaf-spot is caused by *Bacterium lachrymans,* a white, rod-shaped, 1-5-polar flagellate organism with the group number 211.2322123. It may live over the winter on the surface of the seed or in the diseased plants left in the field. During rains it is splashed from the soil to the stems and leaves, and later to the fruit. After infection has once taken place, the organism is transferred readily on the hands and clothing of the pickers. Infection takes place through the stomata and, since they close during the night, most of the entrances by the bacteria take place in the early morning after daylight but before the dew or rain has dried. Inoculation may take place on the fruit just before picking, so that when such cucumbers are shipped long distances much spotting may appear during transportation on fruit that apparently was healthy when picked. When affected cucumbers are saved for seed, the method of fermenting the pulp and seed to separate them is an ideal way for contaminating the seed and inoculating the new crop.

Frequent rains with a temperature of 77° to 86° F. are optimum conditions for the growth of the bacterium. It develops even when the average maximum temperature for five days registers 97° F., and when the heat stamps out bacterial wilt.

Control.

Seed treatment, rotation of crops, and spraying or dusting have proved successful in controlling *Bacterium lachrymans.* The seed is treated by tying small quantities loosely in cheesecloth and dipping for five minutes in a corrosive sublimate solution (1 ounce in 7.5 gallons of water). The solution corrodes metal, therefore should be placed in wooden, glass, or earthenware receptacles. While the seeds are soaking. the bundles should be stirred so that all the air is liberated from the cloth and the liquid permitted to touch the seed. After the seed is treated five minutes, it is removed and rinsed in clean running water, then dried.

The spraying or dusting is conducted in the manner suggested for the control of the bacterial wilt of cucurbits, except that if the angular leaf-spot is the only disease to be combated, the dusts on the seedlings need not be applied.

References

Carsner, Eubanks. Angular leaf-spot of cucumber: dissemination, overwintering, and control. Jour. Agr. Research **15**: 201-220. 1918.

Gardner, M. W. and W. W. Gilbert. Cucumber angular leaf-spot and anthracnose overwintering and seed treatment control. Phytopath. **8**: 79, 80. 1918.

Jones, L. R. and S. P. Doolittle. Angular leaf-spot of cucumber. Phytopath. **11**: 297-298. 1921.

Smith, E. F. and Mary K. Bryan. Angular leaf-spot of cucumbers. Jour. Agr. Research **5**: 465-476. 1915.

Mycosphærella Wilt and Black-Rot of Cucurbits

Caused by *Mycosphærella citrullina* (Smith) Gross.

The wilt and black-rot of cucurbits has been present in the United States at least since 1890 and possibly longer. It has also been reported from many parts of Europe and from Japan. Ordinarily it is rare, but occasionally may appear in epidemic form, especially in the greenhouse where it causes much loss. At Geneva, New York, and in Japan muskmelons were badly wilted and cucumbers shipped from Florida were severely injured by the black-rot. It has been found on watermelon, muskmelon, cucumber, gourd, pumpkin, and squash.

Symptoms.

The disease causes a leaf-spot, a stem-wilt, and a black-rot of the fruit. On muskmelon the stem-wilt seems to be the most common, while on cucumber and squash the fruit-rot is the most prevalent. The spots on the leaves are grayish to brown dead areas of varying size and marked by the presence of black pycnidia partly embedded in the tissue. In a severe attack the leaves turn yellow and shrivel. The fruit-rot begins as a water-soaked or grayish area on the rind, which gradually darkens until in many cases it is almost jet-black. Only a small part of the surface may be included, or the disease may spread over all the fruit even in the case of large pumpkins and squashes. The flesh remains firm until soft-rot organisms enter when the whole fruit will soften and collapse. The rind is marked by great numbers of black pycnidia and perithecia, which aid in differentiating the trouble from other rots.

The infection on the stem begins at the node and usually in the axil of the leaf. The affected tissue takes on a water-soaked or oily green appearance. The invasion of the fungus does not spread far up or down the stem, but causes a girdle that wilts the upper part of the vine. The surface of the lesion may be partly covered with a dark exuded gum. If there is no gum present, the affected part turns gray after the tissue dies. Pycnidia are present, but not in such large numbers as on the fruit.

Cause.

The organism was first described as a Phoma, then changed to *Ascochyta citrullina* (Chester) Smith and to *Diplodina citrullina* (Sm.) Gross. Later the perfect stage was discovered and designated as *Mycosphærella citrullina.* There were some investigators who believed it was identical with the organism causing Ascochyta blight of tomatoes, and that it should be classed with the latter. A comparison of authentic cultures has since shown this to be wrong.

The two-celled, hyaline, asexual spores are borne in almost innumerable pycnidia, from which they are discharged in long tendrils. Scattered among the pycnidia and similar in appearance are the perithecia with their asci and two-celled hyaline ascospores. Both the pycnospores and ascospores may cause infection during the summer, and possibly may live over the winter to serve as primary inoculum in the spring. Evidently the fungus cannot live long in the soil without the presence of the host.

The fungus spreads most rapidly in a warm moist environment.

Control.

Spraying, as recommended for the control of bacterial wilt of cucurbits, has proved effective against the Mycosphærella.

References

(See also Ascochyta Blight of Tomatoes, page 575)

Grossenbacher, J. G. A Mycosphærella wilt of melons. N. Y. (Geneva) Agr. Exp. Sta. Tech. Bull. 9: 196-229. 1909.

Hemmi, Takewo. On the occurrence of Mycosphærella wilt of muskmelons in Japan. Phytopath. 12: 394-397. 1922.

Kauffman, C. H. A black-rot of squash. Rept. Mich. Acad. Sci. 22: 201-202. 1921.

Massee, G. Cucumber and tomato canker. Jour. Bd. Agr. (London) 16: 579-581. 1909.

Meier F. Chas. Drechsler, and Emery Eddy. Cucumber black-rot caused by *Mycosphærella citrullina*. Phytopath. 12: 43. 1922.

Smith, C. O. Leaf-spots on cucurbits. Del. Agr. Exp. Sta. Bull. 70: 1-16. 1905.

Macrosporium Blight of Cucurbits

Caused by *Macrosporium cucumerinum* E. and E.

Since 1894 a leaf-blight has been very prevalent on muskmelons in America and Europe. The same fungus is also found on squash, cucumber, and watermelon, and has been inoculated successfully into potato and tomato, though the last two hosts are probably not affected naturally under field conditions. The disease does much injury in many cucurbit fields in defoliating the vines and thereby diminishing the yield, and in injuring the quality of the melons by ripening them prematurely, especially those harvested late.

Symptoms.

The first symptoms usually appear during the middle of the season and on the leaves nearest the center of the hill. The spots rapidly increase in number, later spreading to the leaves toward the tips of the vines. The affected areas at first are

FIG. 63.—Macrosporium leaf-spot on muskmelon.

small, circular, and somewhat water-soaked in appearance (Fig. 63). The spot enlarges in all directions, ordinarily showing concentric rings and a definite margin on the upper side of the leaf. On the lower side rings seldom occur and the margins are very indefinite. The lesions when single may vary in size from mere points to half an inch or more in diameter, while confluent spots may cover a fourth of the surface of the leaf. The muskmelon foliage is more susceptible to the injury than is that of other cucurbits; therefore, is often found curled downward, or the vines will be almost completely defoliated. The fruit also is injured by depressed spots.

Cause.

By the aid of a hand-lens, a black moldy growth may be observed covering parts of the dead areas on both sides of the leaf. These are the conidiophores with large, brown, muriform spores. Evidently there frequently are two fungi present, one with spores borne singly and the other with spores borne one above the other. Investigators do not agree as to which of these is the causal organism. Some of them say it is the former, and name it *Macrosporium cucumerinum,* and others insist that the parasite has catenulate spores and is named *Alternaria brassicæ* var. *nigrescens* Pegl. It is possible that in different sections of the country the parasites are not identical, but a number of herbarium specimens collected by various pathologists seem to be the same, though labelled differently. The disagreement in nomenclature has no practical bearing as each fungus has the same life history. The parasite lives over as mycelium in old diseased refuse, and possibly also in the soil for a year. The inoculum is spattered by rain to the leaves where infection occurs, a crop of spores are borne, and the simple life cycle is repeated.

The disease spreads more rapidly in warm humid weather. Its optimum temperature is about 86° F., although it will increase at temperatures considerably below and above this point.

Control.

Spraying or dusting as suggested for the control of bacterial wilt of cucurbits has proved effective. If the Macrosporium blight alone is bothersome, the dust on the seedlings need not be applied.

REFERENCES

Brisley, H. R. Studies on the blight of cucurbits caused by *Macrosporium cucumerinum* E. and E. Phytopath. **13**: 199-204. 1923.

Orton, W. A. Spraying for cucumber and melon diseases. U. S. Dept. Agr. Farmers' Bull. 231: 1-24. 1905.

Sheldon, J. L. Diseases of melons and cucumbers during 1903 and 1904. W. Va. Agr. Exp. Sta. Bull. 94: 121-138. 1904.

Sturgis, W. C. A disease of melons (Alternaria sp.). Conn. Agr. Exp. Sta. Ann. Rept. **1895**: 186-187. 1896.

CORYNESPORA LEAF-SPOT OF CUCURBITS

Caused by *Corynespora melonis* (Cooke) Lindau

The leaf-spot, which has caused much injury to melons in England, Holland, France, Denmark, and Germany, has never been reported from America. The first written description was published in England in 1896. It was then becoming serious and since then has done much damage in that country and on the continent. It evidently appears on both greenhouse and field cucurbits.

The spots occur here and there over the leaves as light yellow areas, which turn brown as they enlarge. The lesions average about one-third inch in diameter. The centers of the older spots dry and become gray and are surrounded by a brown zone which is inclosed by a lighter border. Often the spots are bounded by the small veins. The dead tissue of the lower side of the leaf, and to a less extent on the upper side, is covered with a dark downy layer in contrast to the white downy hair of the normal tissue. The trouble is so destructive to the foliage that the term "leaf-fire" is used to describe it in Holland.

The fungus when first discovered was classed with the Cercosporas. Later it was observed that the spores are borne in chains of two or three, therefore the name *Coryespora mazei* Güs. was applied. It has since been pointed out that the original name of the species should have been retained, and that *Corynespora melonis* is correct.

The fungus is characterized by spores varying greatly in size and number of cross walls. They are slightly bent, a little larger at the base than at the tip, have from three to thirty-three cells, and are borne on rather long conidiophores. They are carried by rain, insects, or tools to neighboring plants where they soon cause infection. The whole life history of the fungus is not known, but it is supposed that spores or mycelium are disseminated with the cucurbit seed. A warm humid atmosphere is necessary for the propagation of the pathogene.

Control measures for the disease are based more on theory than practice. Seed treatment is suggested. According to work done in Holland, the seed is soaked twelve hours in one-half per cent of a copper-sulfate solution or four hours in formaldehyde (1 part in 400 parts of water). In either case the seed is rinsed after the treatment and dried. In addition it may be necessary to spray the plants as recommended for the control of the bacterial wilt of cucurbits.

References

Güssow, H. T. Uber eine neue krankheit an gurken in England (*Corynespora mazei* Güssow gen. et spec. nov.). Ztschr. f. Pflanzenkr. **16:** 10-13. 1906.

Höstermann. Blattfleckenkrankheit der gurke. Handelsblatt f. d. Deut. Gartenbau **34:** 336-337. 1919.

Laubert, R. Die Corynespora blattfleckenkrankheit der gurke, ihre verbreitung und bekämpfung. Deut. Landw. Presse **38:** 819-820. 1911.

Quanjer, H. M. Het "Bladvuur" der komkommers veroozaakt door *Corynespora mazei* Güss. Tijdschr. Plantenz. **14:** 78-95. 1908.

Scab of Cucurbits

Caused by *Cladosporium cucumerinum* Ell. and Arth.

Since 1887, when scab was first observed in New York, the disease has been discovered in all the principal cucumber-growing sections of the world, and occasionally causing much damage. It has been observed on cucumbers, muskmelons, and pumpkins.

The affected leaves may have water-soaked spots and be wilted, and the stems have slight cankers, but most of the injury occurs on the fruit. While the fruit is still small, the lesions may appear as gray slightly sunken places about an eighth of an inch in diameter. Frequently a viscid substance is exuded in drops from the lesion as if the trouble were the result of an insect puncture. The exudation is so generally prevalent that in some localities the trouble is referred to as gummosis. The canker grows darker colored with age, and the collapsed tissue sinks farther in until a pronounced cavity is formed. The cavity is lined with a dark green velvety layer of the causal fungus. Occasionally adjoining spots unite forming a lesion one-half inch or more in diameter.

The fungus is composed of greenish colored mycelium, short simple conidiophores, and one-celled spores borne terminally. When the fungus affects the leaf, knots of hyphæ collect in the stomatal cavity, and clumps of the conidiophores, arising from these knots, protrude through the stomata. On the fruit, the mycelium seems to thrive on the gummy exudate in the canker, and with this gum forms a dense layer which lines the cavity. The spores are disseminated by any method in which microscopic bodies may be carried from plant to plant.

The organism apparently requires wet warm weather for optimum growth. The maximum temperature is 95°, the optimum about 77°, and the minimum 52° F.

Spraying with bordeaux mixture or dusting as suggested for the control of bacterial wilt of cucurbits has proved fairly successful when the applications were made repeatedly and before rains. Diseased refuse, especially in the greenhouse, should be burned or buried deeply and the house fumigated with sulfur before another cucurbit crop is planted.

References

Aderhold, R. Cladosporium und Sporidesmium auf Gurke und Kürbis. Ztschr. f. Pflanzenkr. 6: 72-76. 1896.

Arthur, J. C. Spotting of peaches and cucumbers. Ind. Agr. Exp. Sta. Bull. 19: 5-10. 1889.

Doolittle, S. P. Cucumber scab caused by *Cladosporium cucumerinum*. Mich. Acad. Sci. Rept. 17: 87-116. 1916.

Gilbert, W. W. Virulence of different strains of *Cladosporium cucumerinum*. Phytopath. 7: 62. 1917.

Humphrey, J. E. The leaf-blight, *Cladosporium cucumerinum* E. and A. Mass. Agr. Exp. Sta. Ann. Rept. 10: 227-229. 1893.

Fusarium Wilt of Cucurbits

Caused by *Fusarium* sp.

Since the work of Smith in 1899, much has been written about the Fusarial wilt of cucurbits in America. It is extremely important in the South, especially on watermelon. In the North most of the trouble is found on cucumbers and muskmelons grown under glass. All cucurbits, no doubt, are more or less susceptible to the fungous wilt, the watermelon and squash being most often reported as injured, muskmelons and cucumbers are less susceptible, while pumpkins, gourds, and citron show marked resistance.

Symptoms.

The term wilt pictures vividly the symptoms of the disease. At first the vines usually wilt only during the middle of the day, the symptoms becoming more pronounced until the whole plant remains wilted and dies. Sometimes the lack of turgidity is not evident until the vine has been infected for some time. In the meanwhile the vine does not grow as rapidly as it should, the foliage has a yellow color, and there is a pronounced appearance of under-nourishment. After the plant is dead, or even before it has quite reached that stage, a mycelial weft may grow about the base of the vine. After the vine dies, the fungus fruits on the surface of the stem if moisture is present.

Cause.

The wilt organism generally found on watermelon is *Fusarium niveum* EFS. The two additional species, *Fusarium ci-*

trulli Taub. and *Fusarium poolensis* Taub., are accredited with causing a wilt on the same host. The wilt on squash is attributed to *Fusarium cucurbitæ* Taub. Still another species of Fusarium has been reported on cucumber and muskmelon. Each of the above named species is considered to be limited to the one or two hosts, and does not affect any of the other cucurbits or any unrelated plants. They differ morphologically in the presence or absence of sclerotia, and the size and septation of the spores. For instance, *niveum* has large sclerotia, and spores measuring 25-55 x 3.5-5 μ, mostly three-septate; *citrulli* has large sclerotia, and spores measuring 50-80 x 3-4 μ, mostly five-septate; *poolensis* has no sclerotia, its spores measuring 32-35 x 2.5-3 μ, a few being four- or five-septate; and *cucurbitæ* has no sclerotia, its spores measuring 30-80 x 2-3 μ, nearly all being three-septate.

The life history of each parasite is much the same. The fungi not only live on diseased vines but can live indefinitely in the soil while the host is absent. After the mycelium enters the vascular tissue, it grows so luxuriantly in the water-ducts that the cavities are completely filled with the hyphæ. The Fusarium sporulates after the host is killed. The spores and mycelium are disseminated with infested seeds, soil, and manure. It is a common practice to make hay from the wild grass or cover-crops growing in the watermelon fields. In raking up the hay the old dead vines are also collected, so that many of them are stored in the barn and finally are cast from the manger to the manure heap. In this manner fields that have never grown melons become contaminated.

Control of Fusarium wilt.

A soil organism is difficult to control under field conditions. In the greenhouse it is practicable to sterilize the beds with steam or formaldehyde (page 602). In outdoor plantings care should be taken in procuring manure free from contamination. If such is not available, it is better to buy commercial fertilizers to supply the plant-food. Long rotations with immune crops are desirable. If the soil once becomes infected, no melons should be grown there for ten or fifteen years even though such a restraint forces the renting of land on some other farm to use as a melon patch.

As the fungus grows on the fresh pulp attached to the watermelon seed and is also able to survive in a dormant condition while the seed is stored, seed treatment is recommended. Corrosive sublimate (1 ounce in 7.5 gallons of water) may be used as the disinfectant, the seeds being soaked five to seven minutes after which they are thoroughly washed in running water. The seeds may be planted wet or dried quickly in a shallow tray with a screen bottom. Formaldehyde (1 part in 200 parts of water) may also be employed, the seed being soaked one-half hour. The seed in this case is dried before planting.

Resistant varieties offer a solution of the problem. Orton has demonstrated the feasibility of crossing a desirable type of watermelon with a resistant citron to obtain a commercial type of melon that will withstand the wilt. In order to meet the demand for a definite type of melon in the different states, a number of crosses with immune hosts may be necessary; nevertheless, all these difficulties will finally be overcome. On farms on which the squash is seriously affected by the wilt, the "Sugar Through" gourd is suggested as taking the place of the susceptible squash, at least for home consumption.

References

Cook, Mel T. A greenhouse disease of melons. Phytopath. 13: 462, 463. 1923.

Fulton, H. R. and J. R. Winston. Watermelon wilt spread by contaminated seed. N. C. Agr. Exp. Sta. Bien. Rept. 1913-14: 48-51. 1915.

Orton, W. A. A study of disease resistance in watermelons. Science n. s. 25: 288. 1907.

Smith, E. F. Wilt disease of cotton, watermelon, and cowpeas. U. S. Dept. Agr. Div. Veg. Phys. and Path. Bull. 17: 1-72. 1899.

Stone, G. E. Fusarium disease of cucumbers and other plants. Mass. Agr. Exp. Sta. Ann. Rept. 23: 62-68. 1911.

Taubenhaus, J. J. Wilts of the watermelon and related crops. Tex. Agr. Exp. Sta. Bull. 260: 1-50. 1920.

Powdery-Mildew of Cucurbits

Caused by *Erysiphe cichoracearum* DC.

The powdery-mildew of cucurbits is probably present in every country of the world where the hosts are grown. It has long been known in Europe, but was not reported in America until about 1890. It is seldom of economic importance in the field, although it may occasionally become destructive in greenhouses. The disease affects nearly every known cucurbit, having been mentioned especially on cucumbers, squash, pumpkin, and gourd. A fungus with the same name and similar morphologically is found on goldenrod and aster, yet the race of the fungus appearing on the cucurbits will not infect the two unrelated hosts.

The powdery-mildew on the foliage and young stems is first evident as a talcum-like growth on the surface of the plant. During this stage the tissue still appears normal, but finally the spots turn brown and dry. In severe attacks the leaves may be killed. Rarely the perfect stage of the parasite is present. The diagnostic sign then is the number of small, black, globose fruit-bodies on the plant surface and large enough to be seen easily without the aid of the hand-lens. When all the foliage on a vine is attacked, the plant is weakened, has a yellowish color, and much of the fruit must be discarded as culls.

Several powdery-mildew fungi have been reported on cucurbits. The most common, and possibly the only one, is *Erysiphe cichoracearum.* The talcum-like growth on the host is made up of mycelial threads from which arise numerous short conidiophores bearing chains of barrel-shaped conidia. The black fruit-bodies appearing later are perithecia in which are borne the asci and ascospores. The parasite is able to live over winter in the perithecial stage on infected leaves and petioles. Spraying or dusting as recommended for the bacterial wilt of cucurbits will eradicate the fungus. Dusting the plants with sulfur, or fumes from a heated pot of sulfur in the closed greenhouse will effectually stop the growth of the parasite.

References

Humphrey, J. E. The powdery mildew of the cucumber, *Erysiphe cichoracearum* DC. Mass. Agr. Exp. Sta. Ann. Rept. **9**: 222-226. 1892.

Reed, G. M. Infection experiments with the mildew on cucurbits, *Erysiphe cichoracearum* DC. Trans. Wisc. Acad. Sci. Arts and Letters **15**: 527-547. 1907.

Reed, G. M. Physiological specialization of parasitic fungi. Brooklyn Bot. Gard. Mem. **1**: 348-409. 1918.

Salmon, E. S. Specialization of parasitism in the Erysiphaceæ. New Phytol. **3**: 55, 60, 109-121. 1904.

Stewart, F. C. Notes on New York plant diseases, I. N. Y. (Geneva) Agr. Exp. Sta. Bull. 328: 349. 1910.

Soft-Rot of Cucurbits

Caused by *Bacillus carotovorus* Jones

(For a more detailed discussion see Bacterial Soft-Rot of Carrots, page 76.)

The same bacterium that causes a soft-rot of carrots, cabbage, celery, and many other vegetables is present also on the cucurbits. It is sometimes serious on the muskmelon, and has been inoculated successfully into the cucumber and citron. It is not known whether the trouble is the same as the bacterial soft-rot of cucumbers found in Florida and other southern states or whether the two are distinct. The symptoms, however, are much the same.

The organism gains access to the fruit only through wounds. The entrance into the muskmelon is commonly through growth cracks on parts of the melon touching the moist soil. The outer part of the plant is not decomposed, but shrivels as the inner flesh collapses. The tissue becomes soft and slimy, and has a disagreeable odor. The cucumber exhibits similar symptoms.

The organism, when first isolated from muskmelons, was thought to be slightly different from the ordinary strains of

Bacillus carotovorus in that it readily caused decay of beets and potatoes, and had a slightly different reaction towards starch. It is now recognized that the bacillus on carrot is capable of many variations; therefore the organism named *Bacillus melonis* Gid. may rightly be included with the common soft-rot pathogene.

As the bacterium gains an entrance through wounds from the soil it is suggested that when possible the melons be laid on dry bits of board or on small flat stones. Any fruit that shows the beginning of rot will serve as a source of inoculum if it is not removed at once and buried. Thorough cultivation, turning under the cover-crops, and intelligent use of fertilizers will keep the vines growing uniformly, thus aiding in the avoidance of growth cracks. Favorable results may also be obtained by spraying with bordeaux mixture 3-4-50 or dusting with 10-90 copper-lime dust, following the directions offered for the control of bacterial wilt of cucurbits.

Reference

Giddings, N. J. A bacterial soft-rot of muskmelon caused by *Bacillus melonis* n. sp. Vt. Agr. Exp. Sta. Bull. 148: 363-416. 1910.

Bacterial Rot of Cucumber

Caused by *Bacillus* sp.

In some of the southern states, particularly in Florida, a soft-rot of cucumbers has caused much loss since 1911. At one time it was thought to be connected with the angular leaf-spot of cucurbits, but this has since been disproved.

The lesions are small watery spots on the fruit, from the ruptured epidermis of which exudes a viscid liquid that on drying becomes an amber mass of gum. The decay progresses inwardly until it reaches the vascular bundle and from there passes to all parts of the cucumber leaving the fruit a soft watery pulp. The younger the fruit the more easily the tissue is invaded. The disease commonly is present in the field; the greatest loss, however, is sustained in shipping fruit that appears healthy when crated, but which develops the rot before it reaches the market.

Isolations and inoculations have been made, resulting in the finding as the cause a white bacillus with the group number 222.3332113. It dissolves the middle lamellæ of the host cells and thereby produces the soft-rot of the fruit. The organism grows rapidly causing visible infections in five to ten days after inoculation. It probably lives throughout the year in old diseased tissue, and is constantly available to threaten the new crop.

The organism demands an abundance of moisture, and is able to infect more readily the crop with an increased succulence brought on by liberal applications of nitrate fertilizer.

Control measures consist of spraying and care in handling the fruit. As an added precaution it is advised that nitrates be used sparingly. The fruit when picked is easily bruised, so that in harvesting the stems should be cut with a sharp knife, the crates lined with cloth, and at no time should the cucumbers be poured from a height out on the floor or into a deep crate. All the bruised or diseased fruit should be removed in the packing-shed, and those showing bacterial rot buried. The spraying or dusting in the field is the same as outlined for the control of bacterial wilt of cucurbits.

References

Burger, O. F. A bacterial rot of cucumbers. Phytopath. 3: 169-170. 1913.

Burger, O. F. Cucumber rot. Fla. Agr. Exp. Sta. Bull. 121: 97-109. 1914.

Timber-Rot of Cucurbits

Caused by *Sclerotinia libertiana* Fuckel

(See also Lettuce-Drop, page 243.)

The timber-rot is almost wholly, if not entirely, a greenhouse disease. The trouble usually appears about the time the vine begins to bear fruit. Both the stems and fruit may be attacked. The largest number of lesions are found on the nodes near the base of the stem, and are made conspicuous by the aerial white mold of the parasite. At first, even in the presence of so much superficial mycelium, the host tissue remains fresh and plump. Later the stem shrivels and turns yellow while the invaded tissue begins a dry granular decay. Because of the appearance of the stem at this stage, the name timber-rot has been applied. The vine may rot off completely near its base, or remain attached merely by the vascular bundles. During the development of the fungus, large sclerotia may be formed within the tissue or on the outside of the stem. The fruit, when attacked, becomes soft and watery. The same white mold and black sclerotia are found on the fruit as on the stem.

The causal fungus is the same one that affects lettuce, tomatoes, celery, cabbage, and many other plants. It lives in the soil, therefore can be eradicated in the greenhouse by soil sterilization (page 600) or by the use of virgin soil. Ordinarily it is not necessary to sterilize each year if this disease alone is to be combated.

Reference

Humphrey, J. E. Diseases of the cucumber plant. Mass. Agr. Exp. Sta. Ann. Rept. 1892: 212-231. 1893.

Hypochnus Foot-Rot of Cucurbits

Caused by *Hypochnus cucumeris* Frank

A foot-rot of cucumbers has been reported from Germany and England. The first symptoms are the yellowing and dying of the foliage, beginning with the older leaves at the center of the hill and gradually extending to the tips of the vines until the whole plant is dead. Beginning at the surface of the ground or at the crown of the root and extending an inch or more up the stem, and sometimes also at the base of the leaf-petioles, is a loose weft of gray or brown mycelium. Below the weft the host tissue is fresh and healthy, but in certain depressed areas cankers are formed, frequently all the parenchyma being destroyed leaving only the vascular tissue. The white mycelial weft is never found on the dead brown tissue.

When the aerial mass of mycelium is examined with the microscope, it is found to consist of a hymenial layer of thick-standing basidia, each with four sterigmata and oval hyaline spores. The spores germinate in water in twenty-four hours, and in the presence of the host soon cause infection. The fungus lives as mycelium on organic matter in the soil.

The fungus in Germany has been named *Hypochnus cucumeris*. It is probable that the same fungus has been found on cucurbits in America where it is described as Rhizoctonia or *Corticium vagum*. This point may need further investigation.

No control methods have been suggested.

References

Collinge, W. E. Plant diseases due to fungi. Rept. Econ. Biol. **2**: 41-49. 1912.

Frank, A. B. Ueber einige neue und weniger bekannte pflanzenkrankheiten. Ber. d. Deut. Bot. Ges. **1**: 58-63. 1883.

Nectriella Wilt of Cucumbers

Caused by *Nectriella cucumeris* Hanzawa

It is not certain whether Nectriella wilt is present in the United States and known by some other name, or whether it is confined to Japan where it was described in 1913 as occurring in greenhouses.

The plants at first grow normally, but later the base of the stem stops growing in thickness and remains much thinner than do those of the normal vines. This is followed by the turning brown of the leaves especially along the margins and between the veins, on the upper part of the stem. They finally drop from the vine. The discoloration spreads from the foliage to the spindling stem, causing the plant to wilt and die. In the first stages of the disease the fruit remains normal, but that formed later is dwarfed, abnormal in shape and yellow. The roots begin to die from their tips, and are much thinner than are those on the healthy plant.

The fungus has asexual spores, shaped somewhat like those of *Fusarium vasinfectum,* but a perfect stage is connected with the fusarial spores, consequently the fungus has been named *Nectriella cucumeris.* It lives in the soil, and apparently attacks only cucumbers. The mycelium penetrates all parts of the plant, and produces the spores over the fallen dead stems. The perithecia are produced for the most part on the roots and not on the stems.

The trouble should be controlled by soil sterilization (page 600) or changing the soil, and in being careful that the inoculum is not carried to disease-free parts of the house.

Reference

Hanzawa, J. Über das Welken der Gurkensflanzen. Ztschr. f. Pflanzenkr. 23: 65-72. 1913.

Cercospora Leaf-Spot of Cucurbits

Caused by *Cercospora* spp.

A Cercospora leaf-spot has for many years occurred in Europe and America on muskmelon, cucumber, squash, watermelon and other cucurbits. The disease is seldom of economic importance.

On watermelon foliage the spots are small, black, circular and with a grayish center. The spots occur first on the leaves near the center of the hill, and gradually increase in number until all the leaves are spotted. The leaves, when badly infected, die, fall to the ground, or are blown about the field. The spots on cucumber, muskmelon and squash foliage are larger, and of a gray-ochre color. The fruit is reduced in size by the defoliation of the vines.

The Cercospora usually mentioned in American literature is *Cercospora citrullina* Cke. Other Cercosporas that have been described on cucurbits are *Cercospora cucurbitæ* E. and E. on squash and *Cercospora melonis* Cke. on muskmelons. The host range and distribution of each species has not been determined definitely. The only characteristic which seems to have been relied on for the classification is the size of the fungous spores. For all practical purposes the species does not matter, for the life history of each is identical. The genus is characterized by thin long septate spores borne on clumps of short conidiophores, which when plentiful form a faint ashen mold over the surface of the lesion. The spores are blown about either detached or with the leaves. Tools and splashing rains assist in their further dissemination. Apparently the fungus lives over winter in diseased refuse left in the field.

The control measures consist in destruction by burning of the diseased vines, two- or three-year rotation of crops, and spraying or dusting as described for the control of bacterial wilt of cucurbits (page 179).

References

Anonymous. Cucumber and melon leaf-blotch. Jour. Bd. Agr. (London) 9: 196-198. 1902.

Chester, F. D. Report of the mycologist. Del. Agr. Exp. Sta. Ann. Rept. 1901: 36-50. 1902.

Cooke, M. C. A new melon disease. Gard. Chron. ser. 3. 20: 271-272. 1896.

Cooke, M. C. Cucumber leaf disease. Jour. Roy. Hort. Soc. (London) 26: CXLIV. 1902.

Eriksson, J. Die einbürgerung neuer zerstörender gürken-krankheiten in Schweden. Centralbl. f. Bakt. u. Par. II, 44: 116-128. 1916.

Hall, A. D. The cucumber leaf-blotch. Jour. Bd. Agr. (London) 12: 19-21. 1905.

Sheldon, J. L. Diseases of melons and cucumbers during 1903 and 1904. W. Va. Agr. Exp. Sta. Bull. 94: 121-138. 1904.

Septoria Leaf-Spot of Cucurbits

Caused by *Septoria cucurbitacearum* Sacc.

The fungus, Septoria, has been collected on the leaves of muskmelons, pumpkins, and watermelons, and on the fruit of squash. The spots on the foliage are described as being either small gray or whitish spots in some cases, and large brown spots in others. Both types of lesions are circular, rather conspicuous, and frequently bordered by a zone of yellow-colored tissue.

The fungus fruits abundantly on the upper surface of the foliage, producing black pycnidia with thin long septate spores, typical for Septoria. The species of the fungus on most of the cucurbits is known as *Septoria cucurbitacearum*. On watermelon the spores are much shorter than those of the above species; therefore, the fungus on the latter host has received the name *Septoria citrulli* E. and E. In both cases the parasite, no doubt, winters over in the old diseased leaves, and is ready to cause infection in the spring.

No methods for controlling the disease have been suggested. If the leaf-spot ever proves serious, it probably can be held in check by spraying or dusting as recommended for the control of bacterial wilt of cucurbits (page 179).

Reference

Stewart, F. C. Notes on New York plant diseases, I. N. Y. (Geneva) Agr. Exp. Sta. Bull. 328: 364, 365. 1910.

Stemphylium Leaf-Spot of Cucumbers

Caused by *Stemphylium cucurbitacearum* Osner

In 1915 a new leaf-spot was found on cucumbers in Indiana and Ohio. It was then important in only a few fields. Very few reports have been made of it since then, consequently its present distribution is unknown.

The spots on the leaves are much like those described for the Sporodesmium leaf-spot, only that the lesions are smaller. In addition the Stemphylium has not been found on the fruit.

Within three to five days after inoculation the fungus is able to grow into the plant. Infection takes place through the stomata or directly through the epidermis and on either side of the leaf. After the fungus has developed in the tissue for a week or longer, mycelial threads grow over the surface of the spot and from these arise septate conidiophores with globose multiseptate spores. The mycelium is able to survive in the old infected tissue during the winter, and cause new infections the following year. It is not known how long the spores will live.

In controlling the fungus, it is desirable to practice a rotation of crops, to destroy old diseased vines, and to spray with bordeaux mixture 4-4-50 or dust with 15-85 copper-lime dust as described for the control of bacterial wilt of cucurbits.

Reference

Osner, G. A. Stemphylium leaf-spot of cucumbers. Jour. Agr. Research **13**: 295-306. 1918.

Sporodesmium Rot of Cucurbits

Caused by *Sporodesmium mucosum* var. *pluriseptatum* Karst. and Har.

In 1896 in Germany lesions were found on cucumbers and pumpkins from which a Sporodesmium was isolated. The spots on the fruit are circular brown depressions, appearing like finger-prints in soft tissue. After the spots become older they are overrun by Cladosporium. The lesions on the foliage are brown dry spots, lighter colored in the center and with a dark border. Often the lesions are so numerous that they coalesce; the dead tissue then crumbles away leaving the foliage in shreds.

The fungus belongs to the same group as Stemphylium which has been described in America as causing a leaf-spot of cucumber. This together with the fact that Saccardo finds that the two genera are difficult to separate, would seem to indicate that the disease in Germany and the one in the United States may be caused by the same parasite. Further work needs to be done to clear the matter.

The fungus has multiseptate spores with persistent elongated stems. The spores are borne on conidiophores that arise in clusters from the stomata of the host. The detailed life history is not recorded.

Evidently spraying with bordeaux mixture 4-4-50 controls the fungus.

Reference

Aderhold, R. Cladosporium und Sporodesmium auf gurke und kürbis. Ztschr. f. Pflanzenkr. 6: 72-76. 1896.

Phyllosticta Leaf-Spot on Cucurbits

Various species of Phyllosticta have been reported on cucurbits among which are *Phyllosticta orbicularis* E. and E. found on pumpkin, *Phyllosticta citrullina* Chester on watermelon, and *Phyllosticta cucurbitacearum* Sacc. on cucumber. It is possible, and later authors seem to agree, that formerly the imperfect stage of *Mycosphærella citrullina* when present on the leaves may have been mistaken for a Phyllosticta. The present literature does not assure the reader that there is or is not a distinct fungus on cucurbits, and which can be classed under the latter genus.

References

Selby, A. D. Further studies of cucumber, melon and tomato diseases. Ohio Agr. Exp. Sta. Bull. 105. 217-236. 1899.

Smith, C. O. Study of diseases of some truck crops. Del. Agr. Exp. Sta. Bull. 70: 1-16. 1905.

Blossom-Rot of Squash

Caused by *Choanephora cucurbitarum* (B. and Rav.) Thaxt.

In wet weather the blossoms of summer squashes are occasionally attacked by a fungus related to the bread-mold. The flowers, when invaded early, turn soft and usually drop off. Later infections cause the flowers to become brown, prematurely, and if the fruit has begun to develop its tissue is entered and caused to decay. In nearly every case it enters the squash through the blossom-end, but occasional infections may occur through wounds.

The fungus is differentiated from Mucor by its less erect mycelium, by the metallic lustre of its mycelial weft, and by its less compact sporangium. The conidia are borne on a head that is composed of short swollen branches arising from the swollen tip of a conidiophore. These conidia germinate by means of a germ-tube. Spores of a second type are borne in closed sporangia, and are peculiar in that they bear at their ends tufts of hair-like appendages. They germinate in the same manner as do the conidia. Zygospores, typical of the Mucorales, have also been observed and probably aid the fungus in surviving the winter. The various types of spores are disseminated by bees, cucumber beetles, other insects, and by wind.

No control methods are suggested.

References

Clinton, G. P. Diseases of plants cultivated in Connecticut. Conn. Agr. Exp. Sta. Ann. Rept. 27: 359. 1904.

Thaxter, Roland. A New England Choanephora. Rhodora 5: 97-102. 1903.

Wolf, F. A. A squash disease caused by *Choanephora cucurbitarum*. Jour. Agr. Research 8: 319-328. 1917.

Thielavia Foot-Rot of Watermelon

Caused by *Thielavia basicola* (B. and Br.) Zopf

(See also Black Root-Rot of Bean, page 37.)

The foot-rot of watermelon was found several years ago in Utah and Oregon. The margins of the leaves turn yellow and die, followed by similar symptoms over the whole leaf. The disease shows first on the older foliage near the center of the hill, and gradually spreads outward toward the tips of the vines. On examination, the base of the plant from one to four inches below the surface of the ground and sometimes extending upward above the soil surface is found to be affected by a decay, which causes the invaded part of the stem to be dark and roughened. A miscroscopic examination reveals the presence of many chlamydospores of *Thielavia basicola.*

No control measures are suggested.

Reference

McKay, M. B. *Thielavia basicola* on watermelon in Oregon. Phytopath. 12: 445. 1922.

Stem-End Rot of Watermelon

Caused by *Diplodia* sp.

Stem-end rot of watermelons is common and destructive wherever this crop is grown extensively. It is present in the field, but more often enters the cut stem after harvest and causes great loss during transportation. It usually occurs only on the fruit, but Burger reports that it may attack the base of the vine and cause a wilt. It causes infection of the fruit only through a wound. Most often this is at the stem-end, but bruises or cuts on the side caused by rough handling or by insects may furnish place for entrance. Where the disease is common it is not unusual to find it causing a blossom-end rot, although decay at the blossom-end of the fruit ordinarily is due to another fungus.

The rot causes a water-soaked appearance of the rind and a browning of the fruit stem. Even before there are any outward symptoms, the flesh may be softened so that buyers are in the habit of testing the end by pressing on it with the thumb. As the disease progresses the end of the melon shrivels and the flesh becomes soft and slimy (Fig. 64). Under favorable conditions the disease may cover a band one-half to one and one-half inches wide in the development of one day. Finally the whole melon is decayed, changing into a black wrinkled mummy.

The rot is caused by *Diplodia tubericola* E. and E., and probably other species of the same genus. The pathogene is supposed to be identical with the species that causes a stem-

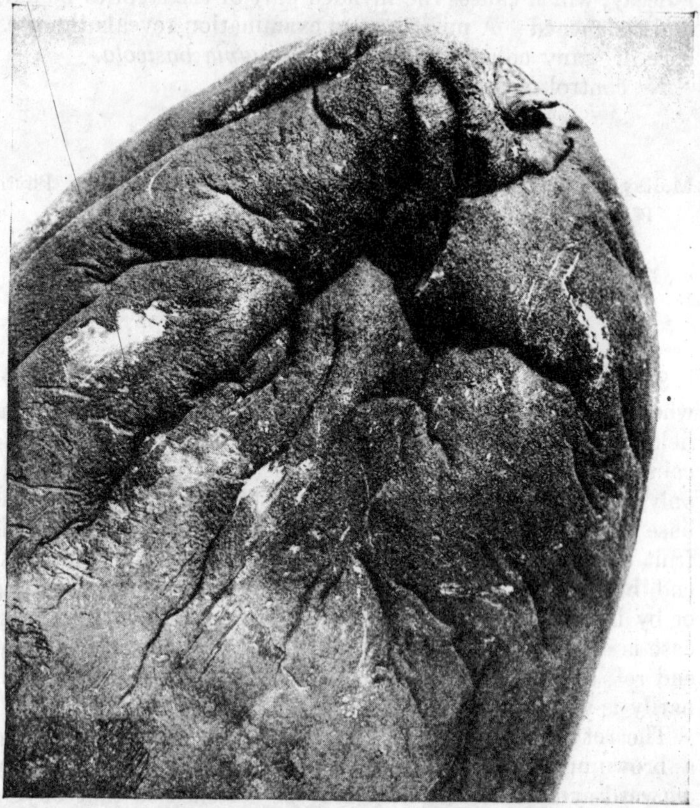

FIG. 64.—Stem-end rot of watermelon.

end rot of citrus fruit, a decay of sweet-potatoes, and a boll-rot of cotton. The occurrence of the fungus on these hosts is significant, as sweet-potatoes and cotton are grown in the same regions as watermelons, and may serve as alternate crops in the rotation.

The spores are borne in great numbers in pycnidia in the rind of the host. When water is present, they ooze out and are carried by wind, rain, or tools to other injured or cut melons. The melons when harvested are often inoculated by handling or are placed in cars infested by previous loads of diseased melons. When the stem of the fruit is cut a drop of sap collects on the cut surface and furnishes an ideal place for the incubation of the spores. The fungus lives over winter on diseased crop refuse or among the trash in the field and among the fence rows.

Control.

Methods for the elimination of the disease have proved very effective. They consist in sanitation, spraying, and stem treatment. All diseased melons should be destroyed as soon as they are found. As they cannot be burned, they may be fed to the hogs or buried several feet under the ground. The fence corners and other places where trash may collect should be burned over during the winter. In spraying the vines the suggestions made for the control of bacterial wilt of cucurbits also hold for watermelons. Bordeaux mixture 4-4-50 is applied. Since the rows are so far apart and the vines grow so long, it is impossible to drive straddle of a row with an ordinary sprayer. Consequently the sprayer is fitted with two or four leads of rubber hose instead of a boom, and a man is assigned to each hose. The outer longer hose may be supported on a long pole extending horizontally some distance from the top of the tank, and thus eliminate injury to the vines by its dragging. When planting the field a roadway is left open between groups of six to eight rows, so that the sprayer may be hauled through the planting without injury to the vines. A traction sprayer cannot be used for the machine must stand still while the vines are being covered. Therefore, a power sprayer is a necessity. It should be large enough to furnish a high pressure, and the men should be careful to cover all the foliage and young fruit.

In order to protect the stem-end of the melon a paste is recommended for treating the cut stem. Three and one-half quarts of water and eight ounces of copper sulfate are heated together in an enamel-ware kettle (a metal vessel is corroded by the copper). Eight ounces of laundry starch are stirred up with a pint of cold water until a milky suspension is obtained. The latter is then slowly poured into the boiling hot copper sulfate solution, being stirred vigorously during the mixing. The paste is boiled for one or two minutes longer or until the starch thickens. This paste should be made up fresh each time it is used. A quart is sufficient to treat a carload of melons.

When the melons are harvested, the longest stems possible are retained. The melons are at once placed in the car one tier at a time, turning the stem-end outward. The stem of each melon is then cut shorter, and the copper paste applied to the fresh surface, care being taken not to daub the rind of the fruit and thereby make it unsightly. The application in the car is more desirable than one in the field, for the fungicide may be rubbed off if the fruit is handled after the stem has been treated.

References

Meier, F. C. Watermelon stem-end rot. Jour. Agr. Research **6**: 149-152. 1916.

Orton, W. A. and F. C. Meier. Diseases of watermelons. U. S. Dept. Agr. Farmers' Bull. 1277: 1-31. 1922.

Blossom-End Rot of Watermelon

Caused by *Pythium* sp.

The decay of watermelons known as blossom-end rot is not always caused by the same organism. The Diplodia that produces stem-end rot has been associated with this trouble. The same is true of *Pythium debaryanum*. Recently a new species of Pythium has been found on the melon fruit. The decay, beginning at or near the blossom-end, progresses toward the stem leaving a brown colored dry-appearing pulp and water-soaked rind (Fig. 65). The disease appears mostly on ill-shaped or wounded fruit.

All the nubbin melons should be removed from the patch, and care taken in not bruising the remaining ones. Spraying, as suggested for the control of bacterial wilt of cucurbits and stem-rot of watermelons, will aid in protecting the fruit.

FIG. 65.—Watermelon showing blossom-end rot caused by Pythium.

REFERENCE

Drechsler, Charles. A new blossom-end decay of watermelons caused by an undescribed species of Pythium. Phytopath. 13: 57. 1923.

LITTLE KNOWN OR UNIMPORTANT DISEASES OF CUCURBITS

The following fungi have been reported on cucurbits; but little is known or at least little has been written about their life histories, or how they may be controlled: *Scolecotrichum melophthorum* P. and D. reported on cucumber and melons from France and Holland where it causes brown spots on stems, leaves, and fruit (possibly identical with *Cladosporium cucumerinum*); *Venturia cucumerina; Acremonium*, a fungus associated with leaf-glaze; *Dendryphium comosum* Wal., the cause of a leaf-spot of cucumber in England; *Cephalothecium roseum* Corda, a cause of bitter-rot of melons; and *Hormodendrum hordei* Bruhne, assigned as the cause of a leaf-blotch in England.

FIG. 66.—Damping-off of cucumber seedlings.

REFERENCES

Anonymous. A cucumber leaf disease. Jour. Bd. Agr. (London) **10:** 166-170. 1903. (*Dendryphium comosum.*)

Collinge, W. E. Plant diseases due to fungi. Rept. Econ. Biol. **2:** 41-49. 1912. (*Hormodendrum hordei.*)

Humphrey, J. E. Another disease of cucumbers. Mass. Agr. Exp. Sta. Ann. Rept. **9:** 227-228. 1892. (*Acremonium.*)

Laubert, R. Bittere melonen. Handelsblatt f. d. Deut. Gartenbau **26:** 601-602. 1911. (*Cephalothecium roseum.*)

Lindfors, T. A new cucumber disease. K. Landthr. Akad. Handl. och Tidskr. **58:** 424-430. 1919. (*Venturia cucumerina.*)

Prillieux and Delacroix. La nuile, maladie des melons, produite par le *Scolecotrichum melophthorum*. Bul. de la Soc. Myc. de France **7**: 218. 1891.

Damping-off of Cucurbits (Fig. 66)

Caused by various fungi

(See Damping-off of Tomatoes, page 546.)

Southern Sclerotium Rot of Cucurbits

Caused by *Sclerotium rolfsii* Sacc.

(See Southern Sclerotium Rot of Sweet-Potato, page 510.)

Root-Knot of Cucurbits

Caused by *Heterodera radicicola* (Greef) Müll.

(See Root-Knot of Tomato, p. 550.)

Verticillium Wilt of Cucurbits

Caused by *Verticillium alboatrum* Reinke and Berth.

(See Verticillium Wilt of Eggplant, page 234.)

References

Lindfors, T. A wilting disease of cucumbers. K. Landthr. Akad. Handl. och Tidskr. **56**: 627-636. 1917.

Van der Lek, H. A. A. Studies on tracheomycoses: Verticilliose of cucumber. Meded. Landbouwhoogsch. (Wageningen) **15**: 1-18. 1918.

Soft-Rot of Cucurbits

Caused by *Rhizopus* spp.

(See Soft-Rot of Sweet-potato, page 519.)

CHAPTER VIII

DISEASES OF EGGPLANT

THE eggplant is a relatively unimportant vegetable, there being less than two thousand acres planted to this crop in the United States. The states having the largest acreages are New Jersey and Florida.

PHOMOPSIS BLIGHT OF EGGPLANT

Caused by *Phomopsis vexans* (Sacc. and Syd.) Harter

The Phomopsis blight, also known as wilt, foot-rot, tip-over, leaf-spot, and fruit-spot, has been known for many years in America, Cuba, and southern Europe. Where the disease occurs, the destruction of the crop may be complete. Every part of the plant above ground is attacked, beginning as a damping-off of the seedlings and continuing on stems, leaves, and fruit throughout the season. In one community where the average annual value of the crop is approximately two thousand dollars, the value in a serious blight year was reduced to four hundred dollars. This is merely an example of what happens in most of the intensive eggplant sections.

Symptoms.

The stems of the seedlings at the surface of the soil discolor and weaken, causing the young plant to fall over as in typical damping-off. The later stages are stem-canker, leaf-spot, and fruit-rot. The foliage is attacked at any time, the lesions usually showing first on the leaves that are near the ground. The spots are clearly defined, circular, gray to brown areas, which have a lighter colored center than is the marginal zone. As the spots become older, they take on irregular shapes and are marked by the presence of numerous small black pycnidia. When the spots attain a large size or are numerous, the leaf may turn yellow and finally die.

The cankers at the base of the stem may appear at any time of the season after the damping-off stage. They may show only as constrictions at the place of infection or as a gray dry-rot of the tissue. Sometimes the bark falls off leaving the woody part of the stem exposed. The inner tissue ordinarily is discolored. Occasionally adventitious roots arise from the stem above the decay and aid the plant in surviving. Frequently the weakened stem is not able to support the plant in a heavy wind, thus producing what is known as tip-over.

The spots on the fruit are pale sunken areas which may finally include the whole fruit (Fig. 67). The lesions are characterized by the presence of numerous black pycnidia. The fruit surface being glabrous while the calyx is much rough-

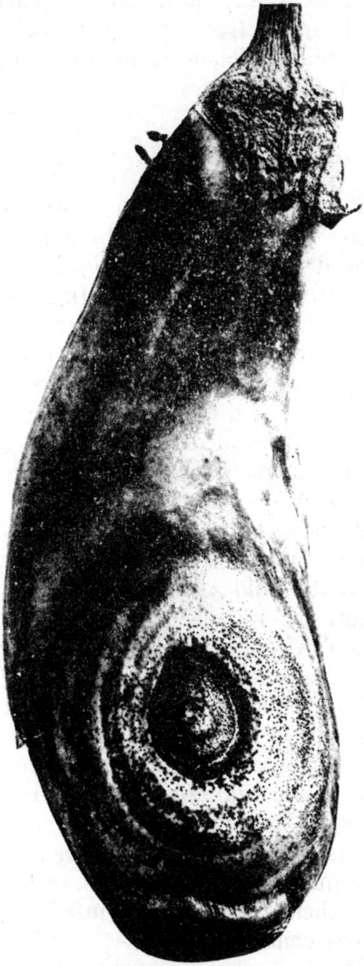

FIG. 67.—Phomopsis rot on eggplant.

ened, the latter seems to be more easily infected than the fruit as it affords a lodging place for spores. The parasite often grows from the calyx into the fruit-pedicel, then into the fruit, which is transformed into a black mummy by the dry-rot.

Cause of blight.

The fungus which has been associated with the different stages of the eggplant blight has in the past been known by several names, such as *Phyllosticta hortorum*, *Phoma solani*, and *Ascochyta hortorum*. In later investigations, the pycnidia were found to bear the peculiar stylospores or pseudoparaphyses characteristic of the genus Phomopsis, therefore the pathogene was referred to *Phomopsis vexans*. It is possible that there is a true Phyllosticta on eggplant, but if so it has not been recorded in the United States.

The fungus lives during the winter as mycelium within the spores, as spores on the seed, and probably both as mycelium and spores in the diseased plant refuse and soil. It is not known how long it can survive in soil without preying on the host; it would seem, however, that as it can attack only eggplant, the fungus could be starved out within a few years. When infected seeds are planted, the seedlings become inoculated, or when young plants are set into infested soil they may contract the disease. In either case the fungus is established for another season. By the formation of numerous pycnidia, an innumerable number of spores, discharged in a gelatinous matrix, are available for dissemination. The spores are carried by splashing rain, on tools, and by insects. When slightly diseased fruits are used for the source of seed, the latter become contaminated by the spores, or the mycelium may have grown into them. From these infected seeds the new seed-bed becomes contaminated.

The parasite grows most luxuriantly in wet weather. Its optimum temperature is approximately 85° F., yet it will grow at slightly higher and considerably lower temperatures.

Control of blight.

The length of rotations with other crops for starving the parasite is not known, but a period of at least three or four years intervening between crops of eggplant is recommended. The rotation, accompanied by the destruction of diseased parts, the avoidance of contaminated manure, the use of disease-free seed, and spraying where there is danger of infection, includes a program that has proved successful when followed carefully.

As the infested seeds cannot be differentiated from the healthy ones, it is necessary to know the source of the seed stock, or the grower must produce his own eggplant seed, if he wishes to be sure of a healthy crop. In making selections for seed, the best-producing most desirable plants are marked, and only the perfectly healthy fruit picked from these selected plants. The calyx is broken from each fruit, which is then treated in corrosive sublimate (1 ounce in 7.5 gallons of water) for twenty minutes. The seed is then extracted and sun-dried for a week, preferably in a shallow box with a screen bottom and a glass top, after which it is stored in a dry room in a tightly corked jar or bottle.

The spores on the seed may be killed by treating, although it is impossible to reach the mycelium under the seed-coat. Nevertheless when the source of the stock is not known, the results of treating are sufficiently beneficial to make the procedure worth while. The seeds are placed in corrosive sublimate (1 ounce in 7.5 gallons of water) for ten minutes, then rinsed in running water for fifteen minutes and planted at once. If any drying is necessary the seeds should be spread thinly on a wire screen, and just enough water removed so that the seeds will not stick together or to the planter. Much drying after treating injures germination.

Spraying and dusting have proved satisfactory in controlling the disease and increasing the yield when the fungicide is applied very thoroughly. About five applications at weekly intervals, beginning when the plants are well established in the field, are recommended. Two similar applications may be made on the seed-bed. The nozzles on the boom of the sprayer or duster are so arranged that one whole row and half of each adjoining row can be treated at one time, four nozzles being trained on each row, particularly when the plants are large. Two of the nozzles are directed either upward or at a ninety-degree angle from the side, while the other two are so directed from the top that every part of the plant will be covered. The fungicide used is bordeaux mixture 4-6-50 or 15-85 copper-lime dust. Only a slight amount of an insecticide is added or the foliage will be badly burned.

References

Edgerton, C. W. and C. C. Moreland. Egg-plant blight. La. Agr. Exp. Sta. Bull. 178: 1-44. 1921.

Geise, F. W., H. H. Zimmerley and H. Spencer. Spraying and dusting vegetable crops in 1922. Va. Truck Exp. Sta. Bull. 41: 237-256. 1922.

Halsted, B. D. Some fungous diseases of the egg-plant. N. J. Agr. Exp. Sta. Ann. Rept. 1891: 277-283. 1892.

Harter, L. L. Fruit-rot, leaf-spot, and stem-blight of the egg-plant caused by *Phomopsis vexans*. Jour. Agr. Research 2: 331-338. 1914.

Sherbakoff, C. D. Egg-plant diseases. Fla. Agr. Exp. Sta. Bull. 139: 232-237. 1917.

Smith, C. O. A leaf-spot disease of egg-plant. Del. Agr. Exp. Sta. Bull. 70: 10-15. 1905.

Verticillium Wilt of Eggplant

Caused by *Verticillium alboatrum* Reinke and Berth.

A wilt of eggplant has been known for many years. In some parts of Europe and in northern United States the disease causes such serious losses that it is questionable whether the crop is profitable on many farms. No varieties of eggplant seem to be immune. Aside from this host, the causal fungus has been reported on potato, tomato, cucumber, melons, cotton, okra, soybeans, raspberry, ginseng, peppers, and many other cultivated and wild plants. There seem to be different races of the fungus, however, for Jagger and Stewart could not infect eggplant or barberry from the form isolated from maple and Haenseler was unable to obtain infection on tomato and pepper with the form causing wilt of eggplant.

Symptoms.

The very young plants usually appear normal; but as the affected plants grow older they may show marked dwarfing. Frequently the stunting is the only manifestation of the disease. Badly affected plants, however, take on a yellow unhealthy color while the lower leaves wilt and drop off. These symptoms become more pronounced each day until the plant dies. When the stems and roots of a diseased plant are cut open, the woody tissue shows a dark brown discoloration, this darkening even extending into the fruit. After the plant is seriously invaded, the roots and the base of the stem may decay.

Cause.

The fungus causing the wilt of eggplant has repeatedly been proved to be *Verticillium alboatrum*. Its mycelial threads that penetrate the cell-walls and often fill the xylem-ducts are very small in diameter. They grow in great profusion throughout the plant from the tips of the small roots to the ends of the branches. It is difficult to find the fruiting stage of the parasite before the death of the plant. The spores are borne in great numbers on slender conidiophores whose branches are arranged verticillately, that is, in whorls. The spores are small, oblong, single-celled hyaline bodies. The fungus, probably as mycelium, is able to live in the soil indefinitely. Infection takes place through roots or the base of the stem. Plants may be infected in their seedling stage, but probably the trouble starts more often in the field after transplanting. At least sterilized seed-beds do not seem to reduce the amount of disease observed later in the field. So far as is known, the organism is not disseminated on or in the eggplant seed. It, however, may be carried on potato tubers or with other hosts, so that Verticillium is distributed over nearly the whole area where the temperature is suitable for its growth.

An interesting fact has been discovered in the study of the response to temperature which races of the parasite exhibit when brought from different sections of the country. A strain isolated in Maine could not develop in a temperature higher than 86°, while a similar strain from Virginia was able to grow in temperatures up to 95° F. The optimum of both strains was 77° and the minimum 41° F. The amount of water in the soil is not so important, as the fungus will cause infection wherever there is enough moisture for the plant to grow.

Control of Verticillium wilt.

There is very little that can be done in controlling the disease. If the farm is large enough to permit long rotations with immune crops such as cereals, grasses, onions, and corn, the disease will never become destructive enough to make the crop unprofitable. When the fields once become badly infested, it is well to discontinue the growing of eggplant for four or five years, or until the amount of inoculum in the soil is much reduced if not eliminated. The fungus is sometimes carried with contaminated manure, consequently manure with vegetable refuse should be placed in fields where susceptible crops are not grown.

References

Haenseler, C. M. Studies on eggplant wilt. N. J. Agr. Exp. Sta. Ann. Rept. 1921: 469-472. 1922.

Jagger, I. C. and V. B. Stewart. Some Verticillium diseases. Phytopath. 8: 15-19. 1918.

Van der Lek, H. A. A. Over de z. g. "verwelkingsziekten," in het bijzonder die welke door *Verticillium alboatrum* veroorzaakt worden. Tijdsch. Plantenz. 25: 17-52. 1919.

Bacterial Wilt of Eggplant

Caused by *Bacterium solanacearum* E.F.S.

(See Brown-Rot of Potato, page 401, and of Tomato, page 564.)

Anthracnose of Eggplant

Caused by *Colletotrichum* sp.

(See also Anthracnose of Potatoes, page 441.)

For many years an anthracnose has been recognized on the leaves and fruits of eggplants. More recently a similar fungous disease has been observed causing a foot-rot and wilt on the same host. The spots on the leaves are at first yellow, turning to brown as the cells die. Small to medium large depressed spots appear on the fruit. A similar brown decay occurs on the roots and stems. On all the affected parts of the plants acervuli with setæ are present after the parasite has once become well established. Following a rain or heavy dew the spores are formed in small pink masses in the acervuli.

Several species of the fungus have been reported on eggplant. At first it was known as *Glœosporium melongenæ* E. and H. Later it was considered the same as the one causing anthracnose of potatoes, or *Colletotrichum atramentarium.* The fungus produces minute sclerotia which survive the winter and produce spores for inoculation in the spring. The spores are splashed by rain or carried by insects and on tools and workmen to the plants which later become infected.

Control measures are not suggested, but careful spraying or dusting as recommended for the control of Phomopsis blight will probably protect the plants from infection by the fungus. A rotation of crops in which tomatoes or potatoes are not alternated with eggplant will also decrease the amount of inoculum, and together with spraying will no doubt result in a clean crop in most cases.

References

Butler, E. J. Pilzkrankheiten in Indien im Jahre **1903**. **Ztsch. f. Pflanzenkr. 15**: 44-48. 1905.
Halsted, B. D. Fungous diseases of various crops. N. J. Agr. Exp. Sta. Ann. Rept. **1890**: 345-366. 1891; **1891**: 233-340. 1892.
O'Gara, P. J. The occurrence of *Colletotrichum solanicolum* O'Gara on eggplant. Phytopath. **7**: 226, 227. 1917.

Cercospora Leaf-Spot on Eggplant

Caused by *Cercospora melongenæ* Welles

A serious spotting of the leaves of eggplant has recently been observed in the Philippine Islands. So far as is known the trouble has not been reported elsewhere. The parasite affects only the one host; furthermore there seems to be a difference in the susceptibility of the varieties of eggplant to the leaf-spot. The native Philippine variety is much more injured than is the one from Siam.

The spots, beginning on the older lower leaves, increase in number until all the foliage may be affected seriously. The lesions are first yellow in color, and observed only on the upper surface of the leaf. The disease quickly passes through the foliar tissue, causing large grayish-brown areas with concentric rings.

The fungus is characterized by its thin, long, septate spores borne on short conidiophores. The parasite probably lives during the resting stage in the old diseased refuse.

Spraying as suggested for the control of Phomopsis blight of eggplant has proved effective in protecting the crop. As it does not now occur in America, care should be taken in making this exclusion permanent.

Reference

Welles, Colin G. Cercospora leaf-spot of egg-plant. **Phytopath. 12**: 61-65. 1922.

Gray Mold-Rot of Eggplant

Caused by *Botrytis* sp.

(See also Gray Mold-Rot of Lettuce, page 253.)

The purple-fruited eggplant may be made conspicuous by a rot which changes the purple of the fruit to a tan color. Large areas of the fruit exhibit a soft-rot, and over the surface is a fruiting layer of the gray mold fungus. Halsted states that the fungus is *Botrytis fascicularis* (Cd.) Sacc.

Only general precautions can be taken in combating the trouble. The plants should not be set too thickly in the field, so that the foliage and fruit will dry quickly after a rain. As many insect and mechanical injuries as possible should be avoided. It may be practicable to control the insects by proper insecticides. When fruits have begun to decay, they should not be left lying on the ground among healthy plants, but should be removed and buried or destroyed in some other manner.

Reference

Halsted, B. D. Report of the Botanist. N. J. Agr. Exp. Sta. Ann. Rept. **1890**: 345-366. 1891.

Rhizoctonia Fruit- and Stem-Rot of Eggplant

Caused by *Corticium vagum* B. and C.

(See Wire-Stem of Crucifers, page 151.)

The Rhizoctonia fruit-rot softens the tissue of the affected part and makes the product unsalable. The same fungus causes a foot-rot at the base of the stem similar to that described as wire-stem of crucifers. Probably the same means of control may be employed, although it has not been shown that corrosive sublimate is perfectly safe to use on eggplant seedlings.

Reference

Wolf, F. A. Fruit-rots of egg-plant. Phytopath. 4: 38. 1914.

Nectria Stem-Rot of Eggplant

Caused by *Nectria ipomœæ* Hals.

A Nectria was reported in 1892 as causing serious injury to eggplant. It was considered at that time to be the same fungus that causes wilt of sweet-potatoes. The latter, subsequently, was proved to be a Fusarium. Recently in Canada the Nectria was again reported as a dangerous parasite of eggplant. According to the description the plants when half grown turn yellow and have a sickly appearance. No fruit is borne on the more seriously affected plants, which finally turn brown and die. A white mold appears on the base of the stem, which gradually decays. The lesions later are distinguished by pink stromæ in which are the perithecia of the fungus. The rot continues to spread from the surface of the ground down into the roots and up the stem until most of the plant is included.

A Nectria, which was supposed to infect sweet-potatoes, has been isolated and named *Nectria ipomœæ*. It is a soil organism and generally considered as a saprophyte. Originally it was supposed to be connected with a Fusarium stage. Until its pathogenicity and relation to other fungi are investigated anew, little can be said of the life history and control of the fungus.

References

Halsted, B. D. The egg-plant stem-rot, *Nectria ipomœæ* Hals. N. J. Agr. Exp. Sta. Ann. Rept. 1891: 281-283. 1892.

Stone, R. E. and J. E. Howitt. Plant diseases and fungi comparatively new or rare in Ontario. Phytopath. 10: 317. 1920.

PHYTOPHTHORA FRUIT-SPOT OF EGGPLANT

Caused by *Phytophthora melongenæ* Sawada

This rot of the eggplant fruit has been reported only from Japan. The parasite causes a decay which later is covered with a white mold. When the fungus was isolated, it was successfully inoculated into tobacco, tomato, potato and several other plants. The fungus reproduces by means of conidia borne on their long conidiophores and by oospores. The conidia give rise to as many as forty zoospores, which after being motile for a brief period cause infection.

No control measures have been suggested. It is hoped that it may be excluded from America.

REFERENCE

Tanaka, Tyozaburo. New Japanese fungi. Notes and translations II. Mycologia 9: 249-253. 1917.

BLACK DOT-ROT OF EGGPLANT

Caused by *Rhabdospora melongenæ* Han.

In 1907 a rot was observed in Japan on eggplant fruits that had been left to ripen on the vine for seed. The affected parts of the fruit lose the purple color and the surface is marked thickly with small black pycnidia. The trouble is not found on the leaves and stems.

The parasite, characterized by irregularly shaped pycnidia in which are borne thin, long, hyaline, one-celled spores, is named *Rhabdospora melongenæ*. The pycnidia have somewhat the appearance of those borne by Phomopsis. The spores, also, are shaped very much like the pseudoparaphyses, which distinguish Phomopsis from other genera. The two fungi, however, may not be the same.

REFERENCE

Hanzawa, Jun. Ueber einen neuen fruchtkrankheitserregenden Pilz (*Rhabdospora melongenæ* sp. nov.) der Eierpflanzen. Trans. Sappora Nat. Hist. Soc. 3: 83-87. 1909-10.

MACROSPORIUM LEAF-SPOT OF EGGPLANT

Caused by *Macrosporium solani* E. and M.

(See Early-Blight of Potatoes, page 360.)

ROOT-KNOT OF EGGPLANT

Caused by *Heterodera radicicola* (Greef) Müll.

(See Root-Knot of Tomatoes, page 550.)

PHYTOPHTHORA BLIGHT OF EGGPLANT

Caused by *Phytophthora infestans* (Mont.) DeBary

(See Late-Blight of Potatoes, page 349.)

REFERENCE

Haskell, R. J. *Phytophthora infestans* on egg-plant in the United States. Phytopath. 11: 504-505. 1921.

SEPTORIA LEAF-SPOT OF EGGPLANT

Caused by *Septoria lycopersici* Speg.

(See Septoria Leaf-Spot of Tomato, page 531.)

REFERENCE

Smith, C. O. The study of the diseases of some truck crops in Delaware. Del. Agr. Exp. Sta. Bull. 70: 1-16. 1905.

CHAPTER IX

DISEASES OF LETTUCE

MORE than twenty thousand acres of lettuce are grown in the United States each year, the largest acreages being in California, New York, Florida, and New Jersey. Lettuce is extremely susceptible to injury resulting from diseases. It is not unusual for growers to lose an entire crop in an unfavorable season.

LETTUCE-DROP

Caused by *Sclerotinia libertiana* Fuckel
and *Sclerotinia minor* Jagger

The organism most commonly associated with lettuce-drop was described in Europe more than fifty years ago, but the disease itself was not recognized until twenty-five years later when it was found in Florida, North Carolina, and Massachusetts. At present it probably is present in every state of the Union, in many of which it causes extremely large losses. In the North most of the losses occur in the greenhouses, but in the southern and central states the fungus attacks also the outdoor crop. The most frequent reports have come from the states bordering on or near the eastern shore. The fungus has been prevalent and destructive in Bermuda and present to some extent in Holland. Strangely it is either not present or unimportant in England and the European countries, although Sclerotinia is reported there as common on many other hosts.

Before adequate control measures were known, the disease was so devastating that many greenhouse men gave up the growing of lettuce. The better methods of combating the fungus have decreased the losses materially, but still there are states in which the average reduction is reported as 10 to 20 per cent of the crop. It is possible that in such estimates the diseases caused by Botrytis, Rhizoctònia, and other organisms are not always separated from the Sclerotinia rot. Even so, the loss is known to be enormous.

The fungus which is found so generally on lettuce, is present on nearly every other vegetable host as well as on other crops and on weeds. It causes a stem- and tuber-rot of potatoes, a field and storage-rot of cabbage, a pink-rot and damping-off of celery, a storage rot of carrots, a pod-rot of beans, a stem- and fruit-rot of greenhouse tomatoes, a blight of parsley, a damping-off of cauliflower, pepper, spinach, carrot and lettuce seedlings, a rot of rhubarb, a stem-blight of eggplant, a rot of onion, a stem- and fruit-rot of cucumber, a head-rot of cauliflower, and causes injury to mustard, caraway, beets, peas, rutabagas, chicory, mangel-wurzel, Jerusalem artichoke, kohl-rabi, kale, marrow-cabbage, and turnip.

Symptoms.

The rot usually begins on the stem near the surface of the soil, and presents a soft water-soaked spot that spreads downward until the roots are decayed, and at the same time passes upward until the bases of the leaves are affected. When the outer lower foliage has its water supply cut off by the attack of the fungus on the petioles, the leaves die, wither, and droop until their tips rest on the ground. The parasite rapidly ascends the stalk, killing the leaves in succession until it reaches the heart of the plant. Each leaf in its turn drops downward until it rests on the one below. The inner ones, however, do not dry out as quickly as do the outer lower ones, and therefore are invaded completely by the mycelial threads, which reduce the tissue of the tender foliage to a wet, slimy, decayed mass. After becoming infected, the entire plant has the appearance of having been crushed or having "dropped." Under moist conditions the fungus grows not only within the host, but produces a white mycelial weft over the entire plant. In this weft and particularly among the collapsed leaves, numerous, black, variously-shaped sclerotia are developed. Usually they are about the size of peas; in rare cases they may be small as mustard seeds, depending on the species of

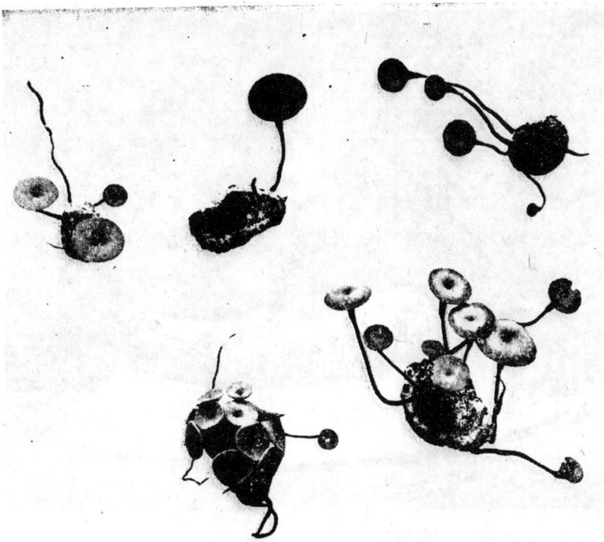

FIG. 68.—Apothecial cups of *Sclerotinia libertiana* growing from sclerotia.

the fungus involved. The presence of a white mass of hyphæ and black sclerotia are signs which differentiate this disease from any other lettuce trouble.

Cause.

The most common cause of lettuce-drop is *Sclerotinia libertiana* or as the English writers prefer to name it,—and rightly so—*Sclerotinia sclerotiorum* (Lib.) Massee. At one time it was supposed to have a Botrytis as its conidial stage but this has since been proved erroneous. The parasite is abundant in most of the older lettuce sections. It lives over-winter by means of the innumerable sclerotia. At any time after the latter are formed, they may germinate by sending out one to many slender hollow stalks, which vary in length from almost nothing to more than an inch, and on

Fig. 69.—Apothecia of Sclerotinia giving off a cloud of ascospores.

the tips of which are formed saucer-shaped discs (Figs. 68, 69). The discs are known as apothecia and are the fruit-bodies that bear the ascus layers. The top surface of the apothecium is covered with the spore-sacs standing as tightly together as is possible to crowd them among the sterile threads or paraphyses. The number of these sacs may be imagined when it is realized that one sclerotium is able to produce almost a square inch of fruiting area, and the asci that cover the space are too small to be seen with the unaided eye. Each ascus in turn bears eight spores and this increases the tremendous number of possibilities for infecting the new crop. The spores are ejected forcibly from the asci, and are carried by air-currents sometimes for a considerable distance. Where the fungus has been well propagated, there is a literal rain of spores, so that every plant is inoculated.

In a wet relatively cool environment the spores germinate and the mycelium, after growing saprophytically for a time in the soil humus, passes into the tissue where it dissolves the middle lamellæ causing the whole leaf to become a soft watery mass. Knots of hyphæ are interwoven to form the new sclerotia, which are white at first, gradually turning black on the outside and having a pronounced wall for a covering. The sclerotia may live for a year or more when the soil is not too wet; but they are readily attacked and destroyed by molds when moisture is abundant. They withstand a deep soil covering but will not fruit unless they are within one and one-half inches of the surface, and are much more prolific still nearer the top of the soil. If the stalks or stipes are broken off near the sclerotium after they have begun to form they do not develop again; if only the tops of the stalks are removed, they grow as readily as ever.

The more rare fungus, *Sclerotinia minor*, differs only slightly in its life history from that of *Sclerotinia libertiana*. Its sclerotia are more numerous and much smaller, being more nearly the size of mustard seed. The fungus is even more virulent in its attack on lettuce than is the species with the large sclerotia, but fortunately is not often present. As its sclerotia do not fruit readily, it was first supposed to be a degenerate strain of the *libertiana*. It is now known to be distinct. Instead of producing apothecia, the sclerotia frequently develop only masses of new hyphal growths that are able to infect the plant without the intervention of spores.

Both species of Sclerotia need a relatively large amount of soil-moisture and air humidity to develop rapidly. It has been observed, however, that the species, *minor*, is able to begin growth more quickly because it reproduces by mycelial threads growing directly from its sclerotia, and therefore responds more readily to a brief rainy period than does the *libertiana*.

Both fungi grow at extremely low temperatures, having been known to increase the length of their mycelium when the thermometer registered a point slightly below freezing. This quality of withstanding cold is of especial benefit to the parasites, for they can continue their development in late autumn on the old lettuce refuse left in the field, and have available in the spring a greatly increased volume of inoculum. Their optimum temperature is from 60° to 75° F., so that both of them can grow at nearly any degree in which lettuce thrives well. The organisms rarely appear in epidemic form in the open field in the middle of the summer, unless it is in some of the cooler northern states.

Control of lettuce-drop.

As the disease in the northern states is almost altogether in the greenhouses, soil sterilization is the most important control measure. Either steam or formaldehyde may be used (see page 600). Krout has shown that formaldehyde 1-50 or 1-100 will kill the sclerotia no matter how large they are, and that in commercial beds where covering the soil with boards or canvas is not possible, the fungicide will act successfully without a covering after the application.

If a few plants are affected in a bed, it is well to remove them without scattering any of the diseased tissue or the sclerotia. The vacant spot may then be soaked with a blue vitriol solution (1 pound in 7 gallons of water). Keeping the top soil well stirred and as dry as good growth of the lettuce will permit will retard the development of the fungus. Similar results may be obtained by subirrigation rather than by overhead sprinkling. If the system of greenhouse management can be so arranged, it is desirable to alternate crops of lettuce with less susceptible vegetables, such as cucumbers, radishes, and tomatoes.

In the field where soil sterilization is out of the question, only precautionary methods can be adopted. Crop rotation is of great value even though few vegetable hosts are wholly immune. Most of the outdoor lettuce is grown on muck, which will also bear good crops of sweet-corn, tomatoes, potatoes, cucumbers, radishes, beets, onions, spinach, and horse-radish, all of which will help in reducing the number of sclerotia. Celery and lettuce should not be grown continuously on the same soil nor alternated with one another without the use of some resistant vegetable in rotation. In intensive gardens, cabbage and lettuce are sometimes alternated in rows. A fairly well-matured head of cabbage is very susceptible to Sclerotinia, so that almost a half pint of sclerotia may form on one plant. Naturally such a host should not be grown in a row adjoining that of lettuce if the parasite is to be eliminated.

Well drained soil, eradication of weeds, care in not crowding the plants, and diligence in shallow cultivation of the soil tend to reduce the surface moisture, and thereby lessen the chances of the fungi to cause infection.

References

Beach, W. S. The "lettuce-drop" due to *Sclerotinia minor*. Penn. Agr. Exp. Sta. Bull. 165: 1-27. 1921.

Jagger, I. C. *Sclerotinia minor*, n. sp., the causes of a decay of lettuce, celery, and other crops. Jour. Agr. Research **20**: 331-334. 1920.

Krout, W. S. Control of lettuce-drop by the use of formaldehyde. Jour. Agr. Research **23**: 645-654. 1923.

Smith, R. E. Botrytis and Sclerotinia: their relation to certain plant diseases and to one another. Bot. Gaz. **29**: 369-407. 1900.

Stevens, F. L. and J. G. Hall. A serious lettuce disease (sclerotiniose) and a method of control. N. C. Agr. Exp. Sta. Tech. Bull. 8: 89-143. 1911.

Stone, G. E. and R. E. Smith. The rotting of green-house lettuce. Mass. (Hatch) Agr. Exp. Sta. Bull. 69: 1-40. 1900.

Wakefield, E. M. On the names *Sclerotinia sclerotiorum* (Lib.) Massee and *S. libertiana* Fuckel. Phytopath. **14**: 126, 127. 1924.

Bottom-Rot of Lettuce

Caused by *Corticium vagum* B. and C.

It is strange that until the work of Dye in 1919 there should be only slight reference in literature to the bottom-rot, which causes the loss of such a large acreage of lettuce each season. The trouble has been present for many years in New York where it is not uncommon to find the entire crop in a field destroyed. It has been reported from neighboring states, and is probably present wherever the Big Boston head lettuce is grown intensively. The Iceberg and Romaine lettuce, because of their more upright habit of growth, are rarely affected

FIG. 70.—Typical leaf-blade rot of lettuce due to bottom-rot.

seriously. When the last two varieties are inoculated, however, the fungus readily causes infection, showing that the plants are disease-escaping rather than immune to the bottom-rot.

Symptoms.

The disease may be present the entire season, beginning when the first crop is half grown, and continuing until the last plants are harvested late in the autumn. The first parts of the plant to be affected are the leaf-petioles that touch the ground. The two most important symptoms which distinguish this disease from other lettuce troubles are rust-colored slightly sunken lesions on the leaf-petioles and midribs and the slimy dark-brown rot on the leaf-blade. The midrib does not disintegrate, but remains whole even after the blade has rotted away. After the lower leaves are infected (Fig. 70) the decay frequently spreads to the immediately adjoining leaves in succession until the entire head has been changed to a dark slimy mass. Later the invaded tissue often dries, leaving an erect, black, mummied plant. Permeating all parts of the head and extending over its surface and even over the soil at the base of the stem are microscopic strands of brown mycelium. Nestled among the infected leaves, and especially at the bases of the petioles, are crust-like brown sclerotia.

Cause.

The fungus, *Corticium vagum*, known generally as *Rhizoctonia solani* Kühn, is the same parasite that causes damping-off of seedlings, besides producing other diseases on many hosts, some of which are mentioned under rhizoctoniose of potatoes (page 367). The parasite lives in the soil, on growing plants, and on diseased plant refuse. Nearly all soils have been inoculated so that the lettuce grown on virgin muck may not be free from the disease. The fungus reproduces by developing mycelial strands from dormant hyphæ or from sclerotia. A sexual spore stage is present, but does not seem to be of much importance in the development or dissemination of the organism. When the humidity of the air is high, this stage appears as a delicate white mycelial weft about the base of the stems of various hosts. On the weft is produced a layer of club-shaped basidia, on the tips of which are four sterigmata each bearing a spore. The spores germinate by means of a germ-tube.

The parasite requires the presence of much moisture and moderately high temperatures in order to develop rapidly.

The disease may begin in epidemic form during a wet period and then disappear almost entirely after the advent of dry weather. The disease is most pronounced on poorly drained soils, yet there are few lettuce fields in New York where the trouble does not occur. Apparently no known fertilizer has any influence on the amount of bottom-rot present.

Control.

Since Rhizoctonia is a soil organism, the only effective means of control are soil sterilization or procuring resistant varieties. Soil sterilization is practicable only in small gardens, being much too expensive to attempt on a large acreage. Rotating lettuce crops with onions, sweet-corn, and other resistant vegetables will help in reducing the amount of the disease, especially if the soil is well drained.

Only the Big Boston head lettuce is seriously affected. Attempts have been made to select a resistant strain of this variety, but none has been found. Dye made considerable progress in procuring a desirable head lettuce by crossing the Big Boston and Romaine with the purpose of obtaining a large yellow head like that of Big Boston on a long stem similar to that of Romaine. This breeding work is still being carried on, and probably will be brought to a successful conclusion. In localities where Iceberg lettuce can be grown profitably, bottom-rot need not be feared. There are many lettuce sections, however, which are not adapted to the growing of this variety.

REFERENCES

Duggar, B. M. and F. C. Stewart. The sterile fungus, Rhizoctonia, as a cause of plant diseases in America. N. Y. (Cornell) Agr. Exp. Sta. Bull. 186: 51-76. 1901.

Dye, H. W. Two diseases of western New York lettuce: the bottom-rot and the stunt. Unpublished Thesis. 1919.

Stone, G. E. and R. E. Smith. A Rhizoctonia disease of lettuce. Mass. (Hatch) Agr. Exp. Sta. Bull. 69: 16-17. 1900.

Gray Mold-Rot of Lettuce

Caused by *Botrytis* sp.

The gray mold-rot is of great importance under certain environmental conditions, but generally is not so serious a disease as either drop or bottom-rot. The fungus is found everywhere and on many hosts. It is usually considered only a weak parasite. Among the diseases which this species of fungus or closely related species causes are root-rots of chicory, carrot, and turnip; fruit-rots of squash, eggplant, tomato, and pepper; leaf-rot of rhubarb, bud-rot of globe artichoke, and pod-rot of bean. In fact, there are few vegetables which are not attacked by the fungus.

Symptoms

The lettuce plant may be affected with gray mold-rot at any stage in its development, in the field, in the greenhouse, or during transportation. When the seedlings are attacked, the disease resembles the common damping-off. It is more common to find the rot on partly or fully developed plants. The first indication of the disease is a soft, dark, decayed area on the base of the stem. The decay spreads rapidly so that it soon includes the lower leaf-petioles, thus causing these leaves to wither. The disease progresses upward through the head until all the inner tender leaves are transformed into a slimy mass. Occasionally the plants collapse before any rot is apparent from the outside, the fungus having rotted away the stem and leaves in the center of the head before it reached the margins of the outer leaves. Another common method of attack is that of causing decay on one side of the plant while the opposite side appears healthy. On seed lettuce infection often takes place on the floral parts and pedicels during and subsequent to the flowering period.

After the lettuce head is thoroughly rotted, black, flat or cylindrical sclerotia may be found among the decayed mass. The parasite also forms its ashen-gray fruiting layer over the surface of the diseased host parts. The layer is more conspicuous on the under sides of the leaves or in other protected places where the moisture is retained. After the plant has fully collapsed, it dries and withers away.

Cause.

The fungus causing gray mold-rot is a Botrytis and is usually referred to as *Botrytis cinerea* Pers. (*Botrytis vulgaris* Fr.). At one time it was considered as the conidial stage of *Sclerotinia libertiana*, but this has been disproved. The fungus is composed of a luxuriant growth of white mycelium, which in certain cases unites to form relatively large flat to cylindrical sclerotia. The conidial layer is made up of branched conidiophores with the small spores borne in fairly compact heads at the tips of the branches. The spores germinate by means of the germ-tube that cannot enter the host directly unless through wounds. The mycelium draws sustenance from the humus in the soil and after it has developed abundantly it seems able to enter the host even though no wound is present. The plant cells are killed quickly, after which the parasite again forms the sclerotia and conidia. The sclerotia have never been known to form apothecia, but always germinate by means of conidiophores and conidia. The mycelium is able to live indefinitely on organic matter in the soil, and the sclerotia can withstand long periods of inactivity. The fungus is disseminated with diseased plant parts, by wind, and with the soil.

The Botrytis is extremely susceptible to changes in its environment. It is most prevalent in greenhouses during the winter when there is much cloudy weather, the humidity of the air is high, and the temperature low. The fungus will grow at temperatures between 35° and 80° F., but its optimum temperature is between 68° and 75° F. According to Newhall, the parasite frequently is destructive on lettuce seed plants in September when the nights are cool and heavy dews are present.

Control of gray mold-rot.

Very little can be done in the field to reduce the amount of gray mold-rot. Newhall obtained beneficial results on seed lettuce by dusting the plants at five-day intervals with 15-85 copper-lime dust. In the greenhouse the trouble can be controlled satisfactorily by regulating the temperature and humidity, and by sterilizing the soil (page 600). The night temperature for lettuce is lower than that at which the fungus grows well; therefore, if at night the greenhouse can be cooled sufficiently, and during the day the plants and top soil can be kept fairly dry, the Botrytis will not be bothersome. All diseased material should be destroyed at once.

REFERENCES

Humphrey, J. E. The rotting of lettuce—*Botrytis vulgaris* Fr. Mass. Agr. Exp. Sta. Ann. Rept. 9: 219-222. 1892.

Peltier, G. L. A consideration of the physiology and life history of a parasite, Botrytis, on pepper and lettuce. Mo. Bot. Gard. Ann. Rept. 23: 41-74. 1912.

ANTHRACNOSE OF LETTUCE

Caused by *Marssonina panattoniana* (Berl.) Magnus.

In 1895 this disease of lettuce was found almost simultaneously in Italy and in Ohio. It has since been observed in England, Australia, Germany, Holland, New York, Michigan, Kansas, Oregon, Utah, North Carolina, Florida, Washington, and probably other countries and states. Very serious losses occasionally are produced by the trouble, but as a rule it is not of great economic importance.

The disease is found on all varieties of lettuce, possibly on endive, and a recent report suggests that it also attacks the common wild lettuce. Because of the characteristic symptoms, the anthracnose is known as shot-hole, and occasionally as rust.

FIG. 71.—Marssonina leaf-spot of lettuce.

Symptoms.

The disease progresses from the older to the younger inner leaves, and in severe cases covers so much of the plant that the latter takes on a sickly yellow caste and becomes stunted. Any part of the leaf may be infected. The spots start as small water-soaked areas that, as they enlarge, become yellow or brown (Fig. 71). They are inclined to be circular in outline, but may be angular when near a vein or may be elongated if on the midrib. The affected tissue shrinks, forming a depression on the vein, or breaks away leaving a hole or a frayed edge in the leaf. If the lesion on the midrib is deep enough, the latter may break permitting the blade to fall over and wilt. In wet weather the sunken spots show faint traces of pink spore masses, but the color is not as pronounced as it is in other anthracnoses on vegetables.

Cause.

The fungus was placed in the genus *Marssonia* and given the specific names *panattoniana* and *perforans*, the former having preference because of priority. Later it was discovered that the name given to the genus had already been applied to some other plant, therefore was void. The difficulty was surmounted by revising the spelling of the genus name, designating it as *Marssonina panattoniana*.

It is not known whether the fungus is disseminated with the seed or is able to live saprophytically for a period in the soil. It is certain that it can survive the winter in diseased refuse left in the beds or in the fields. From this debris the spores are splashed by rain to the growing crop, they germinate in six or more hours, and their germ-tubes enter directly through the epidermis into the host. The mycelium does not spread far in the leaf, but as soon as it becomes well established it begins to form acervuli beneath the host epidermis. As the fruit-body matures the covering bursts and the numerous two-celled spores are available for dissemination. They may be scattered by rain or insects to healthy plants. It may often be observed that the mycelium in a spot on an outer leaf will affect the closely adjoining inner leaf, and continue in this manner until a corresponding lesion is formed on every leaf from the oldest to the youngest in the center of the head. Germination and penetration of the germ-tube occurs in a day.

The parasite requires a moderate amount of moisture and

a comparatively low temperature. It is difficult to procure infection during the warmer summer months. It is much more severe when overhead sprinkling is practiced, than when spattering is avoided by wetting down the soil with a fine mist or a slow stream from a wide open hose.

Control of anthracnose.

If rotation of crops and destruction of diseased plants were practiced, there would be little or no anthracnose, but the lettuce too often follows lettuce without intermission with the result that inoculum is always present. In the greenhouse it is advisable to raise the temperature as soon as the disease appears. The increased temperature may not be ideal for the most rapid development of the lettuce, yet it is better that the crop suffer slightly from heat than much from the fungus. An increase of a few degrees is often sufficient to check the growth of the parasite particularly if the humidity of the air is reduced by repairing leaky water pipes, by applying the water in a fine mist from overhead pipes or by subirrigation, and by keeping the soil moderately dry. Diseased plants should be removed at once. Spraying has been suggested, but it is not practicable for large plants, and therefore is not recommended.

References

Appel, O. and F. Laibach. Ueber ein im Frühjahr 1907 in Salatpflanzungen verheerendes auftreten von *Marssonia panattoniana* Berl. Anst. f. Land. u. Forstwirtsch 6: 28-37. 1908.

Berlesee, Antonio. Un nuovo marciume dell' Insalata (*Lactuca sativa*) Rivista di Pathologia Vegetale 3: 339-342. 1896.

Brandes, E. W. Anthracnose of lettuce caused by *Marssonina panattoniana*. Jour. Agr. Research 13: 261-280. 1918.

Chittenden, F. J. On some plant diseases new to, or little known in, Britain. Jour. Hort. Soc. (London) 37: 531-550. 1912.

Dandeno, J. B. A fungus disease of greenhouse lettuce. Rept. Mich. Acad. Sci. 8: 45-47. 1906.

Ritzema Bos, J. Het andijvierot, veroorzaakt door *Marssonia* (*Marssonina*) *panattoniana* Berl. Tijdschr. Plantenz. 21: 169-186. 1915.

Downy-Mildew of Lettuce

Caused by *Bremia lactucæ* Reg.

The downy-mildew of lettuce is widely distributed in Europe and America. It was observed first about Boston in 1875, and since then has often proved a serious pest, especially in greenhouses. In California, however, it causes much loss in the field. Not only are affected plants dwarfed and discolored, but when slightly diseased stock is shipped, the pathogene, or saprophytic organisms that follow, cause a decay during transportation.

The mildew is found on the cultivated lettuce and the different species of wild lettuce, which grow luxuriantly about compost heaps near greenhouses. Voglino in Italy has reported the disease also on chicory. Some varieties of the cultivated lettuce are much more susceptible than are others, and occasionally might be replaced by the more immune kinds. Usually, however, the market demands a certain type and will accept no other. In some tests the Los Angeles Market (New York Head) proved extremely susceptible, while the Iceberg was almost immune and Big Boston and Hanson only slightly less so.

Symptoms.

The disease ordinarily produces most of the lesions on the older foliage although any exposed part may become infected. The spots begin as yellowish or lighter green areas on the upper surface of the leaves and as the lesions enlarge the fungus fruits with white mycelial wefts in corresponding spots on the opposite side. The tissue may later turn brown, and when the discoloration is on the edge of the leaf, the disease is sometimes known as brown margin. In severe cases the whole plant is dwarfed and yellow. During transportation the affected leaves offer a suitable substratum for the growth of saprophytes and often succumb to a soft-rot.

Cause.

The pathogene, *Bremia lactucæ*, belongs to the downy-mildew family, and for a long time was known as *Peronospora gangliformis* (Berk) DeB. It has well developed mycelium and tall profusely branched conidiophores, which arise from the stomata in clumps of two or three. The characteristic cup-like formation with scalloped edges at the tips of the branches distinguish Bremia from the other related genera. The conidia are borne on the tips of the scallops or sterigmata, and after becoming detached germinate either by germ-tubes or by zoospores. No oospores have been found. The fungus is able to live in the soil during part of the summer months, but cannot withstand the freezing and thawing of winter. It is supposed to live over on wild lettuce, which grows as a winter annual. The seeds of these weeds germinate in the fall, and the plants begin growth before snow arrives. Infection follows. The fungus remains dormant in the weed hosts until spring growth begins, when it develops with the wild lettuce foliage and is soon ready for dissemination.

As in the related genera, the Bremia spores germinate best in a low temperature, the optimum being 43° to 50° F. Abundant moisture is required. The disease is most severe when the differences between the day and night temperatures are greatest and when the days are most cloudy. These conditions usually prevail during the winter when lettuce is grown under glass.

Control of downy-mildew.

If the wild lettuce hosts could be eradicated and rotation of crops practiced, the mildew question could be solved. In many greenhouses one crop of lettuce follows another, so that if there is any contamination the second crop is sure to be inoculated. This statement may suggest that soil sterilization is the solution of the problem, but when sterilization has been practiced the mildew has not been controlled. Light being a factor, it is well to have a house constructed in such a manner that plenty of air space is available and that the sunlight is obstructed as little as possible. Temperatures also are important and for the control of this disease should be three or four degrees higher at night than is recommended under normal conditions. In addition, the watering can be done better by means of a sluggish stream out of an open hose, than when the water is applied by an overhead sprinkling system in which the foliage is covered with moisture.

The very young plants are much more susceptible to the downy-mildew than are those having attained to a greater degree of maturity. Consequently if the small plants are protected with a fungicide during the critical period, the disease can be much reduced if not wholly avoided. Several days before the seedlings are ready to transplant, the upper and lower sides of their leaves are well sprayed with bordeaux mixture 3-4-50 or 4-4-50. Another application is made two or three days after they have been transplanted. If the spray is put on too near the time of transplanting, the seedlings wilt severely and are checked in their growth. A lapse of two or three days, however, is sufficient to avoid any unfavorable results.

References

Erwin, A. T. Controlling downy mildew of lettuce. Iowa Agr. Exp. Sta. Bull. 196: 307-328. 1921.

Link, G. K. K. Downy mildew: a transit disease of lettuce. Phytopath **12**: 48, 49. 1922.

Maynard, S. T. Lettuce mildew. Mass. (Hatch) Agr. Exp. Sta. Bull. 4: 11-14. 1889.

Milbrath, D. G. Downy mildew on lettuce in California. Jour. Agr. Research **23**: 989-994. 1923.

Erwin, A. T. Control of downy mildew of lettuce. Proc. Amer. Soc. Hort. Soc. **17**: 161-168. 1921.

Melhus, I. E. Bremia on hothouse lettuce. Phytopath. **11**: 54. 1921.

Noffray, E. La Lampsane commune principale nourricière du meunier des laitues. Jour. Agr. Prat. n. s. **22**: 781-783. 1911.

Schneider, Numa. Traitement du meunier des laitues et romaines et du mildew des épinards. Rev. Hort. (Paris) **84**: 493-494. 1912.

Schweizer, Jean. Die spezialisation von *Bremia lactucæ* Regel. Schweiz. Verhandl. Naturf. Gesell. **99**: 224. 1918.

Damping-off of Lettuce (Fig. 72).

Caused by various fungi

(See Damping-off of Tomatoes, page 546.)

Fig. 72.—Showing the effect of damping-off of lettuce in unsterilized soil. Sterilized soil in the pot on the right.

Lettuce Stunt

Caused by *Pythium* sp.

In recent years a stunt disease or dwarfing has been found on lettuce plants in Michigan, Missouri, and New York. It resembles in outward appearance the rosette caused by Rhizoctonia and some of the lettuce diseases resulting from bacteria, and for this reason may have been confused with these troubles. The disease ordinarily does no great damage, although fields have been reported where the reduction in yield was greater than 10 per cent, and the crops in some greenhouses have been almost a total loss. Stunt, so far as is known, affects only lettuce. All varieties of this vegetable are susceptible.

Symptoms.

At first the only outward symptom on the affected plant is that of dwarfing (Fig. 73). Evidently infection may take place and check growth as soon as the plant is through the ground, or at any time until the lettuce is mature. There are no necrotic lesions on the leaves and stems, but the foliage becomes darker green than normal, and has a lusterless appearance. The central leaves may not develop, resulting in a

FIG. 73.—Pythium stunt of lettuce. A diseased plant on the right compared in size with a healthy one.

flattened plant. As the disease progresses, the outer leaves die and shrivel, followed by the death of the remainder of the plant. Previous to death the plant may exhibit the symptom of wilting. If the stem and roots are split longitudinally, a blackening of the vascular bundles is very evident.

Cause.

Both in Michigan and New York the trouble has been proved to be caused by a Pythium. No oospore stage has been discovered either in culture or under field conditions, therefore the species of the fungus has not been determined. The life history seems to be the same as that given for *Pythium debaryanum* (page 546). The fungus persists in the soil, but apparently is not carried on the seed. As soon as the lettuce seed germinates, infection may take place through the young root. The conducting tissue of the plant is invaded and gradually killed, which accounts for the slow death of the plant.

In Ohio a similar disease known as rosette has been attributed to *Rhizoctonia solani.* The chief difference in symptoms is the presence of surface lesions typically like those caused by Rhizoctonia on the stems of potatoes and other hosts, while the attack of Pythium on lettuce causes no superficial lesions.

The stunt is abundant in the field only during periods of wet cool weather, such as occur in early spring or late in the fall.

Control.

No methods for controlling the fungus in the field are known. Good drainage and long rotations of crops will aid in reducing the amount of disease. In the greenhouse the fungus can be killed by soil sterilization (page 600).

References

Coons, G. H. Lettuce diseases. Mich. Acad. Sci. Rept. 17: 129-131. 1916.

Coons, G. H. Lettuce stunt. Mich. Agr. Exp. Sta. Rept. 30: 314. 1917.

Dye, H. W. Two diseases of western New York lettuce: the bottom-rot and the stunt. Unpublished Thesis. 1919.

Selby, A. D. Studies in potato rosette II. Ohio Agr. Exp. Sta. Bull. 145: 27. 1903.

Bacterial Wilt of Lettuce

Caused by *Bacterium vitians* Brown

If the number of bacterial organisms on lettuce and their corresponding literature increase as the present rate of nomenclature threatens, a separate science will finally be necessary to deal with this subject. Future work, however, may simplify the problem. Judging from symptoms alone, it is possible that some of the bacterial diseases that are now considered as distinct will later be combined.

It is not known just how prevalent the bacterial wilt is, as pathologists generally are not sufficiently acquainted with the symptoms to make a definite diagnosis. Reports based on appearances in the field and not on microscopical or cultural examinations, indicate that the disease is common and destructive in many lettuce districts of this country. In individual fields there may be almost a total loss. Actual counts have shown infections ranging from 15 to 98 per cent.

Symptoms.

Lesions may appear either on the stem or leaves. At first the affected host has an unhealthy yellow cast, often followed by wilting. In some cases the center leaves decay and in others the outer ones are first to be destroyed. In slight attacks there may be only spots of dark dead tissue on the foliage, without any extended rot. Invaded plants are not firmly rooted, and the stem is brittle enough to be snapped off easily near the surface of the ground. In the early stages of the attack a longitudinal section of the stem reveals a blue-green streak through its center; in later stages a brown cylinder is formed of the inner stem tissue. Occasionally these inner symptoms may be accompanied by hollowness of the stem, but the latter is more the result of irregular growth on the part of the lettuce than any reaction toward the parasite.

Cause.

The organism is a short rod-shaped bacterium having the group number 211.3332523 and bearing the name *Bacterium vitians*. Where it has been proved to be present in the tissue, the plants had previously been weakened either by freezing or by poor soil conditions. Evidently it is only weakly parasitic. A detailed life history has not been recorded.

Control.

No direct methods for the eradication of the organism are available unless the trouble occurs in the greenhouse. Soil sterilization is then advisable (see page 600). In the field the only precaution that can be taken against the disease is to grow sturdy plants. Well-rotted manure, soil favorable for the crop, and careful cultivation tend toward the production of such a plant.

Reference

Brown, Nellie A. Some bacterial diseases of lettuce. Jour. Agr. Research 13: 367-388. 1918.

Bacterial Rosette Disease of Lettuce

Caused by *Aplanobacter rhizoctonia* Thomas

The rosette is still another bacterial disease of lettuce. In outward appearance it resembles the injury due to Rhizoctonia, and probably was mistaken for the latter during a number of years. The bacterial trouble was not recognized as a distinct disease until it was studied carefully in 1919, and after it had caused much injury in greenhouses in Ohio. Apparently the malady is present in several states but up to the present time the symptoms are not familiar enough to the pathologists for any detailed reports regarding its prevalence and severity.

The chief symptom above the ground is the dwarfing of

the plant and a slight wilting of the leaves during the middle of the day. The root system bears the most severe injury (Fig. 74). The smallest roots are attacked first. The killing and discoloration extends upward until the larger roots are invaded. As the infected roots cease to function, the parts

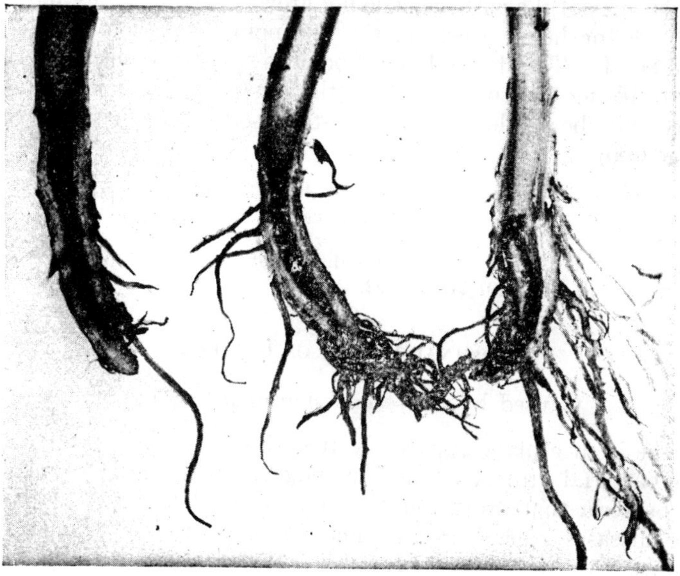

FIG. 74.—Longitudinal section of lettuce stem showing internal browning caused by *Aplanobacter rhizoctonia.*

above ground decrease in the rapidity of their growth. When a much dwarfed plant is pulled up almost the whole root system has disappeared. If the stem is cut longitudinally a yellow streak is found to extend up the center from the crown of the root to an inch or more above the ground. There is never any rot or spotting of the stems or leaves.

The organism is a small rod-shaped yellow bacterium, with the group number 211.3333523, and because of its non-motility is known as *Aplanobacter rhizoctonia.* Evidently it remains alive in old roots or even in the soil during the periods in the rotation when lettuce is not planted. Infection takes place through the root-hairs or the finer roots. The bacteria quickly fill the invaded xylem-ducts. The parasite is disseminated from one bed to another with particles of soil clinging to tools or clothing.

Cultures of the pathogene do not grow at temperatures above 100° F. or below freezing, the optimum being between 77° and 80° F. The thermal death point is approximately 125° F.

Sterilizing the soil eradicates the organism. The sterilization may be made either with steam or formaldehyde (see page 600).

Reference

Thomas, R. C. A bacterial rosette disease of lettuce. Ohio Agr. Exp. Sta. Bull. 359: 197-214. 1922.

Marginal Blight of Lettuce

Caused by *Bacterium marginale* Brown

The lettuce plant apparently furnishes an ideal substratum for bacterial growth if one can judge from the large number of bacteria that are listed as parasitic on this host. Among the diseases caused in this manner is a trouble found on lettuce by Melchers in Kansas. How widely it may be spread is not known. Where it was first found the pathogene exhibited a predilection for certain varieties, for example, Black-Seeded Simpson, Improved Hansen, and Big Boston, while it was much less severe on Early Curled Simpson, and Vaughan's All Season, and did not attack at all the variety Grand Rapids.

The first symptoms may appear as a wilting and drying of localized areas along the margin of the leaves, or as a darkening of the veins near the foliage tips. In either case it usually affects simultaneously all the leaves in a single whorl of the head. The lesions are small at the beginning and far apart, but as they enlarge they join on either side until a large part of the leaf edge is included. The disease does not progress much farther than two or three centimeters back from the margin, and in no case does it cause a soft-rot. The tissue becomes dry and papery, and may turn brown, tan, reddish or even black, thus destroying the appearance and salability of the plants.

The pathogene, *Bacterium marginale,* is a rod-shaped organism with one or two, rarely three, flagellæ at each pole. It bears the group number 211.2323123. Evidently it hibernates in the soil, and from there is splashed to the tips of the leaves where drops of water often collect. It enters the unwounded tissue, and in an optimum temperature of 77° to 80° F. develops rapidly.

Good ventilation and watering the soil without splashing the plants is usually sufficient to protect the crop, but if the organism still persists in causing injury, it may be well to sterilize the soil (see page 600).

Reference

Brown, Nellie A. Some bacterial diseases of lettuce. Jour. Agr. Research **13:** 367-388. 1918.

Bacterial Rot of Lettuce

Caused by *Bacterium viridilividum* Brown

A soft-rot of lettuce has frequently been observed by gardeners and pathologists, and usually has been attributed to Sclerotinia. Therefore, it is not known how widespread or how serious the bacterial rot really is. Judging from the few authentic reports on record, the malady may become epidemic in nature both in the field and during transportation.

Symptoms.

The infection has its inception at the margins of the leaf, usually following tip-burn or some other injury. The decay spreads rapidly inward, always showing a distinct line of demarcation between the healthy and diseased tissue. In slight cases there may be only isolated dead spots on the leaves, or streaks mostly on the leaf-blade, although occasionally along the midribs; ordinarily the disease progresses toward the center of the plant until the whole head is included. The stems and roots are not affected.

Cause.

The parasite is *Bacterium viridilividum*, a white organism that liquefies gelatine slowly and has a thermal death point of about 120° F. Its life history is not recorded.

Another organism, described in Italy and mentioned in work done in Florida, causes a similar rot of lettuce, yet seems to differ from *Bacterium viridilividum* in that it does not liquefy gelatine and its colonies later exhibit a rose color. It has been named *Bacillus lactucæ* Voglino.

Control.

No definite control measures are known. It was at one time suggested that the disease might be held in check by spraying the plants with formaldehyde (1 pint diluted in 30 gallons of water). Later trials with this fungicide did not give satisfactory results.

REFERENCES

Brown, Nellie A. A bacterial disease of lettuce. Jour. Agr. Research **4**: 475-478. 1915.

Brown, Nellie A. Some bacterial diseases of lettuce. Jour. Agr. Research **13**: 367-388. 1918.

Burger, O. F. Bacterial rot. Fla. Agr. Exp. Sta. Ann. Rept. **1913**: LXXXVII-LXXXVIII. 1914.

Floyd, B. F. Bacterial rot. Fla. Agr. Exp. Sta. Ann. Rept. **1908**: LXXX-LXXXVII. 1909.

Levin, Ezra. Control of lettuce rot. Phytopath. **7**: 392-393. 1917.

Voglino, P. Sulla batteriosi delle lattughe. Annali R. Accad. di Agric. di Torino **46**: 25-33. 1903.

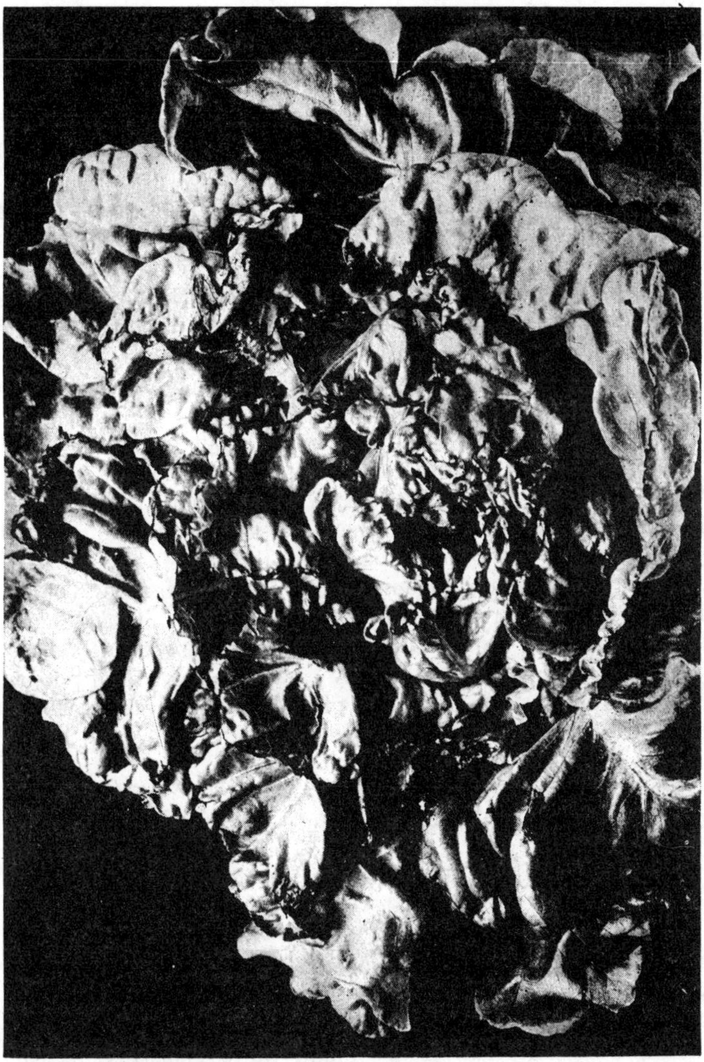

FIG. 75.—Tip-burn of lettuce.

Tip-Burn of Lettuce

Physiological

(Refer also to Tip-Burn of Potato, page 357, Black-heart of Celery, page 106, and Blossom End-Rot of Tomato, page 580.)

Tip-burn is frequently a very serious trouble on lettuce grown in the field. On affected plants, the edges of the leaves die and turn brown (Fig. 75), making the crop less salable and in severe cases interfering with growth. The hard-heading varieties are in general more subject to the disease than the loose-heading and leaf varieties.

Various combinations of climatic conditions resulting in too rapid loss of moisture from the lettuce leaves are responsible for the trouble. It is usually brought about by two or three warm moist days, resulting in a soft watery growth, followed by a warm bright day with rapid evaporation. The disease is most troublesome during the hottest part of summer and seldom injures crops maturing in the fall.

Under glass the conditions causing tip-burn can be avoided. In the field, control measures are unknown, although good cultural methods are of value. Deep and frequent cultivations when the soil is packed by heavy rains, sparing use of potash, and medium applications of phosphate have reduced the amount of tip-burn in certain fields.

Reference

Newhall, A. G. Unpublished reports. 1921-1923.

Mosaic of Lettuce

Cause undetermined

The mosaic is one of the minor troubles of the lettuce crop, yet cases have been observed, especially late in the summer, when aphids are plentiful, in which 15 or 20 per cent of the plants were affected. It is found quite generally in this country and on most varieties of lettuce.

The symptoms are typically those of a mosaic disease, including the yellowish unhealthy color of the foliage, the dwarfing of the plant, the green and yellow mottling of the leaves, the ruffling of the leaf-blade, and in head lettuce the absence of the head when severely infected. The mottling is most pronounced when the lettuce is grown in cool cloudy weather. There is some difference in manifestations of the symptoms on the various types of lettuce. On the head lettuce the general yellow discoloration is very pronounced but mottling is rare. The mottling or yellow blotching is common on the Romaine lettuce. The leaf lettuce is much corrugated. When partly grown plants are inoculated, the leaves formed before infection usually show no signs of the trouble. It is the younger foliage that becomes abnormal. Occasionally a plant may show the effects of the disease, then apparently recover and grow almost to normal size.

The virus is present in the sap of the infected plant. It is transmitted in the field by means of aphids, and possibly by other sucking insects, or in any other manner in which sap may be conveyed from a diseased to a healthy plant. It has been proved that the virus also lives in the seed and when thus carried is the source of primary infection of the new crop. When seed is obtained from infected plants, the progeny will in many cases be affected similarly.

If the disease ever becomes serious enough to make heavy inroads on the yield of the crop, the trouble may be avoided by procuring clean seed and isolating the fields from those having infected plants. It may be possible in some cases to destroy the insect carriers by the use of insecticides. No doubt the seed producers have already had their attention called to the matter and will, if the demand is insistent enough, grow isolated healthy seed plots for their commercial stock.

References

Jagger, I. C. A transmissible mosaic disease of lettuce. Jour. Agr. Research **20:** 737-739. 1921.

Newhall, A. G. Seed transmission of lettuce mosaic. Phytopath. **13:** 104-106. 1923.

Rio Grande Lettuce Disease

Cause undetermined

In the Rio Grande Valley in Texas this disease of lettuce has been very severe during certain seasons. The plants have the appearance of being under-nourished, the older leaves turn red, particularly at their tips, the younger central leaves blanch and do not grow normal in size, numerous lateral adventitious shoots develop, and the roots die and become dried. The most interesting symptom, however, is the formation of small dark-colored blister-like projections along the edges of the blanched leaves. These are formed by the exudations of the milky lettuce sap. The severity of the disease depends on the age when the plant first comes under the influence of the abnormality. The very young plants soon die, half grown plants may not succumb entirely but show the typical blanching of the center leaves, and the blistered leaf borders; old plants are not much changed in their appearance.

Fig. 76.—White-heart or rabbit-ear of lettuce.

It has been suggested that the trouble in Texas is the result of a root disease or injury. In New York a disease with identical symptoms and known as "rabbit-ear" or white-heart (Fig. 76) because of the arrangement of blanched central leaves, causes much injury in dry seasons. It is being studied by Newhall, who does not agree with the Texas findings. It may be possible later to suggest the cause and a possible remedy.

REFERENCE

Carpenter, C. W. The Rio Grande lettuce disease. Phytopath. **6**: 303-305. 1916.

SEPTORIA LEAF-SPOT OF LETTUCE

Caused by *Septoria lactucæ* Peck.

The leaf-spot of lettuce has been known for many years both in America and Europe, and occasionally it does much damage. It is more destructive on some varieties than on others. The lesions are irregularly shaped reddish spots, marked sparingly with black pycnidia (Fig. 77). The fungus has been named *Septoria lactucæ* by Peck and also by Passerini and *S. consimilis* by Ellis and Martin. Judging from the description, the two species are the same. According to Newhall, the life history of the fungus, including the dissemination with the seed, is the same as that of the Septoria on celery.

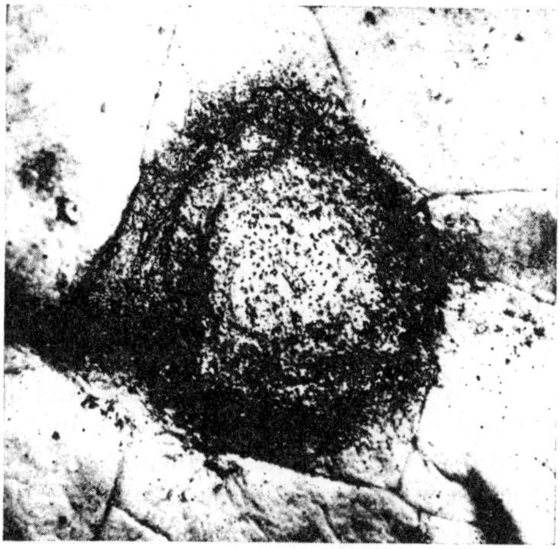

FIG. 77.—Enlarged Septoria spot on lettuce, showing pycnidia.

No attempt has been made to control the fungus.

REFERENCES

Ellis, J. B. and Geo. Martin. New Florida fungi. Jour. Mycol. **1**: 97-101. 1885.

Halsted, B. D. Report of the Botanist. N. J. Agr. Exp. Sta. Ann Rept. 1898: 291-370. 1899.

Saccardo. Sylloge Fungorum **3**: 551. 1884. **10**: 368. 1891.

Cercospora Leaf-Spot of Lettuce

Caused by *Cercospora lactucæ* Stevenson

Recently this leaf-spot of lettuce was discovered in the Philippine Islands. It has not been reported from any other country. The lesions appear on the edge of the leaf as small water-soaked spots, which as they enlarge turn brown. The disease injures only the leaves on which the spots appear. The dead tissue is covered with tufts of short conidiophores on which are borne the thin long septate spores.

Control measures are unnecessary.

Reference

Welles, C. G. Leaf-spot of lettuce. Phytopath. **13**: 289. 1923.

Fusarium Root-Rot of Lettuce

Caused by *Fusarium* sp.

In Kentucky a lettuce disease was observed, the symptoms of which were manifested by dwarfing of the midribs, corrugation of the leaf-blades, browning of the veins, and mottling of the green surface. When the plants were removed from the soil, the roots showed numerous lesions on their sides or had decayed tips. A Fusarium was isolated from the diseased areas. No further work has been reported regarding the trouble.

Reference

Anonymous. Green-house lettuce disease. Ky. Agr. Exp. Sta. Ann. Rept. **1919**: 45-46. 1920.

CHAPTER X

DISEASES OF ONIONS

ALTHOUGH onions are grown in each of the forty-eight states, the important sections are widely separated. The states in the order of their acreages are California, New York, Texas, Ohio, Illinois, Massachusetts, and Indiana. These states produce more than half of the total crop which is planted on about sixty-five thousand acres.

ONION SMUT

Caused by *Urocystis cepulæ* Frost

The disease known as smut, or rarely as black-rot, is the most destructive malady found on onions. It was present in the Connecticut Valley as early as 1860, and from there spread to all the northern intensively-growing onion sections. It has been reported from nineteen states, most of which are north of latitude thirty-six and extending from Massachusetts to Oregon. It has been present for many years in France, Germany, Denmark, and Belgium. Strangely it has not been found in England until recently. The origin of the fungus is not known.

The disease is known to occur only on species belonging to the genus Allium. In America it is of economic importance only on the cultivated onion, but in Europe it is sometimes destructive where leeks are grown extensively. It also has been found on shallots, chives, and the Welch onion. In a test made with twenty-nine varieties of the common onion, all were about equally susceptible. In a similar test with eighteen species of Allium, there was a gradation of infection from

Allium Cepa (common onion), one of the most severely infected, to *A. Oreoprasum* and *A. Moly* (golden garlic), both of which are immune.

Losses from smut are confined largely to sections where onions are grown from seed and are produced intensively without rotation. When green sets are employed, the injury is confined wholly to the seed-bed. Onions grown for the production of seed or from dry sets are never attacked. The loss is due to thinning of the stands on infected soil, or by the shriveling or decaying of diseased bulbs before the crop is cured and ready to store. It is not a storage disease. The percentage of loss varies for different sections, but it is not uncommon to see whole fields plowed up because of the poor crop. The owners of many muck farms no longer attempt to grow onions, since their previous plantings resulted in annual losses. Eighty-one counts were made of diseased plants in three counties of New York in 1920 and the average percentage of infected onions was nearly 30 per cent. The reduction in yield was not that great because growers, anticipating the poor stands, used almost twice the amount of seed that otherwise would be necessary.

Symptoms.

Readily visible dark streaks show within the tissue of the leaves (Fig. 78), leaf-sheaths, and bulbs. These streaks are filled with a dark brown powdery mass, very similar to the smut masses on corn. The lesions first appear in the one-leaf seedlings and may be seen at that time by transmitted light. They continue to appear in new parts as they develop throughout the life of the plant. Some abnormal bending and twisting of the earlier diseased leaves often occur. The plants usually are much stunted and may die slowly through a gradual drying out process at any stage in their life. The smut is sometimes followed by a soft-rot due to other organisms, which have gained entrance through the bursted pustules. Not all the seedlings die which show disease in their first leaves, for these

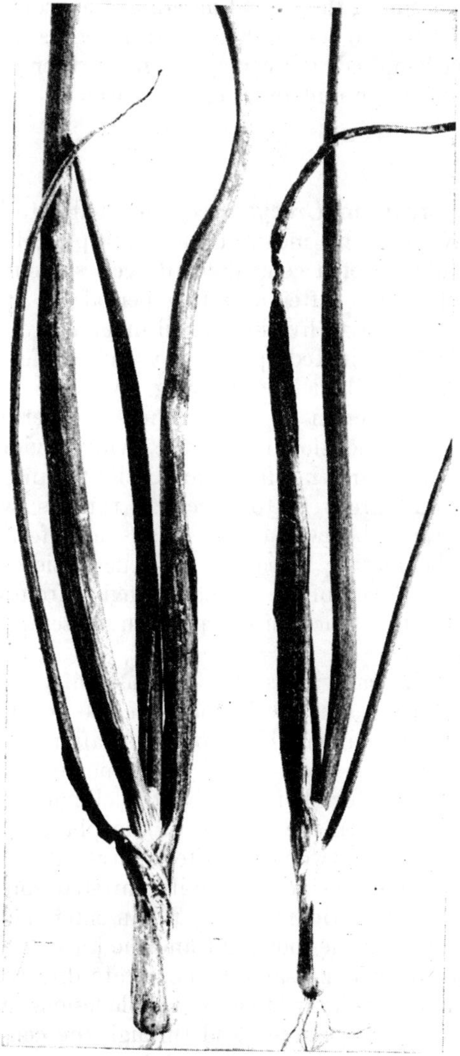

FIG. 78.—Smut bursting the epidermis of young onion leaves.

leaves are shed during the second month of the growing season, and when the infected foliage has dried up or the pustules on the bulb have emptied, there may be no further advance of the disease during the remainder of the season.

Cause.

The causal organism, *Urocystis cepulæ,* is able to live in the soil for many years by means of its multiple-celled spores. These are made up of a central fertile cell surrounded by a layer of sterile cells. After a resting period the spores may germinate or continue to live in the soil indefinitely. If onions are present the germinated spores may cause infection, or if there is no host available the growing mycelium may thrive on the organic matter in the soil. The spore germinates by putting out a hyaline globular vesicle, which evidently is a type of basidium. Growing from this basidium and extending in all directions are septate mycelial threads with short branches emerging below the cross-walls and forming right angles with the mother threads. No true basidiospores are produced, but each cell of the hyphæ springing from the basidium may take the place of a spore in dissemination and reproduction.

Fortunately the parasite is able to cause infection only for a brief time during the development of the seedling. This period extends from about the second day after the germination of the seed until the seedling is in the first leaf. This period is lengthened or shortened by the climatic conditions, which retard or hasten the growth of the plant. Ordinarily the time extends over twelve or fifteen days, after which the host is safe even in the most heavily infested soil. During the brief susceptible period, the fungus can enter the epidermis at any point between the root-joint and the knee of the cotyledon, but apparently is not able to penetrate directly the true leaf. About five days after infection visible lesions are present. The mycelium passes between and through the cells where it feeds by means of variously shaped haustoria until it has reached the true leaves. The fungus, when well established, forms fruiting pustules just below the host epidermis, and which later are broken open permitting the spores to drop into the soil. They are then disseminated by running water, by tools, on the feet of men and animals, on the roots of transplanted vegetables, or in any other manner in which particles of soil may be conveyed from one place to another.

The smut of onions is confined almost wholly to the states with a cool summer climate. Even though diseased sets are sold each year to growers in the South, they are not bothered with the disease, indicating that the soil temperature is the limiting factor for smut. Experimental work has shown that the fungus will develop at temperatures too low for good growth, but that a soil temperature of 85° F. will inhibit almost entirely the growth of the pathogene. The optimum for infection is 72° F. Soil-moisture has no direct effect on the amount of the disease, but by delaying the growth of the seedling and cooling the soil, an excessive amount of water favors the development of smut.

Control of smut.

Since onion sets are immune to smut, it is possible in a limited way to use these instead of sowing seed. However, the planting of onion sets will never become popular among large groups of growers. Some will sow the seed each year until the inoculum in the soil becomes too plentiful for a profitable yield, then will turn to some other crop. No doubt the growing of some non-susceptible host on infested land for a period of years will tend to lessen the number of spores present if flooding or other means of recontamination are avoided.

The only direct method of control is that of applying formaldehyde (1 pint in 16 gallons of water) in the furrow with the seed (Fig. 79). Approximately two hundred gallons of the solution are needed for an acre, or sixteen gallons for three thousand feet of row. Such a large amount of water is difficult to apply when the soil is already moist, for the wet soil sticks to the rear wheel of the seeder. Consequently many growers have reduced the number of gallons applied to an acre and have made the solution more concentrated. They

FIG. 79.—Yield of two onion rows (at right) treated with formaldehyde for the control of smut, as compared with two untreated rows.

have often been successful in this change, although the greater concentration sometimes results in considerable injury to the seed. In Massachusetts the results are said to be very good on loam soil when one gallon of the solution is diluted with fifty gallons of water and applied according to the condition of the soil. If the soil is very dry, fifty to sixty gallons are sufficient for one acre; ten more gallons are added if the soil is medium moist, and twenty more gallons if the soil is moist and heavy. The muck soil requires much more liquid than does the loam.

Tanks made of tin or galvanized iron and varying in capacity from one to five gallons are fastened on the seeder in such a manner that the lead outlet pipe discharges the solution into the furrow with the seed. With the larger amount of solution (200 gallons), this pipe should have an inner diameter of at least five-sixteenths of an inch (slightly larger if a stand-pipe is attached), and be supplied with a stop-cock, which may be opened or closed by manipulating a wire that extends up to the handle of the seeder. The outlet pipe may be supplied with a union into which washers of different sizes may be slipped at will to govern the flow of the solution. About three sizes of washers are all that are needed: one size when the soil is wet, another when it is dry, and a third when the soil is medium moist. If the opening at the top of the tank where the liquid is poured in is not closed air-tight during the operation, the solution runs out faster when the tank is full than when it is nearly empty (Fig. 80). To overcome this difficulty, a tank has ben devised which has an air-tight screw top, the air in the receptacle being supplied through a stand-pipe soldered on near the bottom of the tank

FIG. 80.—An old type of formaldehyde tank without the stand-pipe. The fuller the tank, the faster the liquid flows.

and with its open top rising above the water level as shown in Fig. 81.

The tanks can be made to accommodate any type or size of seeder. It may be necessary when growing onion sets, for which the seed is sown thickly and in wide furrows, to have a fish-tail opening at the mouth of the delivery tube, so that the whole width of the row will be covered with the disinfectant.

Fig. 81.—Onion drill with formaldehyde tank attached. The tank is supplied with a stand-pipe so that air enters near the bottom and insures a uniform flow of liquid.

Lime and sulfur, paraformaldehyde, and other dry materials have been employed, placing them directly in the box with the seed. They have not proved satisfactory. Lime often gives an increased yield of onions, but it is because the acidity of the soil is reduced and not because of any fungicidal value that the lime may have.

References

Anderson, P. J. Development and pathogenesis of the onion smut fungus. Mass. Agr. Exp. Sta. Tech. Bull. 4: 99-133. 1921.

Anderson, P. J. Susceptibility of species of Allium to onion smut. Phytopath. 14: 26. 1924.

Anderson, P. J. The relation of soil moisture to formaldehyde injury of onion seedlings. Phytopath. **13**: 392-403. 1923.

Osmum, A. V. and P. J. Anderson. An improved formaldehyde tank for the onion drill. Phytopath. **13**: 58. 1923.

Sturgis, W. C. Transplanting as a preventive of smut upon onions. Conn. Agr. Exp. Sta. Ann. Rept. **19**: 176-182. 1895.

Thaxter, R. The "smut" of onions (*Urocystis cepulæ*). Conn. Agr. Exp. Sta. Ann. Rept. **1889**: 129-153. 1890.

Walker, J. C. Experiments upon formaldehyde drip control of onion smut. Phytopath. **10**: 323-327. 1920.

Walker, J. C. and L. R. Jones. Relation of soil temperature and other factors to onion smut infection. Jour. Agr. Research **22**: 235-262. 1921.

Downy-Mildew of Onions

Caused by *Peronospora schleideni* Unger

The onion mildew, conspicuous because of its destructiveness, was observed as early as 1841 when it was described in England. It is now present on, and frequently very injurious to, the common onion in Europe, Bermuda, and America, and has been collected in China. It has also been reported on the potato or multiplier onion and on shallots. It is difficult to determine the actual loss resulting from the disease as it does not attack the bulbs directly. The decrease in yield is brought about in the reduction of green leaf surface. Furthermore, the mildew makes it possible for weak parasites of onions to gain an entrance into the tissue and cause destruction of the bulbs. The successive killing of the leaves tends to delay the maturity of the bulb, and when such stock is put into storage it may be damaged by neck-rot and other similar troubles. In California and Louisiana the injury resulting from mildew is sometimes very severe on seed onions.

Symptoms.

The disease may be present one season, then fail to appear for several years, or there may be a continuous epidemic for successive years. The mildew, however, is present more often than the average grower is aware. Coming so late in the season leads the grower to believe that the dying of the tops is nothing more than early maturity. It seldom occurs early in the growing period of the plant; the first symptoms on the leaves are ordinarily not discernible until the bulbs begin tc form. The infection commonly starts on one or a few isolated plants which serve as centers from which the inoculum is disseminated. Even before there is any discoloration of the foliage tissue, the fungus, on wet mornings or during a rain, forms a purplish mold over the surface of the leaf. As the infected spot enlarges the whole leaf takes on a water-soaked appearance, then later turns yellow, and on drying may become white. Soon after infection takes place, the leaf, because of the weakened tissue, may drop over, shrivel and become black with mold. New leaves usually are produced after the old ones have been killed, and may continue to develop during the remainder of the season if the weather is dry. More frequently, however, the new foliage is also attacked, the infection persisting until the bulb no longer puts out shoots. By this process the onion is much dwarfed in size, and left either shriveled or immature at the end of the season.

Cause.

The pathogene is one of the downy-mildews, and bears the name *Peronospora schleideni.* It lives over winter as mycelium in the onion bulbs and sets, and as oospores in the old diseased foliage. If infected sets or large bulbs for seed are planted, the mycelium grows up with the foliage, and when favorable weather conditions prevail the fungus produces numerous long branched conidiophores, which protrude through the stomata of the host epidermis either singly or in groups of two or more. On the branched tips of the conidiophores, which resemble somewhat the open beak of a bird, are borne the numerous violet-colored lemon-shaped conidia. The oospores while in the soil also germinate producing conidia similar to those growing on the foliage. In the presence of water on the leaves, the spores which may be splashed or blown on the plant, germinate by means of a germ-tube that enters the stomata and invades the tissue. The internal mycelium may soon produce the conspicuous purple fruit-layer or remain invisible in the plant for weeks or even months without doing any apparent damage.

The parasite, being a downy-mildew, requires plenty of moisture with rather cool nights and relatively warm days. When such a combination of weather occurs, an epidemic may be expected wherever the fungus is present.

Control of downy-mildew.

Only fairly satisfactory control measures are known for the Peronospora. Inasmuch as the parasite lives in the old onion refuse, a rotation of crops extending over three or four years is the first recommendation. The second is that of having well-drained soil, eliminating or avoiding wind barriers as woods, hills, and weeds, and in every other way aiding the tops of the plants to dry during the day. The third is the burning of diseased onion refuse. The fourth recommendation is the obtaining of disease-free sets or seed bulbs. This may not be possible at present, but if greater care is taken in growing sets, and especially if each consumer knows the source of his sets or grows them himself, he may be able to obtain disease-free stock.

An examination of the seed has revealed an admixture of oospores. Consequently seed treatment has been suggested. No careful work has been done on this phase of the subject. The application of formaldehyde to the soil as for the control of onion smut would probably prove as effective as soaking the seed in the disinfectant before planting.

Much has been written regarding the spraying of onions. At best it is a difficult task, and only under certain conditions is it profitable. Even though the disease does not occur each year, it will be necessary to apply the spray every season, for it is impossible to predict the coming of an epidemic until it has arrived. It is then too late to apply a fungicide. Another danger is that when sets are used, the disease may be systemic and therefore the fungus be inaccessible to any poison. Nevertheless, when rotations are practiced and the crop is grown from seed, it may be advisable to use bordeaux mixture 4-4-50. To this mixture are added three pounds of resin fish-oil soap as a sticker, otherwise the glabrous surface of the onion leaf will not retain enough of the copper for effective protection. In order to avoid the trampling down of many onions, the sprayer must be equipped with a very long boom and many nozzles, or preferably with two or more leads of hose, the longer ones of which are partly supported on a long bar laid across the top of the spray-tank. A man at the end of each hose directs the spray to every part of the plant, taking three or four rows at a time or even more if the spray-pump is run by motor. In this manner a dozen or more rows can be treated at one time. The applications must be made at least once a week, and if the mildew is imminent it will be necessary to put on the bordeaux mixture every three or four days. The applications are begun before the Peronospora is expected to appear. When the violet mold once is present, no further remedy for the affected plants is available. The fungus in the tissue cannot be killed by any fungicide.

References

Dudley, W. R. The onion mold. N. Y. (Cornell) Agr. Exp. Sta. Bull. 15: 193-195. 1889.

Hearst, W. H. Experimental work with plant diseases. Ontario. Ont. Min. Agr. Ann. Rept. **1917**: 60-62. 1918.

Jones, L. R. The onion mildew in Vermont. Vt. Agr. Exp. Sta. Ann. Rept. **9**: 113-115. 1895. **10**: 61-62. 1896.

Murphy, P. A. The presence of perennial mycelium in *Peronospora schleideni* Unger. Nature (London) **108**: 304. 1921.

Thaxter, R. The onion mildew, *Peronospora schleideni* Unger. Conn. Agr. Exp. Sta. Ann. Rept. **1889**: 155-158. 1890.

Trelease, W. The onion mold. Wis. Agr. Exp. Sta. Ann. Rept. **1883**: 38-44. 1884.

Whetzel, H. H. Onion blight. N. Y. (Cornell) Agr. Exp. Sta. Bull. **218**: 139-161. 1904.

NECK-ROT OF ONION

Caused by *Botrytis allii* Munn

The literature dealing with rots of onions extends over many years and composes a long bibliography. The descriptions, however, are frequently so lacking in detail that it is impossible to say when neck-rot became serious or how it compares in economic importance with white-rot, Fusarium bulb-rot and other troubles with which it has been confused. It is now known to be serious on mature onions and on sets, and probably in every part of the world where onions are stored. In poor storage places the loss may comprise the total crop. In some years the fungus also causes blasting of the onion flowers or injures the viability of the seed.

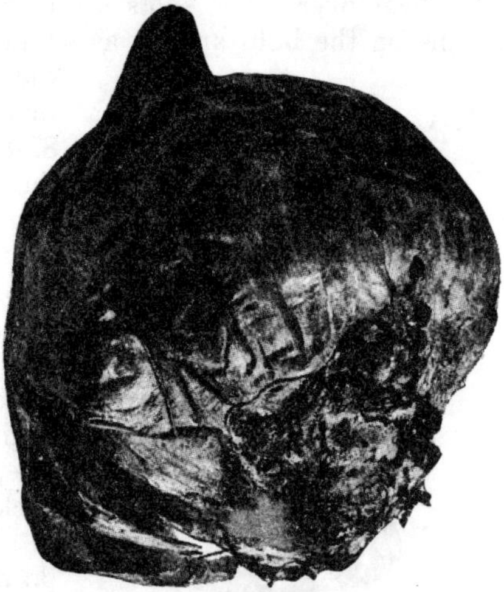

FIG. 82.—Botrytis rot at base of onion bulb.

Because of its general prevalence and the length of time it has been known to growers, the disease has been designated by a long list of names, as gray-mold, storage-rot, onion-rot, bulb-rot, dry-rot, black neck-rot, and dry neck-rot, all of which are in a measure descriptive of the malady.

The neck-rot is confined to varieties of the common onion. The white onion is much more susceptible than is either the yellow or red type. The onion with a thin neck which dries easily is less susceptible than another of the same variety that is thick-necked.

Symptoms.

The disease seldom occurs on the bulbs while they are in the ground, although the fungus may attack injured leaves even when the plants are quite young. The lesions on the foliage have their origin at some thrips injury or where the leaves have bent over. The spots are elongated yellow areas. The lesions on the bulb appear as sunken dried out places about the neck, which finally may involve the whole bulb (Figs. 82-84). In a cross-section the scales appear as if they had been cooked. The tissue is soft and of a brownish color. Between the scales is a gray mass of mycelium, and on the dried parts of the scales small black sclerotia begin to form. On the outside, particularly about the neck, the sclerotia may form in a solid crust. Onions that do or do not have superficial sclerotia often have part of their surface covered with the characteristic ashen-gray fruit-layer of the parasite. The roots may be included also in the decayed area, especially if the bulb is attacked at its base. Occasionally soft-rot bacteria follow the neck-rot, causing the onion to become soft, watery, and foul smelling. The neck-rot, when unaccompanied, results in a dry-rot.

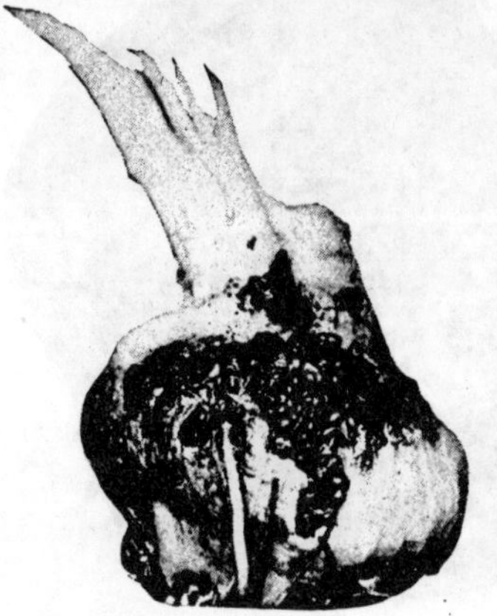

FIG. 83.—Botrytis neck-rot of onion.

The stalk and head of seed onions may be attacked. The fungus usually enters at the base of the stalk where the leaf-sheath clasps the stem, or invades the head directly. In either case the head may be killed and no seed produced, or if the trouble comes too late to hinder seed production, the resulting seed often will not germinate.

Cause of neck-rot.

The classification of the species of the genus Botrytis is in such a muddled condition that no one knows whether a description given by one author fits a specimen found by some other investigator. To make matters more complicated, there are two or three different species of this genus parasitic on onions, to say nothing of other sclerotial forms that are not related to Botrytis. About all that may be said conclusively is that the neck-rot is caused by a Botrytis of the cinerea type and named by Munn *Botrytis allii.* Other names which have been used are *Botrytis cana* (Pers.) Fr., *Botrytis parasitica* Cav., *Sclerotinia bulborum* (Wakker) Rehm., *Sclerotinia libertiana* Fckl., *Botrytis cinerea* Pers. and several species of Sclerotium.

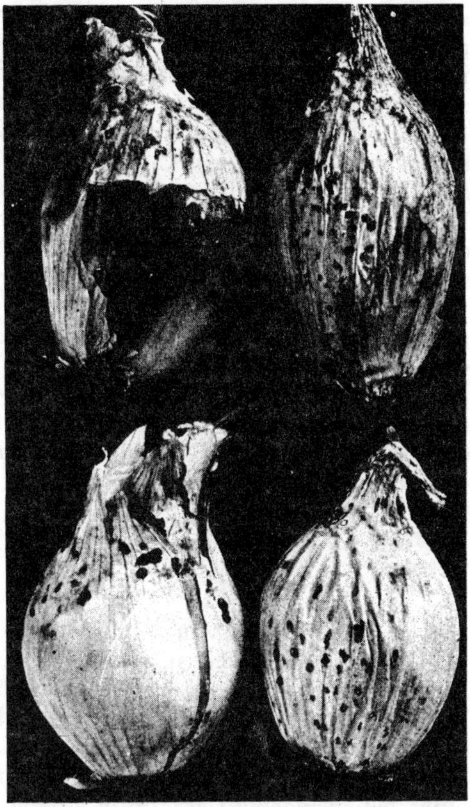

FIG. 84.—A second type of Botrytis on onions.

Infection takes place through wounds at the neck of the

bulbs during or following harvest, the invaded area continuing to enlarge while the onions are curing and after they are housed. The fungus seldom passes from bulb to bulb in storage unless the adjoining tissue is exposed by open wounds. On a normal onion the dry outer scales form a complete barrier against the fungus. It is only when the cut stems or necks are not dried or the onions are kept wet because of the humidity in the room that the parasite can enter. Immature onions and scallions are especially susceptible.

The Botrytis survives the winter by means of the sclerotia, and as mycelium in the decaying host. Both the old mycelium and the sclerotia form new hyphal threads on which spores are formed in small bunches. The spores are blown by wind or carried on tools and clothing to the flower-head, the leaf-sheaths, the inner dry scales, and at harvest time to the cut tops and roots. If enough moisture is present, germination takes place, and after the formation of an appressorium, the germ-tube enters the host. No rot may show until after the onions are put away for the winter, but nearly all the infection takes place before the crop is crated. After the bulb is thoroughly invaded by the mycelium, conidia and sclerotia are formed in great numbers between the scales and on the outer surface.

Control of neck-rot.

If the fact is kept in mind that well dried well stored onions are seldom affected, protection will seem fairly simple. First, the onions should be so grown that they will mature readily. Late applications of fertilizers, downy-mildew, weeds, and wind barriers should be avoided. At harvest the tops should be cut very close, and every attempt made to dry the stubs before placing the crop in storage. It has been shown that mature bulbs or sets may be dried artificially to good advantage. They are placed in shallow trays and held at 90° to 120° F. for forty-eight to seventy-two hours. In addition, all thick-necked bulbs or scallions are to be removed as their slow drying qualities cause them to succumb quickly to the disease.

The onions should be stored in slatted crates, which are so piled that air can circulate all about them. The temperature should be kept as nearly 32° F. as possible, and the humidity of the air as low as the arrangement of the room will permit. A steeply pitched roof with ventilators at the peak will aid in removing the warm moist air. If the doors are then kept tightly closed on damp misty days and opened only when the air is cool and dry, the amount of shrinkage from rot in an onion crop can be much reduced.

Spraying as for the control of onion mildew (see page 287) has shown beneficial results in the keeping quality of the crop. Fumigation of the bulbs with formaldehyde has been attempted. The fumes do not kill the organism but do injure the onions.

References

Humbert, J. G. The neck-rot of white onions. Mon. Bull. Ohio Agr. Exp. Sta. **1**: 176-180. 1916.

Munn, M. T. Neck-rot disease of onions. N. Y. (Geneva) Agr. Exp. Sta. Bull. 437: 363-455. 1917.

Walker, J. C. Control of neck-rot and anthracnose of onion sets. Phytopath. **8**: 70. 1918.

Smudge of Onions

Caused by *Colletotrichum circinans* (Berk.) Vogl.

Since 1851 when it was first discovered, smudge has been described in literature under various names both for the disease and the parasite. Some of the other terms which have been applied in these descriptions of the trouble are anthracnose, vermiculariose, black-spot, scab, mold, and surface-rot. It is present in many parts of Europe, and in most, if not all, of the onion districts in the United States. The disease occurs on the common field onion, onion sets, shallots, and leeks. The injury is most pronounced on the white onion and on sets, and consists in marring the appearance, thereby reducing the sale value, in shrinkage of the bulbs, especially of the sets during storage, and in causing an increase in the premature sprouting of the sets. Where red or yellow onions are grown, the loss is very slight.

Symptoms.

Smudge may appear at any time in the life of the onion plant, and in storage or during transportation as well as in the field. Under certain conditions, especially when the soil is warm, the seedlings may develop damping-off. In this type of injury the outer leaves become shriveled at the surface of the soil, and from this point the pathogene spreads inward until enough of the inner leaves are affected to kill the plant. The damping-off stage is rare in America, since most of the seed is planted early in the spring when the soil is relatively cold.

FIG. 85.—Smudge on white onion bulb.

The most common symptom is the dark green or black smudge on the bulb or neck of the onion (Fig. 85). In some cases the surface of the lesion may be uniformly black, but more commonly the stromata and visible mycelium of the parasite are arranged in concentric circles, the outer one of which may be nearly an inch in diameter. The black portion when examined with the hand-lens will be covered with stiff bristles. Occasionally the black discoloration follows the veins of the scales, or crosses the veins transversely, producing the beginning of a narrow zone of discoloration about the bulb. In severe cases the fungus grows through the outer dry scales into the turgid living tissue causing a collapse of the fleshy scales. On colored onions the fungus is confined almost wholly to the neck of the bulb, where the flattened leaves are colorless. It seems that the pigment of the outer scales of colored onions is toxic to the spores of the parasite. The symptoms on shallots and leeks are similar to those on the onion.

Cause of smudge.

The fungus has received as many names as has the disease. It first was known as *Vermicularia circinans* Berk., and later as *Volutella circinans* (Berk.) Stev. and True. A perfect stage was also described as *Cleistothecopsis circinans* Stev. and True. The common name now employed is *Colletotrichum circinans,* since it has a fruiting-body composed of a black stroma covered by stiff, black, septate setæ, and between which are the conidiophores with their cream-colored conidia. An unusually large stroma is developed below the fruit-body.

The fungus over-winters on the onions, on sets and in the soil. When the spores are formed, they produce mycelium which first penetrates the outer scales. It is only under the most favorable conditions that the organism can enter actively growing tissue. In five or six days the mycelium becomes established sufficiently in the host to form acervuli, with spores. The fungus in the field is disseminated by wind, in splashing water, and on tools and clothing. The parasite is transported long distances on bulbs and onion sets, and may possibly also be carried on the seed. Most of the infection takes place in the field. It is conceivable that the inoculum may spread and infection occur while the onions are in storage; but as the spores need considerable moisture for germination, it is probable that in the average storage-house the conditions are unfavorable for the development of new lesions.

The Colletotrichum develops best at 75° to 80° F., and at the same time requires a relatively high soil-moisture. It does not attack the plants during the early spring when the soil is cold, but does most of the damage just before harvest.

Control of smudge.

No satisfactory control measure has been devised. The usual precautions, however, are helpful. It has been shown that the fungus lives over winter in the soil; therefore, rotations of crops are recommended. In Ohio there was a reduction in the amount of disease when formaldehyde was applied at seeding time as suggested for the control of onion smut (see page 281), although it is difficult to understand why a fungicide applied so early in the season has any effect if infection is as late as some of the investigators suggest.

Careful harvesting of the crop and thorough drying before placing in storage have proved helpful. When the crop is not too large, artificial drying, as suggested for neck-rot, will aid greatly in reducing the loss while the sets and mature bulbs are in storage. A proper storage-house is essential. When all other control measures fail it may be necessary to change from white onions to growing resistant colored ones, or cease growing sets.

References

Stevens, F. L. and E. Y. True. Black spot of onion sets. Ill. Agr. Exp. Sta. Bull. 220: 507-532. 1919.

Thaxter, R. The onion Vermicularia (*Vermicularia circinans* Berk.) Conn. Agr. Exp. Sta. Ann. Rept. 1889: 163-165. 1890.

VanPelt, W. Onion diseases found in Ohio. Ohio Agr. Exp. Sta. Mon. Bull. 4: 70-76. 1919.

Walker, J. C. Onion smudge. Jour. Agr. Research 20: 685-722. 1921.

Walker, J. C. Disease resistance to onion smudge. Jour. Agr. Research 24: 1019-1040. 1923.

The White-Rot of Onions

Caused by *Sclerotium cepivorum* Berk.

The white-rot of onions frequently has been reported from different places in America, but the collectors evidently have confused it with neck-rot or some other disease, for the only authentic specimens in the United States were discovered in Virginia and Oregon. According to Walker, however, it is destructive to onions in the British Isles, Holland, and France, and to garlic in Spain and Italy. It may attack also leeks and shallots.

Symptoms.

The trouble is found mostly in the field, seldom causing injury in storage. In England few infections take place after the first of August. Because of the characteristic basal rot, in which the tissue is covered with a white weft of hyphæ, the disease is known as white-rot, moldy-nose, and dry-rot. The number of plants that show infection depends on the amount of inoculum in the soil. No matter how many diseased plants are present, they are nearly always scattered promiscuously about the field and seldom or never comprise a continuous part of any row. The leaves of an invaded plant decay at the base, turn yellow, wilt and fall over, the older ones being the first to collapse. The roots of such plants are usually badly rotted, so that the plant may be pulled up easily. It is possible that the roots are attacked first, and that the fluffy mycelial

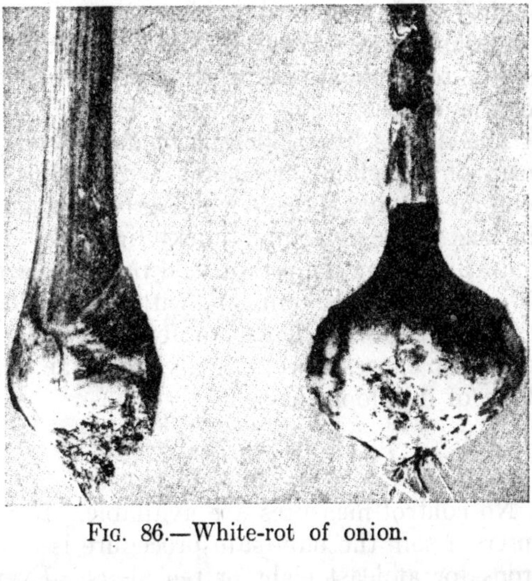

FIG. 86.—White-rot of onion.

growth passes from them to the lower half of the bulb which, as it is invaded, decays and shrinks (Fig. 86). In the meanwhile numerous sclerotia about one-half millimeter in diameter are formed throughout the tissue and on the surface of the onion. They are not as large nor as nearly flat as are those of Botrytis.

Cause.

The parasite was described in England in 1841 and named *Sclerotium cepivorum.* It since has often been confused with Sclerotinia and Botrytis, even though all the available data point to the fact that there is no relationship. Voglino, working in Italy, announced the discovery of a conidial stage, which he designated as *Sphacelia allii.* As other investigators were unable to duplicate his work, the name Sclerotium is retained for the present.

The fungus lives indefinitely in the soil, presumably in the form of mycelium or sclerotia. The latter, on germination, send out hyphal threads which grow through the soil and enter the host. The mycelium quickly invades all parts of the tissue, and after becoming established produces a new crop of sclerotia, which are disseminated with the soil or onion refuse.

The parasite evidently prefers a warm humid climate, and even then causes infection only when the bulbs are at a certain stage of development.

Control of white-rot.

No control measures are available. If the Sclerotium once enters a soil, the only safe procedure is that of growing other crops for at least eight or ten years. Even this may not be long enough. In small gardens it may be possible to sterilize the soil with formaldehyde (see page 610).

References

Cotton, A. D. and M. N. Owen. The white-rot disease of onion bulbs. Jour. Min. Agr. Gt. Brit. 26: 1093-1099. 1920.

Voglino, P. Sur parassitimo e lo sviluppo dello *Sclerotium cepivorum* Berk. nell *Allium sativum* L. Staz. Sper. Agr. Ital. **36:** 89-106. 1903.

Walker, J. C. Occurrence of white-rot of Allium (*Sclerotium cepivorum* Berk.) in Europe and America. Phytopath. **14:** 26. 1924.

PINK-ROOT OF ONION

Caused by *Fusarium malli* Taub.

For many years there has been a root-rot and dwarfing of onions in America, which was considered to be the result of poor soil, wrong fertilizer, or other unfavorable environmental conditions. It was not until 1919 that it was shown to be caused by a soil organism. It is present in many of the onion districts in the United States and Canada, and is very destructive in Bermuda. The disease attacks common onions (Fig. 87), multiplier onions, shallots, and garlic.

Symptoms.

The term pink-root is suggestive of the most striking symptom. The roots one by one or sometimes simultaneously turn pink, shrivel, and die. The bulb continues to produce new roots until frequently a nipple-like growth is formed at its bottom. The lack of feeding roots together with the effort of continuously producing a new root system so depletes the sustenance in the bulb that it fails ever to become of marketable size (Fig. 88). The disease may affect the onion at any stage in its growth, although it is seldom severe enough in the seed-bed to cause any reduction in the size of the green sets. The trouble often follows a weakened condition of the plant, such as that resulting from heat injury, cold injury, smut, or nematodes.

Cause.

Several Fusaria have been isolated from diseased roots. Only one of these seems to cause infection and has been named

Fusarium malli. The suggestion has been made that since this species when alone does not cause infection readily, the concomitant fungi may in some manner aid the pathogene.

The original source of the fungus is not known; it is supposed to have been transported on onion sets shipped from

FIG. 87.—Pink-root reducing stand of onions.

Bermuda to Texas. It is by means of sets that the parasite has now become rather widely disseminated in the United States. When it once gains entrance into the soil, the inoculum gradually increases as onions are grown year after year until the crop is no longer profitable. When the invaded soil is moved about by flooding or on tools, the fungus is scattered still farther.

The Fusarium has a predominance of three-septate spores,

with a few four-septate ones. Microconidia and chlamydospores are also formed. Infection with all these forms takes place very slowly.

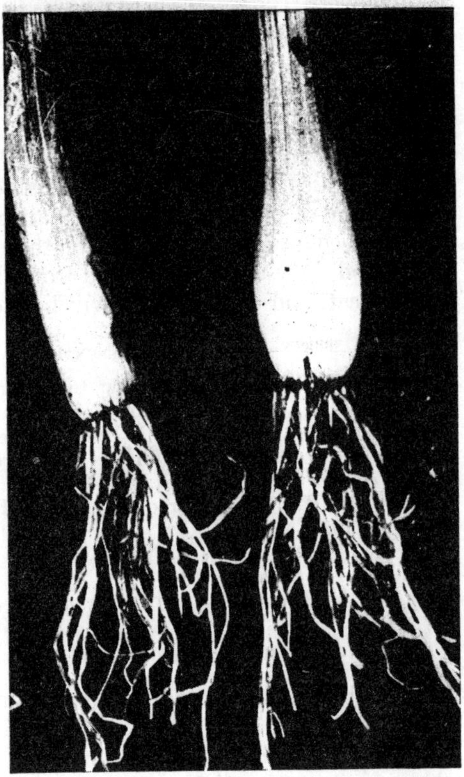

FIG. 88.—Pink-root of onion, causing root discoloration and dwarfed bulbs.

The organism is susceptible to many environmental influences. For example, in poorly leveled fields the disease is more severe in the higher dry soil where the plants suffer from drought. Nitrogenous fertilizers which tend to hasten the growth of the host aid in reducing the amount of loss, as the plant is able in a great measure to outgrow the deleterious effects of the fungus. High temperatures favor the development of the Fusarium, particularly if at the same time they retard the development of the onion.

Control of pink-root.

Attempts have been made to treat onion sets with fungicides, but every chemical which has any effect on the parasite has a similar injurious effect on the host. Therefore, in order to avoid inoculating clean soil, it is necessary to transplant healthy sets. Such sets may be produced in virgin soil or in soil that has been sterilized (see page 600). After a field once becomes infested, nothing can be done to lessen infection. If, however, the plants are grown rapidly, they will be able to overcome the injury to a great extent. The fast growth can be attained by using soil rich in plant-food and humus, and by the application of a quickly available nitrogenous fertilizer. If the time of planting can be so managed that the onions are not exposed to excessively hot weather in their rapidly forming period, they will be protected from the weakening influence of the heat, and consequently will be more resistant.

References

Howitt, J. E. Two diseases new to Ontario. Phytopath. **13**: 291 1923.

Taubenhaus, J. J. Pink-root of onions. Science **49**: 217-218. 1919.

Taubenhaus, J. J. and F. W. Malley. Pink-root disease of onions and its control in Texas. Tex. Agr. Exp. Sta. Bull. 273: 1-42. 1921.

Whetzel, H. H. The pink-root of onions. Bermuda Dept. Agr. Bull. July, 1922, pp. 4-6.

Fusarium Bulb-Rot of Onions

Caused by *Fusarium cepæ* Hanzawa

Very little is known regarding the prevalence and severity of bulb-rot of onions. A number of rots have been mentioned in literature, although the rather meager descriptions do not always aid in identifying them. In some parts of Spain and Japan and in several states in America as Washington, Louisiana, and Illinois the malady has caused injury to sets, seed-stalks, and large bulbs.

Symptoms.

The disease may affect the onion in any stage of its growth and continue to develop in storage. The diseased plants are recognized by a marked discoloration of their tops, which are browned in streaks or throughout the whole length. The infected tops incline toward their more discolored side and finally topple over. The roots are attacked, becoming discolored and thin; consequently the plants rest loosely in the soil and may be pulled up easily. The bulb itself is attacked either on its side or bottom; when attacked on the side, the healthy part grows faster then normal while the diseased side shrivels, thus producing a crescent-shaped bulb. At first the affected tissue is as that of the normal onion but it gradually becomes soft and brown, frequently being covered with aerial mycelium. Maggots may be present in the collapsed flesh. As the bulb succumbs, the aerial part of the plant wilts and dies.

The disease does not occur uniformly over the whole field at once, but diseased plants are located promiscuously, and apparently without injury to the adjoining bulbs. In fact, it has been observed that where two or more bulbs are growing from the same plant, one may be affected and the remaining ones healthy.

Cause.

The fungus in Japan has been named *Fusarium cepæ,* and although it has not been described in great detail, it probably is the same parasite that is found in the United States. The fungus is light yellow or orange color in culture. The three usual Fusarium types of spores are present; one- to three-septate crescent-shaped spores, microconidia, and chlamydospores. The spores germinate from each cell in twenty-four hours and soon thereafter may cause infection. Their germ-tubes can enter only through wounds, often following maggot injury. The parasite probably lives over winter in the soil or on diseased refuse.

Another fungus, *Fusarium allii sativi* Allesch. with three- to five-septate spores and rose-colored mycelium, has been reported from Germany on garlic. It is not the same as *Fusarium cepæ,* differing in color and spore measurement.

The range of temperature at which the bulb-rot organism (*Fusarium cepæ*) grows is about 57° to 90° F., with an optimum of about 79° F. The amount of moisture present does not seem to have any influence on the severity of the disease.

Control.

No definite control measures are available. Sanitation will aid in reducing the loss, and consists in destruction of affected onions, rotation of crops, eradication of the maggot if possible, and care in not injuring the bulbs when cultivating.

References

Boyle, L. W. Fusarium rot of onion. Phytopath. **13**: 510. 1923.

Edgerton, C. W. Onion diseases and onion seed production. La. Agr. Exp. Sta. Bull. 182: 3-21. 1921.

Hanzawa, Jun. *Fusarium cepæ,* ein neuer zwiebelpilz Japans, sowie einige andere pilze an zwiebelpflanzen. Mycol. Centbl. **5**: 11-12. 1915.

Stone, G. E. Onion rot. Mass. Agr. Exp. Sta. Ann. Rept. **21**: 46-47. 1909.

Tims, E. C. and J. C. Walker. A Fusarium bulb-rot of onion. Phytopath. **14**: 26, 27. 1924.

Brittle of Onion

Caused by *Fusarium* sp.

This is a third disease of onion caused by Fusarium, differing from pink-root and bulb-rot in many respects. The latter disease becomes prevalent late in the season, while brittle is primarily a trouble affecting seedlings. The very young plants may be damped-off soon after they come through the soil. The tips of the roots on slightly older plants show slight irregular swellings and their tissue is more brittle. They have no abnormal discoloration. The tops of these plants are stunted, the leaves may be unevenly thickened and be marked with yellow spots. The two most characteristic symptoms, however, are the extreme brittleness of the leaf tissue and the curling of the foliage. In severe cases the leaves are drawn into spiral coils.

Mycelium is present in the affected tissue, so that when diseased plants are placed in a moist chamber, a Fusarium grows out.

Formaldehyde applied at the time of sowing the seed and in the manner suggested for the control of onion smut (see page 281) reduces the amount of disease and increases the yield.

References

Cook, M. T. Report of the plant pathologist. N. J. Agr. Exp. Sta. Ann. Rept. **1912:** 523. 1913.

Clinton, G. P. Report of the botanist. Conn. Agr. Exp. Sta. Ann. Rept. **1905:** 270-273. 1906; also Ann. Rept. **1906;** 332-335. 1907.

Macrosporium Rot of Onions

Caused by *Macrosporium parasiticum* Thümen

For many years a black mold has been observed on the older leaves and flower-stalks of the common onion. A similar symptom has been found on shallots, leeks, chive, and garlic. The fungus that causes the blackening was considered a saprophyte, as it was present in great abundance when downy-mildew or any other disease had injured the plants. Lately it has been shown that the organism is able to enter living tissue and cause considerable injury. This is true particularly on the flower-stalks, which are weakened by the attack until they break over and thus fail to bear seed. A species of the fungus also causes bulb-rot. The infection begins at the necks of the topped onions or through other wounds, and as it progresses downward in the flesh causes a soft decay with a yellow or red discoloration of the invaded tissue.

The fungus is characterized by its large, brown, muriform spores, which on germination can enter the host only through wounds. Clumps of conidiophores push through the stomata, and are so plentiful that with their spores they make a black sooty covering. The parasite probably winters on the old diseased refuse. The inoculum is plentiful everywhere, so that it is almost impossible to find a weakened leaf which has not been attacked.

Different writers seem to have vied with each other in thinking of new names for the fungus, with the result that the list of species has become quite extended. Among the more common names are *Macrosporium sarcinula* Berk., *M. porri* Ellis, and *M. alliorum* Cke. and Mass. There may be a basis for separating some of the species, but as the information on the subject is meager, they are usually all considered as synonymous. A perfect stage known as *Pleospora herbarum* has recently been verified. Still another fungus, *Mystrosporium alliorum* Berk., has been reported from England and South Africa. It is very closely related to Macrosporium, and possibly may be identical.

Warm wet weather and heavy fogs are favorable to the fungus.

The weak parasite is ordinarily not sufficiently injurious to demand attention. When the loss is serious enough to justify the application of bordeaux mixture, it may be possible to hold the disease in check. The application must be begun before the organism has an opportunity for causing infection, and continued as suggested for the control of downy-mildew of onions. Crop rotation and burning of diseased refuse are further aids in combating the Macrosporium.

References

Edgerton, C. W. Onion diseases and onion seed production. La. Agr. Exp. Sta. Bull. 182: 3-20. 1921.

Hanzawa, Jun. *Fusarium cepa,* ein neuer zwiebelpilz Japans, sowie einige andere pilze an zwiebelpflanzen. Mycol. Centralbl. 5: 4-13. 1915.

Miyabe, K. On the life-history of *Macrosporium parasiticum* Thüm. Ann. Bot. 3: 1-26. 1889.

Shipley, A. *Macrosporium parasiticum*. Ann. Bot. 3: 268-271. 1889.

Teodora, N. G. A study of a Macrosporium disease of onions. Philipp. Agr. Rev. 16: 233-275. 1923.

Thaxter, R. The onion Macrosporium. Conn. Agr. Exp. Sta. Ann. Rept. 1889: 158-162. 1890.
Walker, J. C. A Macrosporium rot of onions. Phytopath. 11: 53. 1921.

Black-Mold of Onions

Caused by *Sterigmatocystis nigra* v. Tiegh.

The black-mold of onions was first brought to the attention of pathologists in 1917 when it was reported from Ohio and Virginia. The disease has since been found in a number of other states. It evidently has been present many years but seldom has done much damage. Under certain conditions the trouble results in injury not only to the stored onions but to seeds and sets. The causal fungus may be found nearly anywhere on decaying vegetable matter and heretofore has been considered a saprophyte.

Symptoms.

The black-mold evidently is not able to injure the young growing onions. It is mostly a storage trouble, appearing first as small brown or black circular spots on the surface of the upper half of the bulb. The areas of the black moldy growth enlarge until they become confluent, and may under favorable conditions cover the entire bulb. The invaded area usually includes only the two or three outer layers of scales, which die and shrivel. The value of the affected onions is decreased very much, especially since the large white Spanish onion is most susceptible, and on the surface of which the mold shows most conspicuously. Other organisms are able to follow the disease causing soft-rots of the bulbs, and producing much shrinkage in storage.

Cause.

The fungus, one of the Plectascales, is here named *Sterigmatocystis nigra,* although there are workers who prefer the term *Aspergillus niger.* v. Tiegh. The pathogene is placed under the former genus because of its branched sterigmata. The spores are borne in large compact black heads on the ends of conidiophorcs, which have swollen tips. The swelling is completely covered with closely standing sterigmata, some of which are branched from one to eight times. The dark globose spores are borne in chains on each sterigma, and are so abundant that they form in clouds. They float about in the air like particles of dust until they chance to settle on the ground or possibly on some vegetable matter, or they may be carried with the seeds or sets. If moisture is present they germinate. They can gain entrance into the onion only through wounds. An ascerigerous stage has rarely been found. The fruit-body is composed of loosely woven wefts of hyphæ inclosing the asci. This stage is not of importance so far as the disease on onions is concerned.

The Sterigmatocystis thrives best in a high temperature. Some of the species require 95° to 120° F. for optimum growth. The exact temperature at which onions are most easily infected is not known.

Control.

As the spores are carried on the seeds and sets, it is recommended that where much loss from black-mold has been incurred formaldehyde be applied when drilling the seed as suggested for the control of onion smut (see page 281). The sets also may be treated. According to the result of work done in Ohio, the sets can be soaked successfully for six hours in formaldehyde (1 pint in 30 gallons of water) or in a solution of calcium hypochlorite or ordinary bleaching powder (one-half pound in one gallon of water).

For the protection of mature onions the storage place should be kept as nearly 32° F. as possible, the air fairly dry, and only the best bulbs stored. If the onions cannot be cured thoroughly by ordinary methods, artificial drying will prove beneficial.

References

Taubenhaus, J. J. A study of the black and yellow molds of ear corn. Tex. Agr. Exp. Sta. Bull. 270: 1-38. 1920.
Van Pelt, W. Black mold of onions, a new fungous disease causing serious damage in storage houses. Ohio Agr. Exp. Sta. Mon. Bull. 2: 152-156. 1917.
Van Pelt, W. Onion diseases formed in Ohio. Ohio Agr. Exp. Sta. Mon. Bull. 4: 70-76. 1919.

Rust on Onions

Caused by *Puccinia* sp.

The rust has been reported on fifteen species of Allium, but is most prevalent on chives. It rarely is found in America, although it is rather common and sometimes destructive in Europe.

Both the red and black-pustuled stages are present on the foliage. The sori are much like those of rust on cereals, and cannot be mistaken when once recognized. The seriously affected leaves turn yellow and die prematurely.

There are at least two separate species of rust on Allium. The one which has been observed in America has some of its teliospores two-celled and many of them one-celled, there being much variation in ratio between the two in different localities. Therefore, opposing authors place the fungus in two separate genera. The one-celled form when most abundant is classed as *Uromyces ambiguus* (DC) Fckl., while the two-celled form is known as *Puccinia porri* (Sow.) Wint. There is a second rust, distinct from the one above, the pathogene of which is known as *Puccinia allii* (DC) Rud. Detailed life histories of the two parasites are not recorded, nor are any means of control suggested.

References

Clinton, G. P. Notes on plant diseases of Connecticut. Conn. Agr. Exp. Sta. Rept. of Botanist 1909-1910: 726. 1911.
Saccardo. Sylloge Fungorum 7: 543, 605, 606. 1888.

Cercospora Leaf-Spot of Onions

Caused by *Cercospora duddiæ* Welles

Recently a new leaf-spot was found on onions and garlic in the Philippine Islands. The spots on the leaves begin as small circular areas, which as they enlarge turn brown in the center. The lesions are more numerous near the tip of the foliage than at the base, so that by the coalescing of the spots the extremities of the leaves may be killed and turn brown or gray in color. As the garlic has flat leaves, the spots are visible on both epidermal surfaces. After the fusion of the spots on the latter host, the disease rapidly spreads downward until the entire leaf is killed. In severe cases when all the leaves are attacked, the death of the plant follows.

The fungus is characterized by the long, thin, septate spores borne on short conidiophores that break through the epidermis in bunches or arise from the stomata. Not much of the life history has been recorded.

Where the disease occurs it may be necessary to attempt spraying with bordeaux mixture as suggested for the control of onion mildew. The leaf-spot has not yet appeared in America, consequently care should be taken in excluding it from the country.

Reference

Welles, C. B. A new leaf-spot disease of onions and garlic. Phytopath. 13: 362-365. 1923.

Phytophthora Blight on Welsh Onion

Caused by *Phytophthora allii* Sawada

This disease, which does not affect the common onion, has been observed only in Japan. The trouble is made conspicuous by the white cottony mycelial mass which forms on the leaves and flower-stalks of the affected host, and which may kill the top of the plant. The fungus is composed of non-septate mycelium that penetrates the leaf tissue and finally fruits by pushing through the host stomata bunches of slightly branched conidiophores. The conidia germinate by means of fifteen to sixty zoospores or occasionally by means of a germ-tube. In addition to the conidia, oospores are produced within the tissue, and evidently keep the fungus alive during the winter.

The only means of control necessary is that of excluding the parasite from the country. Such exclusion seems very simple, but may be difficult to put into effect, since an absolute quarantine is almost impossible to enforce.

Reference

Tanaka, Tyozaburo. New Japanese fungi—Notes and translation II. Mycologia **9**: 249-253. 1917.

Rhizoctonia Rot of Onion

Caused by *Corticium vagum* B. and C.

(See Rhizoctoniose of Potatoes, page 367.)

This rot was found on onions when they followed potatoes in a rotation on heavily infested soil.

Reference

Heald, F. D. Some new hosts for the Rhizoctonia disease. Phytopath. **11**: 105. 1921.

Violet Root-Rot of Onion

Caused by *Rhizoctonia crocorum* (Pers.) DC.

(See Violet Root-Rot of Potato, page 439.)

Reference

Eriksson, J. Einige studien über den wurzeltöter (*Rhizoctonia violacea*) der möhre mit besonderer rüchsicht auf seine verbreitungsfähigkeit. Centralbl. f. Bakt. u. Par. II **10**: 721-738, 766-775. 1903.

FIG. 89.—Lightning injury in center of onion field.

Bacterial Soft-Rot of Onion

Caused by *Bacillus carotovorus* Jones

(See Bacterial Soft-Rot of Carrot, page 76.)

Sclerotinia Rot of Onion

Caused by *Sclerotinia libertiana* Fuckel

(See Lettuce-Drop, page 243.)

Reference

Massee, G. An onion disease. Gard. Chron. III. **16**: 160. **1894**.

CHAPTER XI

DISEASES OF OKRA, PARSLEY, AND PARSNIP

THESE three vegetables are classed together because of their alphabetical order, and not because of any morphological relationship which they may have, although parsley and parsnip belong to the same family. Less than a thousand acres is grown of each crop in the United States, consequently the diseases of the hosts have not received much attention.

FUSARIUM WILT OF OKRA

Caused by *Fusarium vasinfectum* Atk.

A very serious wilt of okra has been mentioned in literature for many years. It seems to be found wherever the crop is grown extensively. It was not known until comparatively recently, however, that there are two distinct types of okra wilt, the one caused by Fusarium and the other by Verticillium, both diseases appearing much alike and being equally destructive. The same Fusarium occurs on cotton and a variety has been suggested as the cause of pea wilt.

The Fusarium disease is a typical wilt, beginning with a yellowing and stunting of the plant, followed by wilting and rolling of the leaves as if the roots were unable to supply enough water. After the disease has run its course, the death of the plant results. If a diseased stem, root, or petiole is split longitudinally, the vascular bundles appear as dark streaks. In aggravated cases, nearly the whole stem is blackened by the disease.

The *Fusarium vasinfectum* has mostly three-septate spores. It is a soil organism that enters the host through the roots.

The parasite is disseminated in any way in which soil is transferred from one field to another. Contaminated manure may also aid in spreading it. When the inoculum once enters a field, it slowly increases in quantity until cotton or okra crops are unprofitable.

There are no direct methods for combating the okra wilt. The crop may be grown on a farm as long as the yield is profitable. When the disease is destructive, it is advisable to obtain soil free from the fungus even if such a plan involves the renting of additional land. If immune crops are grown on the infested soil in the meanwhile, a rotation of ten or twelve years will probably reduce the disease to such an extent that okra can be grown profitably.

References

Carpenter, C. W. Wilt diseases of okra and the Verticillium wilt problem. Jour. Agr. Research **12**: 529-546. 1918.

Orton, W. A. Fungus genus, Verticillium, in its relation to plant disease. Phytopath. **4**: 40-41. 1914.

Stevens, F. L. and G. W. Wilson. Okra wilt (fusariose) *Fusarium vasinfectum* and clover rhizoctoniose. N. C. Agr. Exp. Sta. Ann. Rept. **1911**: 70-73. 1912.

Wilson, G. W. Fusarium or Verticillium on okra in North Carolina. Phytopath. **3**: 183-185. 1913.

Verticillium Wilt of Okra

Caused by *Verticillium alboatrum* Reinke and Berth.

(See Verticillium Wilt of Eggplant, page 234, and Fusarium Wilt of Okra, page 314.)

The Verticillium wilt of okra is in nearly every respect like that caused by Fusarium. The Verticillium, however, grows at slightly lower temperatures, and perhaps is found somewhat farther north than is the Fusarium.

The suggestions for growing a healthy okra crop are the same in both cases.

FIG. 90.—Botrytis rot on okra pod.

ASCOCHYTA POD-SPOT OF OKRA

Caused by *Ascochyta abelmoschi* Harter

In 1908 specimens of okra pod- and stem-spots were collected in New York. In 1916 a similar disease was found on collections of okra grown in Maryland and obtained from Russia, Greece, and India. It is not known whether the disease was imported previous to 1908 or whether it is indigenous to the United States. It apparently is not widespread, but may cause considerable loss in certain plantings.

The disease affects the leaves only slightly, the stems moderately, and the young pods severely. The old pods are as nearly immune as is the foliage. The first indication of infection is a small, dark, water-soaked area that enlarges slowly. The tissue turns brown and dies, and about this time the parasite produces an extremely large number of black pycnidia over the infected tissue. They may or may not be arranged in concentric rings. The spots originally are oval to oblong in shape, but after the pods or stems are killed the pycnidia may form indiscriminately over the whole surface.

The causal fungus is *Ascochyta abelmoschi*, characterized by rather large pycnidia in which are borne either one- or two-celled spores. The mycelium in the pod grows through the wall into the seed, and may even produce pycnidia there. In this manner the fungus is introduced with the seed into a new locality.

No demonstrations for the control of the disease have been reported.

Reference

Harter, L. L. A hitherto unreported disease of okra. Jour. Agr. Research 14: 207-212. 1918.

Texas Root-Rot of Okra

Caused by *Ozonium omnivorum* Shear

(See Texas Root-Rot of Sweet-Potato, page 512.)

Root-Knot of Okra

Caused by *Heterodera radicicola* (Greef) Müll.

(See Root-Knot of Tomato, page 550.)

Gray Mold-Rot of Okra (Fig. 90).

Caused by *Botrytis* sp.

(See Gray Mold-Rot of Lettuce, page 253.)

Unimportant Parasites of Okra

Leaf-spots of okra may be caused by *Macrosporium hybiscinum* Thüm., *Cercospora hibisci* T. and Earle, and *Phoma okra* Cke. A blossom-blight may be produced by *Choanephora cucurbitarum* (B. and Rav.) Thaxter.

Bacterial Wilt of Parsley

Caused by *Bacillus nelliæ* Welles

A bacterial wilt of parsley has been found in the Philippine Islands. Very little has been written either regarding the disease or the organism.

Reference

Welles, C. G. Identification of bacteria pathogenic to plants previously reported from the Philippine Islands. Philipp. Jour. Sc. 20: 279-285. 1922.

Downy-Mildew of Parsley and Parsnip

Caused by *Plasmopara nivea* (Ung.) Schroet.

A downy-mildew is present on many of the umbelliferous plants among which are carrots, chervil, parsley, and parsnips. The disease is characterized by yellow spots on the upper surfaces of the foliage, and white mycelial wefts covering the corresponding parts on the lower side of the leaves. The spots turn dark brown with age, and when the disease occurs in epidemic form much of the foliage may be killed.

The fungus, *Plasmopara nivea,* is one of the downy-mildews. The white mycelial mass is composed of numerous branched conidiophores bearing globose conidia that give rise to zoospores. Oospores are also produced, and probably assist the fungus in surviving the winter.

The downy-mildew of parsnips is present usually only when the plants are crowded so closely together that the foliage does not dry after a rain or heavy dew. If the rows are planted far enough apart for circulation of the air, and shading by tall vegetables or by trees is avoided, the disease will not cause enough injury to need attention. It is probable that spraying with bordeaux mixture will protect the plants, but no careful demonstrations regarding this point have been reported.

References

Cotton, A. D. Diseases of parsnip. Roy. Bot. Gard. Kew. Bull. Misc. Inform. 1: 8-21. 1918.

Sorauer, P. Handbuch der pflanzenkrankheiten 2: 162. 1908.

Root-Canker of Parsnip

Caused by various fungi

In England the parsnip-growers have lost a large percentage of their crop because of the root-canker. The lesions usually have their origin near the top of the root and are characterized by a reddish-brown decay. In severe cases the whole root is destroyed.

In many cases a species of Phoma is undoubtedly the cause of the rot, but as frequently there are other organisms which seem to be responsible for the trouble. After some investigation it was found that the parsnip does not form a protecting layer of wound cork when it is injured by the carrot-fly, or when the epidermis is split by growth cracks. Consequently any organism that happens to be present has access to the inner tissue of the root.

The control measures which have been suggested and which apply to the conditions in England are the avoidance of soil which is too rich, planting the seed as late as possible, liming the soil, and proper rotation of crops. Adding to the soil five hundred pounds of salt for each acre has also been beneficial.

Reference

Cotton, A. D. Diseases of parsnips. Roy. Bot. Gard. Kew. Bull. Misc. Inform. 1: 8-21. 1918.

Phyllachora Leaf-Spot of Parsnip

Caused by *Phyllachora pastinacæ* (West.) Rostr.

The only compilation of parsnip diseases is that made by Cotton in 1918, and among the troubles which he lists is the Phyllachora leaf-spot. Brown dead areas occur on the upper sides of the leaf, and after coalescing form large brown blotches. On the lower side of the leaf are glossy black fruiting bodies that at first are barely visible to the unaided eye, but which later may reach two millimeters in diameter. When the disease is severe both the leaves and petioles are killed.

The common stage of the parasite is named *Cylindrosporium pastinacæ* Lind. It has also been known as Septoria. The perfect stage is *Phyllachora pastinacæ*. Evidently the fungus is propagated on the wild cow parsnip and then spreads to the cultivated crop. The fungus is characterized by hyaline, strongly curved, one-septate spores borne in acervuli. In the perfect stage the asci are borne in cavities in the black stromæ.

No methods of control have been suggested.

REFERENCE

Cotton, A. D. Diseases of parsnip. Roy. Bot. Gard. Kew. Misc. Inform. 1: 8-21. 1918.

RAMULARIA LEAF-SPOT OF PARSNIP

Caused by *Ramularia pastinacæ* Bubák

A Ramularia leaf-spot of parsnip is rather common in some parts of Europe. The fungus causes small angular brown spots on the foliage, and narrowly elliptical ones on the petioles. When the parasite is fruiting, a faint white weft is visible over the dead areas on the under side of the leaf. The unbranched knobby conidiophores of the fungus emerge from the stomata, and bear one-septate spores.

No control measures are suggested, except not to crowd the plants in the rows.

REFERENCE

Cotton, A. D. Diseases of the parsnip. Roy. Bot. Gard. Kew. Bull. Misc. Inform. 1: 8-21. 1918.

CERCOSPORELLA LEAF-SPOT OF PARSNIP

Caused by *Cercosporella pastinacæ* Karst

In Europe a Cercosporella leaf-spot is found occasionally and is then usually associated with the spot caused by Ramularia. The same disease has been observed several times in the United States. The lesions on the leaves are circular, very small, at first brown throughout, later with a white center and a brown border. The parasite is characterized by long, slender, septate, hyaline conidia borne on short exposed conidiophores.

No control measures are necessary.

REFERENCE

Cotton, A. D. Diseases of parsnip. Roy. Bot. Bard. Kew. Bull. Misc. Inform. 1: 8-21. 1918.

FIG. 91.—Common scab on parsnip grown in heavily infested soil.

Scab of Parsnip (Fig. 91)

Caused by *Actinomyces scabies* (Thax.) Güssow

(See Scab of Potatoes, page 380.)

Powdery-Mildew of Parsnip

Caused by *Erysiphe polygoni* DC.

(See Powdery-Mildew of Peas, page 333.)

Bacterial Soft-Rot of Parsnip

Caused by *Bacillus carotovorus* Jones

(See Bacterial Soft-Rot of Carrot, page 76.)

CHAPTER XII

DISEASES OF PEAS

The pea is one of the most important vegetables grown in this country as is shown by the plantings which aggregate more than one hundred thousand acres. The states growing the largest quantities of this crop are Wisconsin, New York, California, Michigan, and Maryland.

Ascochyta Blight of Peas

Caused by *Mycosphærella pinodes* (B. and Bl.) Stone

The spotting of the pea pods is so conspicuous that the attention of mycologists was drawn to the disease many years ago. Its general presence in all the countries where peas are grown resulted in many written discussions of the trouble. On account of the great similarity not only of the symptoms but also of the life history of the parasite with that of bean anthracnose, the blight has frequently been confused with the anthracnose in these discussions. It is for this reason that the lists of susceptible hosts, which have been recorded, may not always be authentic. The fungus attacks all varieties of peas with about equal severity, and in addition has been reported on several closely related legumes, as sweet-pea, Egyptian pea, alfalfa, vetch, and a few unrelated hosts.

The disease may be inconspicuous for several years, then suddenly appear in epidemic form over wide areas. A number of such epidemics have been recorded in literature.

Symptoms.

The blight may begin its attack at any time in the season and on any part of the plant. When infected seeds are sown, they may not germinate, or if they do, the tip or stem of the seedling is invaded so early that it soon dies. On older stems the sunken cankers may be numerous enough to girdle the plant, thereby causing its death, or by forming on the growing tip stop further growth unless a secondary shoot is started. In lighter cases of infection, the single lesions are small, brown, deep cankers, which may be present in great numbers on any part of the stem. Even the crown of the root becomes infected. Small black pycnidia are often present in the dead tissue.

FIG. 92.—Ascochyta pea blight.

The spots on the leaves are usually poorly defined. They are inclined to be circular with a light gray center and brown border, the spot being marked by the presence of a few pycnidia. The lesions on the foliage may be numerous enough to kill the individual leaves. The spots on the pods are more easily discernible (Fig. 92). If infection occurs when the pods are young, deep cankers are formed, which frequently extend through the pod to the seeds. On older pods there is no noticeable depression, but the brown pycnidia are borne in indefinite grayish areas. The same is true of the older stems and petioles.

Affected seeds are not always discolored. When severely affected, they shrivel and darken. When only slightly infected, they may show a faint discoloration or none at all. Invaded seeds placed in a moist chamber will often show the presence of the parasite in the formation of brown pycnidia on the seed-coat.

Cause.

The asexual stage of the fungus has long been known as *Ascochyta pisi* Lib. A perfect stage was also recognized for many years, but the two were not supposed to be connected until recent work proved the relationship. The pathogene is now named *Mycosphœrella pinodes.* It remains alive during the winter in the pea seed and in the infected plant refuse, particularly in the stubble left in the field. When infected seed is planted, the moisture of the soil stimulates the production of pycnidia on the seed-coat. Similar fruit-bodies appear on the old stubble. The spores ooze from the pycnidia in long tendrils, and after being splashed by the rain to aerial parts of the host cause infection. The ascospores produced in perithecia on old stems may also serve as inoculum in the spring. When the weather conditions are favorable, secondary infections appear in great numbers, some of which are sure to occur on the pods. The mycelium grows through the pod wall into the seed where the fungus hibernates until the seed is planted.

The fungus is favored by much air and soil-moisture. Its optimum temperature is 77° F., growth being inhibited at about 92° F. and death resulting from a temperature of 100° F.

Control of blight.

In order to destroy the inoculum on the stems and leaves, the harvested vines should be placed in the silo where the fermentation kills the fungus. The stubble should be plowed under as deeply as possible and as soon as the crop is harvested.

If peas can be grown in three-year rotations with cereals or other immune crops, and healthy pea seed sown each year, the disease can be eliminated. The chief difficulty is that of obtaining disease-free seed. Wherever possible it would seem advisable to grow the seed on the farm where it is to be used, for then only the healthy pods need to be selected and this would give disease-free seed. Since home-growing cannot always be practiced, pressure should be brought to bear on the commercial seed-grower that he practice every precaution in obtaining clean stock.

Seed treatment with chemicals and heat, and spraying with bordeaux mixture 4-4-50 have been tried with little or no success. Where peas are grown for seed, spraying may be beneficial when the fungicide is applied often and with high pressure.

References

Halsted, B. D. Failure of pea seed to grow. N. J. Agr. Exp. Sta. Ann. Rept. **14**: 359-362. 1894.

Hiltner, L. Ueber die durch *Ascochyta pisi* Lib. hervorgerufene wurzelkrankheit der erbsen. Centralbl. f. Bakt. u. Par. II **1**: 881-886. 1895.

Krüger, F. Ungewöhnliches auftreten von *Ascochyta pisi* Lib. an erbsenpflanzen. Centralbl. f. Bakt. u. Par. LI **1**: 620-624. 1895.

Stone, R. E. The life history of Ascochyta on some leguminous plants. Ann. Mycol. **10**: 564-594. 1912.

Van Hook, J. M. Blight and powdery mildew of peas. Ohio Agr. Exp. Sta. Bull. **173**: 231-249. 1906.

Vaughan, R. E. *Mycosphaerella pinodes,* the ascigerous stage of *Ascochyta pisi.* Phytopath. **3**: 71, 72. 1913.

Root-Rot of Peas

Caused by *Fusarium martii* var. *pisi* Jones

The root-rot is probably coextensive with the cultivation of peas. It has long been known in literature under various names; in northern Europe where it appears in the middle of the summer the trouble is known as St. John's disease, and in America it has recently been described as wilt. On both continents the root-rot causes much injury during certain seasons, and particularly in fields where rotation of crops is not practiced.

Symptoms.

The disease ordinarily is serious only in definite areas in the field, but in well infested soil the wilt occurs generally. The parasite enters the base of the plant, and when favored by the correct temperature and moisture produces a decay through the entire thickness of the stem. When the temperatures are higher than the optimum, the vascular bundles are discolored and the plants wilt. The roots also may be invaded. On certain plants, the bottoms of which have rotted, roots may develop from the tissue above the lesion, thus aiding the host in surviving the attack of the parasite. More often, however, the leaves turn yellow followed later by the death of the whole plant.

Cause.

In Europe the parasite, *Fusarium vasinfectum* var. *pisi* van Hall, is accredited with causing the injury. In America the greater part of the trouble is supposed to be produced by an undescribed species of Aphanomyces and *Pythium debaryanum* Hesse. Much injury is caused also by *Fusarium martii* var *pisi*. A small percentage of the loss results from the attack of *Corticium vagum*.

The variety of Fusarium on the pea has mostly three-septate spores. It is a soil organism that apparently is able to live indefinitely in the absence of the host. The spores and mycelium are disseminated with soil or contaminated manure. When plants are inoculated, visible infection occurs twenty or more days later, depending on environment. The optimum temperature lies between 75° and 85° F. Soil-moisture does not seem to have any pronounced influence on the amount of infection, although there are slightly more diseased plants in wet soil which is high in organic matter.

Control.

As the different pathogenes causing root-rot of peas live in the soil, there is little that can be done to eliminate the trouble after the fields become infested. Long rotation of crops and good drainage aid in reducing the loss, but the final control measure must be the procuring of a resistant strain. This is difficult since all varieties of peas are susceptible. Some suitable selections have been made among the Early Alaska pea races which probably will later be on the market. Until that time comes, no satisfactory control measures can be recommended.

References

Jones, F. R. Stem and root-rot of peas in the United States caused by species of Fusarium. Jour. Agr. Research **26**: 459-476. 1923.

Schikorra, G. Fusarium-krankheiten der Leguminosen. Arb. K. Biol. Biol. Anst. Land. u. Forstw. **5**: 157-183. 1906.

Van Hall, C. J. J. Die sankt-Johanniskrankheit der erbsen, verursacht von *Fusarium vasinfectum* Atk. Ber. Deut. Bot. Gesell. **21**: 2-5. 1903.

Bacterial Stem-Blight of Peas

Caused by *Bacterium pisi* Sack.

In 1915 a bacterial stem-blight was discovered doing much damage to garden and field peas in the San Luis Valley of Colorado. Later the trouble was reported in Montana, Delaware, Maryland, Indiana, North Dakota, Pennsylvania, South Carolina, and Wisconsin. The disease evidently does not affect any other crop, but in severe cases causes as much as 25 per cent loss of the peas.

The disease is found on all parts of the plant above ground. The affected tissue of the stem assumes an olive-green to olive-brown color, and the stipules and leaflets become yellowish or water-soaked. The trouble may show at any time in the growth of the vine. If the infection takes place before the peas are three to four inches in height, the plants die and wither, leaving only missing spaces in the drill row. In older plants the infection has its source an inch or more above ground, the lesion gradually extending upward on the stem until the top wilts and finally the whole plant may succumb. Frequently only one or two of the main branches are affected, causing the plant to produce other branches, which though weaker than the original ones are able to survive and bear a crop. It is not unusual, however, to have in turn the second group of branches attacked, resulting in the death of the plant.

The organism is a short rod-shaped polar-flagellate bacterium, to which has been given the name, *Pseudomonas pisi*, or following Smith's classification, *Bacterium pisi*. Apparently it lives over winter and is disseminated in or on the seed. It may enter the stomata of the host. In the high altitudes where the disease is found, the strong winds blow particles of soil with such force that the plant is injured, and the organism enters through such wounds. When the field is harrowed after the plants are through the ground many particles of soil are loosened, thus aiding the wind in its destructive wounding. In the eastern states, the disease has been observed following late spring frosts.

No direct control measures are suggested. Rotation of crops and care in cultivation to avoid wounding are helpful. The most satisfactory method, if possible, is that of using resistant varieties. The different varieties have been tested; the Alaska was shown to be one of the most susceptible. Several varieties were found that gave promise of withstanding the disease sufficiently to produce a profitable crop. No doubt seed from these varieties will soon be available.

References

Jennison, H. M. Observations upon the bacterial blight of field and garden peas in Montana. Phytopath. **11**: 104. 1921.

Sackett, W. G. A bacterial stem-blight of field and garden peas. Colo. Agr. Exp. Sta. Bull. 218: 1-43. 1916.

Anthracnose of Peas

Caused by *Colletotrichum pisi* Pat.

The anthracnose of peas is a rather rare disease, having been reported only from Ecuador, Japan, and Wisconsin, but in the few places where prevalent it has demonstrated its capability of causing serious loss. The report has been made that the causal fungus is parasitic also on sweet-peas, but inoculation experiments do not uphold the statement.

The lesions are much like those caused by Ascochyta. They occur on all the parts of the plant above ground. The leaves have few to many, irregularly shaped, gray to brown spots. The same kind of lesions occur on the pods, except that on the latter they are nearly circular. The spots on the stems are elongated and ashen-colored. On mature stems in wet weather, large rusty areas may partly cover the surface.

The fungus was named *Colletotrichum pisi* in 1891 when discovered in South America. It is characterized by the acervulus type of fruit-bodies, which are so thickly lined with setæ that these stiff bristles may sometimes be seen merely by the aid of a hand lens. The spores are washed by water, splashed by rain, and carried by animals or on tools to healthy plants. In the presence of moisture they germinate and quickly cause infection. It is not known definitely how the parasite survives the winter, but it is thought possible that the organism may hibernate in the living seeds and be disseminated with them.

The most favorable temperature for the growth of the Colletotrichum is between 70° and 85° F.

No experiments that deal with the control of the fungus are reported. It is probable that long rotations and procuring disease-free seed are sufficient to hold the anthracnose in check.

References

Hemmi, T. Nachträge zur kenntnis der gloeosporien. Jour. Coll. Agr. Hokkaido Imp. Univ. **9:** pt. 6. 309-311. 1921.

Jones, F. R. and R. E. Vaughan. Anthracnose of the garden pea. Phytopath. **11**: 500-503. 1921.
Patouillard, N. and G. de Lagerheim. Champignons de L. Equator. Bull. Soc. Mycol. France **7**: 180. 1891.

Septoria Blight of Peas

Caused by *Septoria pisi* Westend.

The Septoria blight of peas is fairly common and sometimes destructive in the northern part of Europe and the United States. In the latter country it has been reported from fourteen states, nearly all of which are in the northern half.

The disease, because of its similarity in appearance, is sometimes confused with the Ascochyta blight. The latter is recognized by the lesions with ashen-gray centers, while the spots caused by Septoria are indistinct. The infection commonly begins at the edge of the leaf as a yellowish area which darkens and spreads until the whole leaflet is included. The disease passes from the leaflet to the nodes which are discolored and shrunken. On young plants the lower third of the stems is affected, and marked by the presence of a few pycnidia in the darkened tissue. More pycnidia are found on the leaves and at the base of the stem. The young plant may be killed.

The fungus, *Septoria pisi*, develops rapidly, being able after inoculation of the plant to cause infection and bear pycnidia within two weeks. If the weather is favorable, the parasite soon destroys the young plant and in time causes the death of older vines. The spores are forced from the pycnidia in long yellow tendrils, and blown by wind or possibly carried by other methods, are disseminated over the field. The pycnospores cannot live over winter and neither does the mycelium which affects the pods grow into the seed. Consequently the form in which the parasite lives from one season until the next is not known. It may have a perfect stage, although such a form has not yet been discovered.

The fungus is partial to wet cool weather and is checked in its growth during dry hot months.

Recommendations for the control of the fungus have not been made. Using the same precautions as suggested for the control of Ascochyta on peas (page 325) will probably prove helpful.

References

Melhus, I. E. *Septoria pisi* in relation to pea blight. Phytopath. **3**: 51-58. 1913.

Plant Disease Survey Office. Pea diseases. Plant Disease Surv. Bull. Suppl. **26**: 158, 159. 1923.

Potebnia, A. Beiträge zur Micromycetenflora mittel-Russlands. Ann. Mycol. **8**: 70. 1910.

Bacterial Pod-Spot of Peas

Caused by *Bacterium leguminiperdus* von Oven

In 1905 a disease of pea pods was observed in Germany where it caused much injury by shriveling the seed, which destroyed its viability, and in killing the seedlings. The same parasite causes a similar disease on lupines, and a slight injury on garden beans. It probably has not been observed elsewhere.

The affected pods are conspicuously smaller and ripen earlier. While the normally matured peas have dry brittle pods, those that are diseased feel damp and can be distinguished by their darker color. The disease begins very early in the development of the pod, so that it is possible to find all degrees of infection. The original small, sunken, watery, dark-colored spot spreads until the whole pod is included. Finally the tissue dries, causing the veins to stand out prominently. No infection can be found in the roots or stems excepting when the plants are in the seedling stage.

Although von Oven named the organism Bacillus, according to Smith's classification it is a Bacterium, since the flagellæ are polar. The organism is characterized by rod-shaped white spore-bearing bodies which liquefy gelatine. The parasite is not supposed to pass through the pod into the seed, but by oozing out of the diseased tissue is disseminated by rain and insects to other parts of the field. It enters the tissue through stomata or directly through the epidermis. No mention is made of the manner in which the parasite survives the winter. Presumably it may live in the soil, for the pods that touch the ground are the ones that usually are infected.

The control measures which have been suggested are rotations of crops, and planting varieties of peas that do not fall over easily, or staking the vines. The same effect as staking may be obtained in the field if the peas are grown with dwarf rye or other cereals which hold the vines erect.

References

Von Oven, E. (Translation) A new bacterial disease on the pods of legumes. Centralbl. f. Bakt. u. Par. II **16**: 67-74. 1906.

Powdery-Mildew of Peas

Caused by *Erysiphe polygoni* DC.

The powdery-mildew of peas has been known for many years, and in all countries where peas are grown extensively. The disease usually is not of great economic importance, but occasionally may cause much injury. It is characterized by the white powdery or talcum-like mold on the leaves, stems, and pods. Later in the season, the white mold may be interspersed by numerous small black perithecia. The vines dwarf, the tissue dries, and if infection is severe, the plants may die.

The fungus, having been recognized by the early mycologists, has been given a long list of names. The name here used is *Erysiphe polygoni*, although another common binomial is *Erysiphe communis* (Wallr.) Fr. The parasite has given reported on many other legumes as well as on a large number of unrelated hosts.

The white mold on the plant surface is made up of short stalks on which bead-like conidia are borne in chains. The black globose perithecia, characterized by thin long simple appendages contain few to many four-spored asci. The conidia germinate at once, causing secondary infection, but the ascospores in the perithecia live over winter and serve as inoculum for the primary infection.

In Ohio it was found that spraying with bordeaux mixture 4-4-50 controlled the mildew satisfactorily. One or two applications are made, the first one beginning as soon as the fungus is visible on the plants. No doubt dusting with sulfur will also prove effective. It is only rarely, however, that spraying or dusting are justified on the pea crop. Rotation of crops, storing the harvested vines in the silo, and turning under the stubble at once, are precautions that may well be taken by any grower.

REFERENCES

Halsted, B. D. Report of the botanist. N. J. Agr. Exp. Sta. Ann. Rept. **1893**: 357. 1894.

Van Hook, J. M. Powdery mildew of the pea. Ohio Agr. Exp. Sta. Bull. 173: 247-248. 1906.

DOWNY-MILDEW OF PEAS

Caused by *Peronospora viciæ* Berk.

The downy-mildew of peas is a widely distributed disease in many countries, but is of little economic importance. The symptoms of the disease are the water-soaked appearance of the affected tissue and the white downy fungous growth that may be found on any aerial part of the plant.

The fungus on peas attacks also different species of vetch. The mycelium remains alive in the stems of the perennial form, *Vicia Sepium*, on which it fruits early in the spring. From this host it may be disseminated to peas.

No control measures are necessary.

REFERENCE

Melhus, I. E. Perennial mycelium of species of Peronosporaceæ related to *Phytophthora infestans*. Jour. Agr. Research **5**: 59-70. 1915.

Black-Pitting of Peas

Cause undetermined

In Holland a peculiar black-pitting or black-heart has affected peas. The seeds appear perfectly normal on their outer surface, unless the discoloration is so pronounced that the darkening can be observed through the epidermis. The latter is never affected. When the seed is cut in halves the inner part reveals a blackened area which may be no larger than a mere point or include most of the inner tissue. At times the discoloration shows as numerous black stipples. In the more advanced stages, the plumules and growing tip of the embryo are darkened so that they will not grow. Ordinarily, however, the affected seeds germinate as well as do those which are normal. The main loss results from the fact that buyers are reluctant to receive stock from a community where black-pitting is prevalent.

Attempts have been made to isolate an organism, but the effects have not been successful. Theories which have been suggested to account for the trouble are soil deficient in some element, or the effects of sea water on plants. Nothing can be done to eliminate the injury until the cause has been discovered.

Reference

Von der Lek, H. A. A. Over de zoogenaamde "kwade harten" of "zwarte pitten" der erwten. Tijdsch. over Plantenz. 24: 102-114 1919.

Damping-off of Peas

Caused by various fungi

(See Damping-off of Tomatoes, page 546.)

Black Root-Rot of Peas

Caused by *Thielavia basicola* (B. and Br.) Zopf

(See Black Root-Rot of Beans, page 37.)

CHAPTER XIII

DISEASES OF PEPPER

ONLY about fifteen thousand acres of peppers (capsicums) are grown in the United States most of which are in New Jersey, California, and Florida. Like the tomato and the eggplant, the pepper is susceptible to several destructive diseases.

ANTHRACNOSE OF PEPPERS

Caused by *Colletotrichum nigrum* E. and H.

The anthracnose of peppers has been known for many years and is present in all countries where the host is grown. The varieties of both sweet and pungent peppers are susceptible, and the ripened fruit in each case is more easily infected than are those which are green. The disease causes much injury when it appears in epidemic form, as it occasionally does.

The lesions on the fruit begin as small water-soaked or darkened, soft, depressed areas, which increase rapidly in size, and change to a lighter color as they grow older. The fruiting of the parasite in the lesions forms a black fungous layer over the affected tissue. Usually before the anthracnose covers much of the surface, soft-rot organisms gain an entrance and produce a decay of the whole fruit.

The fungus, *Colletotrichum nigrum,* may not always exhibit setæ or spines in the fruiting bodies, therefore it has also been classified as *Glœosporium piperatum* E. and E. Stoneman, and later Taubenhaus and Dastur have connected the acervulus with a perfect stage, and the fungus has consequently been referred to *Glomerella cingulata* (Stoneman) Sp. and V. Sch. which is the same pathogene that causes bitter-rot of apples.

Still another fungus which bears setæ and causes lesions similar to those produced by Colletotrichum has been reported as *Vermicularia capsici* Syd. The latter is especially injurious in India, where it causes a canker on the stems as well as on the fruit.

The anthracnose fungus bears great numbers of pinkish spores in the acervuli lining the depressed lesions. The spores are splashed by rain or carried on tools and clothing to healthy plants. The fungus lives over winter on contaminated seed, and probably in old diseased peppers left in the field.

The control of the fungus depends on the destruction of diseased fruits, rotation of crops, spraying and seed disinfection. The latter is a delicate operation, since the seed is injured so easily by fungicides. The treatment which has given the best results consists in soaking the seed in water six to fifteen hours, then draining and soaking in a solution of copper sulfate (1 part in 80 parts of water) for five minutes, after which the seed is dusted with air-slaked lime and planted at once.

Spraying with bordeaux mixture 3-6-50 at weekly intervals has been recommended although the results are not always satisfactory and the fruit is spotted by the fungicide.

References

Bancroft, C. K. and R. L. Hunte. A fungus disease of peppers (Capsicum spp.). Journ. Bd. Agr. Brit. Guiana **7**: 139-140. 1914.

Dastur, J. F. *Glomerella cingulata* (Stoneman) Spauld. and V. Schr. and its conidial forms. *Glœosporium piperatum* E. and E. and *Colletotrichum nigrum* E. and H. on chillies and *Carica papaya*. Ann. Appl. Biol. **6**: 245-268. 1920.

Halsted, B. D. Report of Botanist, N. J. Agr. Exp. Sta. Ann. Rept. **1890**: 345-366. 1891; **1891**: 330-333. 1892; **1895**: 297-299. 1896.

Higgins, B. B. Seed treatment in the control of pepper diseases. Phytopath. **13**: 57, 58. 1923.

Jackson, H. S. Anthracnose of cayenne pepper. Del. Agr. Exp. Sta. Ann. Rept. 1904-06; 76-77. 1907.

Ridley, H. N. A new pepper disease. Agr. Bull. Straits and Fed. Malay States **101**: 320-321. 1911.

Sturgis, W. C. Diseases of peppers. Conn. Agr. Exp. Sta. Ann. Rept. **1899**: 277-282. 1900.

Phytophthora Blight of Peppers

Caused by *Phytophthora capsici* Leon.

In 1918 the Phytophthora blight was discovered causing injury to Chile pepper plants in New Mexico. The disease is characterized by cankers on the stems and pods. If the stem is affected, the parasite may girdle it and cause the death of the whole plant. More frequently the girdle is at the base of one or more of the branches which wilt and die. The lesions seldom elongate much on the woody tissue of the stems or branches. On the succulent pod the progress is different. There the parasite rarely or never encircles the pod, but it is common to find the diseased area extending from one tip to the other of the fruit. At first the infection on the fruit is marked by a small water-soaked spot, which later bleaches to a straw color. If the lesion is not over-run by Alternaria, the diseased pods are difficult to distinguish from the healthy ones, and their market value is only slightly impaired. The Alternaria entering as a saprophyte causes a dark discoloration and fills the inside of the fruit with a mouse-gray mycelium. The pathogene grows through the pod wall into the seeds, which may wither and die, or which may retain their normal appearance and viability even when harboring the mycelium. The root is seldom invaded.

The fungus, *Phytophthora capsici*, has a life history resembling that of the Phytophthora causing late-blight of potatoes, or of the one causing downy-mildew of lima-beans. The oogonia have basal antheridia and the conidia, on germination, give rise to zoospores. The fungus is disseminated in infected seed and in infected soil, in both of which it may winter. Rain, also, may spatter the conidia from one point to another.

No control measures have been investigated. Reasoning from what is known of the life history of the parasite, it may be held in check by rotation of crops, obtaining seed from fields where the disease has not occurred, and spraying the plants weekly with bordeaux mixture, 3-6-50, beginning soon after the seedlings have become well established in the field.

Reference

Leonian, L. H. Stem and fruit blight of peppers caused by *Phytophthora capsici* sp. nov. Phytopath. 12: 401-408. 1922.

Bacterial Spot of Pepper

Caused by *Bacterium vesicatorium* Doidge

(For a detailed discussion, see Bacterial Spot of Tomato, page 555.)

The bacterial spot of pepper was mentioned as long ago as 1912, although a detailed study was not made until recently. It is present in nearly all the southern states, and may also occur in foreign countries where peppers are grown.

The spots on the foliage appear on the lower side of the leaf as small, circular, pale green pimples, with a corresponding depression on the upper side. On old leaves, instead of the pimples, there are often water-soaked lesions. As the spot enlarges, the center dies, leaving a straw-colored area of one to ten millimeters in diameter bordered by a zone of water-soaked tissue. Large killed areas of the leaf are never included in a single lesion, but frequently there are so many small brown spots that the whole leaf turns yellow and drops off.

The pimples on the stem are somewhat elongated and inconspicuous. On the fruit the spots are very prominent, being raised as on the leaf, but are further marked by a cracked and roughened surface and brown discoloration. Saprophytic soft-rot organisms enter the cracks and cause a decay of part or all of the infected fruit.

Treating the seed as suggested for the control of anthracnose may prove beneficial. Corrosive sublimate is a better disinfectant than the copper sulfate but injures the seed. It has been shown that dipping the seed for two minutes in corrosive sublimate (1 ounce in 7.5 gallons of water), then washing thoroughly in running water and planting before drying, will kill most of the parasitic bacteria without injuring the viability of the seed very much.

Spraying with bordeaux mixture 3-6-50 at weekly intervals after the young plants become well established in the field has proved fairly successful in protecting the plants from the disease.

References

Gardner, M. W. and J. B. Kendrick. Bacterial spot of tomato and pepper. Phytopath. **13:** 307-315. 1923.

Higgins, B. B. The bacterial spot of pepper. Phytopath. **12:** 501-516. 1922.

Sherbakoff, C. D. Bacterial spot of pepper. Fla. Agr. Exp. Ann. Rept. **1916-17:** 79R-83R. 1918.

Fusarium Wilt of Chile Peppers

Caused by *Fusarium annuum* Leo.

The Fusarium wilt of chile peppers, also known as blackleg and chile blight, was first observed in New Mexico in 1908, and since that time has become so generally present in that state that the chile or red pepper crop is threatened. The trouble has not been observed with certainty elsewhere.

Symptoms.

The disease begins as a drooping of the lower leaves. At this stage the disease has advanced so far at the base of the stem that the wilting of the plant is very rapid. In a few hours the foliage and the younger shoots may wilt, and later die and turn brown. The infections begin at the point where the lateral roots are attached to the base of the stem. Dark brown sunken cankers are formed on the lower part of the stem, which gradually girdle the plant. The roots also are invaded, becoming soft and having a water-soaked appearance. The tissue of both the stems and roots turn dark, the bark of the affected stem rubbing off easily exposing the smoke-colored wood below. In wet soil the base of the stem may be covered with white or bluish-green fruit layers of the parasite. If the vines die after the pods have attained almost full growth, the latter may be sold for green peppers; but those fruits that are small when the disease kills the stem, shrivel and fall to the ground.

Cause.

In making isolations, a fungus has been obtained which is named *Fusarium annuum.* It belongs to the section Martiella of the genus Fusarium and has the usual three types of spores; microconidia, chlamydospores, and macroconidia. The latter are mostly three-septate, although four- or five-septate ones may be numerous under field conditions.

The fungus lives indefinitely in the soil, and is disseminated by irrigation water, or with loose soil in heavy wind storms. Infection takes place through the epidermis at the base of the stalk. The time intervening between inoculation and visible infection depends on weather conditions. When the disease progresses rapidly, the complete wilting of the plant may be accomplished in two weeks, while under less favorable circumstances for the fungus, the wilting may require two or three months.

The Fusarium is very susceptible to changes in temperature and humidity. The optimum temperature apparently is 77° to 81° F. The greatest influence, however, is exercised by the soil-moisture. The wilt does not occur in dry upland soil, but is very serious in low undrained fields.

Control.

In order to avoid too much soil-moisture, which favors the disease, the peppers should preferably be planted in light soils. Artificial drainage may also prove beneficial. In irrigating the soil, only enough water should be applied to keep the plants in a healthy condition. If the plants are set on ridges from eighteen to twenty-four inches in height, the stems and roots cannot come in direct contact with contaminated irrigation water, which must reach the roots by capillary attraction in the ridges. By this method the fungus is filtered out before it can harm the host. Rotation of crops is of no special value, since the wind storms blow the fungus over all the fields in any given community.

Reference

Leonian, R. H. Fusarium wilt of chile pepper. N. Mex. Agr. Exp. Sta. Bull. 121: 1-32. 1919.

Sclerotium Disease of Pepper

Caused by *Sclerotium bataticola* Taub.

(See also Charcoal-Rot of Sweet-Potatoes, page 527.)

The same Sclerotium disease which injures sweet-potatoes has recently been found in the fruit of peppers. Viewing the pepper pod from the outside, the symptoms are so slight that it is sometimes difficult to detect the trouble. The epidermis may be shriveled and blackened. When the fruit is cut open numerous small black sclerotia are embedded in the flesh of the fruit, and adhering to the seeds.

No control measures are suggested.

Reference

Martin, W. H. *Sclerotium bataticola,* the cause of a fruit-rot of peppers. Phytopath. 7: 64, 308-312. 1917.

Pythium Fruit-Rot of Pepper

Caused by *Pythium debaryanum* Hesse.

(For life history, see Damping-off of Tomatoes, page 546.)

During the rainy season of 1919, a soft-rot of pepper fruit was observed in North Carolina. Only the pods near the ground are affected. The lesions begin as small water-soaked areas, which as they enlarge become depressed and the tissue soft and watery. The decay soon reaches the inner part of the fruit, and from the locule immediately below the diseased area the parasite spreads to all parts of the pod, causing a soft-rot as it progresses. The time consumed in the decay of a pod is four or five days. The epidermis remains whole, while the soft misshapen fruit still clings to the plant.

The fungus, which was isolated from the lesions, proved to be *Pythium debaryanum*.

No control methods are suggested.

References

Lehman, S. G. Soft-rot of pepper fruits. Phytopath **11:** 85-87. 1921.

Wolf, F. A. and S. G. Lehman. Soft-rot of pepper fruits (*Pythium—debaryanum*). N. C. Agr. Exp. Sta. Ann. Rept. **43:** 56, 57. 1920.

Choanephora Blight of Pepper

Caused by *Choanephora cucurbitarum* (B. and Rav.) Thaxter

The Choanephora blight of pepper has appeared lately in India, and in seasons when the humidity is high does much damage.

Infection begins at blooming time, the parasite having gained entrance through the blossoms. The flower-buds turn black and rot. The disease is next made evident by the drooping of the leaves and tips of the branches followed by a dying back of the twigs. The invaded tissue exhibits a pronounced wet rot and is soon covered with a silvery layer of fungous fruiting bodies. The infected part has a pronounced water-soaked appearance and the bark peels off in shreds. The decay spreads down the stem, so that in severe cases the whole plant may be killed.

The fungus, *Choanephora cucurbitarum*, formerly was considered as a saprophyte on the blossoms and dead stems of peppers. In 1920 investigational work in India showed that it may become a destructive parasite when weather conditions are favorable. It had previously been reported as a pathogene on flowers and fruit of squash. The fungus is closely related to the bread mold, and produces zygospores in the manner characteristic of this group of lower plants. In addition it forms masses of striated spores on peculiarly shaped conidiophores. The latter end in a swollen tip from which arise numerous sub-heads, all of which are swollen and some of which are branched. The sub-heads are covered with sterigmata on which the spores are borne. Each spore is provided with a hyaline appendage. The spore germinates by means of a germ-tube. A third type of spores is borne in an inclosed sporangium. These latter spores are equilateral with clusters of fine cilia at both ends. The fungus probably lives over winter as mycelium or zygospores in the old diseased tissue on the ground.

No control measures have been suggested.

Reference

Dastur, J. F. *Choanephora cucurbitarum* (B. and Rav.) Thaxter on chillies (Capsicum spp.). Ann. Bot. 34: 399-403. 1920.

Cercospora Leaf-Spot of Pepper

Caused by *Cercospora capsici* Heald and Wolf

The Cercospora leaf-spot of pepper evidently is present wherever the crop is grown in the United States, and in rainy seasons may cause considerable injury.

The spots at first are water-soaked in appearance, but when the affected tissue dries they turn white with a margin of dark brown. The lesion varies in size from one-eighth to one inch in diameter. When many spots occur on a leaf, it turns yellow and drops. When the disease is epidemic, many plants may be defoliated. Occasionally spots are found on the fruit-pedicels and the stems. The parasite cannot grow through the cuticle of the fruit, but it may grow down into the pedicel and cause a stem-end decay of the fruit.

The fungus is characterized by the long separate spores borne on short knobby conidiophores. The mycelium growing from the germinating spore enters the stomata and causes infection. It is possible that the spores live over winter on the seed, as infection may be found on young seedlings.

The control measures for leaf-spot are the same as those recommended for anthracnose of pepper.

Reference

Higgins, B. B. The diseases of pepper. Ga. Agr. Exp. Sta. Bull. 141: 48-75. 1923.

The Black-Spot of Peppers

Caused by *Alternaria* sp.

Recently a fruit-spot of peppers has been reported in America and Italy. It begins as a small water-soaked or yellowish area which resembles scald. As the spot enlarges it becomes depressed, and finally is covered with a black fungous growth. The lesions may appear on any part of the fruit, and serve as a center from which soft-rot organisms spread.

In the United States an Alternaria is considered as the causal organism. In Italy where a similar fruit-spot occurs, the Alternaria is looked on as a secondary fungus which penetrates the lesions made by *Acrothecium capsici* Turconi. The Alternaria is present in either case and is characterized by the large, brown, muriform, catenulate spore. The Acrothecium has long, separate, erect conidiophores, the tips of which are supplied with small sterigmata on which are borne the three-septate olivaceous spores. The pathogene probably lives over winter in diseased refuse and on contaminated seeds.

The control measures are the same as those suggested for anthracnose of peppers.

References

Melchers, L. E. and E. E. Dale. Black-spot of pepper. Phytopath. 7: 63. 1917.

Turconi, M. Un nuovo parassita dei peperoni. Riv. di Pat. Veg. 9: 131-133. 1919.

Mosaic of Peppers

Cause undetermined

(See Mosaic of Tomato, page 542.)

It has not been proved that pepper mosaic is carried in the seed, although this method of dissemination may be possible.

References

Lewis, A. C. Mosaic disease of peppers. Ga. State Bd. Ent. Bull. 55: 1-31. 1919.

Higgins, B. B. The diseases of pepper. Mosaic. Ga. Agr. Exp. Sta. Bull. 141: 56. 1923.

Phoma Rot of Pepper

Caused by *Phoma destructiva* Plowr.

(See Phoma Rot of Tomato, page 584.)

Reference

Higgins, B. B. The diseases of pepper. Ga. Agr. Exp. Sta. Bull. 141: 48-75. 1923.

Brown-Rot of Pepper

Caused by *Bacterium solanacearum* EFS.

(See Brown-Rot of Potato, page 401.)

Gray Mold-Rot of Pepper

Caused by *Botrytis* sp.

(See Gray Mold-Rot of Lettuce, page 253.)

Bacterial Soft-Rot of Pepper

Caused by *Bacillus carotovorus* Jones

(See Bacterial Soft-Rot of Carrot, page 76.)

References

Bennett, C. W. Soft-rot of pepper caused by *Bacillus carotovorus*. Mich. Acad. Sci. Rept. **20**: 351-352. 1918.
Sherbakoff, C. D. Soft-rot of pepper fruit. Fla. Agr. Exp. Sta. Ann. Rept. **1916**: 91R-92R. 1917.

Southern Sclerotium Rot of Pepper

Caused by *Sclerotium rolfsii* Sacc.

(See Southern Sclerotium Rot of Sweet-Potato, page 510.)

Reference

Higgins, B. B. The diseases of pepper. Ga. Agr. Exp. Sta. Bull. 141: 48-75. 1923.

Slightly Known or Rare Diseases of Pepper

A circular ashen-gray leaf-spot caused by a species of Phyllosticta has been observed in New Jersey.

The fungus *Phomopsis capsici* (Mag.) Sacc. occurs on dried pods in the Philippine Islands.

References

Halsted, B. D. Fungous diseases of various crops. N. J. Agr. Exp. Sta. Ann. Rept. **1895**: 360. 1891.
Reinking, Otto. Philippine plant diseases. Phytopath. **9**: 114-140. 1919.

CHAPTER XIV

POTATO DISEASES

THE potato crop in the United States has a greater money value than all the other vegetables put together. More than four hundred million bushels are grown each year. The six states producing the largest quantities of the crop are Maine, New York, Pennsylvania, Michigan, Wisconsin, and Minnesota. No other state produces half as many as any one of those named above, although the following six additional states are important in potato production: Virginia, New Jersey, California, Colorado, Ohio, and Idaho.

The average yield of potatoes in the country did not increase until a few years ago. It is possible that the increase is due in part to better cultivation, and growing the crop on soil better suited for potatoes than formerly, but no doubt the greatest factor in increasing the yield is the use of disease-free high-yielding seed. The statements in recent bulletins from nearly any experiment station praise the certified seed and show how it increased the yields of potatoes in local communities.

There are about forty diseases of potatoes, a dozen, at least, of which are extremely important. According to the Plant Disease Survey Bulletin, the total reduction in yield is estimated at 15 to 20 per cent, which means a loss of approximately seventy-five million bushels each year. Not all of the injury can be avoided, but many of the diseases will be controlled when all growers use inspected seed which has been treated, and when the vines are thoroughly sprayed or dusted.

There are so many diseases of potatoes that no satisfactory sequence for their presentation has been discovered. The maladies of some of the other vegetables may be taken in the order of their importance, but there are eight or ten diseases of potatoes which are almost equal in their destructiveness, or which change their relative positions of importance in different localities. Having these difficulties with which to contend, a crude key is presented which has for its basis of separation both the method of control and the plant parts affected. The index at the end of the volume is more serviceable to the reader than this is meant to be for finding the pagination of any particular disease.

Late-Blight of Potatoes

Caused by *Phytophthora infestans* (Mont.) De Bary

Late-blight is the most spectacular and destructive disease of potatoes. It has been fairly definitely proved that it originated in South America where the potato is indigenous, and that sometime between 1830 and 1840 it was introduced into Europe where it at once began to attract attention. At about the same time or a few years later it was brought into the United States. It evidently spread with startling rapidity, for in 1844 it was present in most of the potato-growing countries, and the following year caused tremendous losses, especially in Ireland where a severe famine resulted. Since that time there have been numerous epidemics of late-blight. Fearing that the potato industry would be exterminated, several countries appointed commissions to study the disease. It is now known in all the potato-producing countries of the world, but is most severe in areas where the nights are relatively cold and the days warm and humid, as in Canada and the northeastern part of the United States. When it occurs in the South it is on the winter crops or in the higher altitudes of the mountain sides.

The disease is most commonly known as late-blight, but is locally also called black-blight, downy-mildew, Irish-blight, potato-blight and brown-rust. The rot on the tubers is sometimes designated as dry-rot, which really should be applied to another disease, late-blight-rot and the mahogany rot.

Late-blight not only affects potatoes but has also been found on tomatoes, eggplant, and other hosts belonging to the genera Solanum and Petunia.

There is seldom a disease of plants which causes such large economic losses. Many writers have enumerated individual cases in which more than half of the crop was destroyed. During epidemics it is not uncommon in New England and in New York state to find that the average reduction in yield is fifty bushels an acre. This is often augmented by later rot in storage. In 1922 late-blight killed the vines early in New York so that in one county where growers before had been harvesting two hundred to three hundred bushels an acre they were fortunate if they obtained one hundred bushels. The loss was incurred almost entirely through the death of the vines, since there was only a trace of tuber-rot.

Symptoms.

The fungus causing late-blight attacks tubers, leaves, petioles, stems and blossom-pedicels. The lesions usually first appear on the lower leaves, although fields have been observed in which the first apparent infection showed on the flower-pedicels. The leaflet when infected shows a water-soaked indefinite spot along the margin. The spot enlarges rapidly until the whole leaflet is included (Fig. 93). It may then dry, blacken and shrivel up, or if there is plenty of moisture in the air, it may rot away entirely, and while doing so emit the characteristic potato odor by which the late-blight may be detected in the field. As the lesion is enlarged recurring wefts of white mold appear adjacent to the healthy tissue on the under side of the leaflet. This is the important diagnostic symptom, inasmuch as no other potato disease organism produces a similar white mold on the under side of the leaf. The lesions on the stems, pedicels, and petioles are similar to those on the leaves.

FIG. 93.—Late-blight on potato leaflets.

The rot on the tuber begins as a minute spot, slightly shrunken and purplish-black in color (Fig. 94). This spot increases in size until the whole tuber is involved. If enough moisture is present a soft-rot may follow the late-blight and cause the complete disintegration of the tuber. If, before a saprophytic organism has invaded the tissue, the tuber is cut open through a discolored spot, the flesh just below the peeling will be marked with granular brick-red blotches (Fig. 95) This, too, is characteristic for late-blight alone. After other fungi or bacteria start the secondary rot, the tissue may become a soft watery mass or turn black and be indistinguishable from that of other storage rots.

Cause of late-blight.

The fungus causing the disease belongs to the downy-mildew group, and is named *Phytophthora infestans.* It lives over

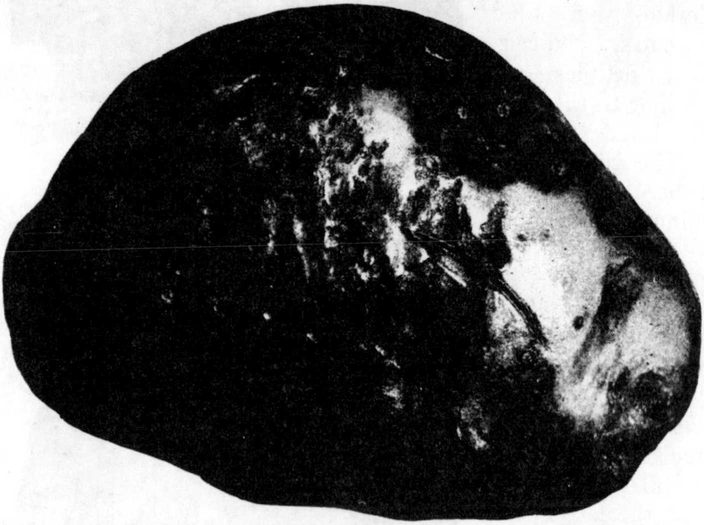

FIG. 94.—Tuber showing purplish sunken lesions caused by late-blight.

winter as mycelium in the tuber, and possibly as oospores, which rarely have been found until recently. The mycelium grows between the host cells, killing the plasma membrane and living on the food thus liberated. It may even send short branch-like haustoria into the cells to aid in procuring its nourishment. When an infected tuber is planted in the spring the parasite may produce fruit-bodies on the cut surface of the seed piece, or the mycelium may grow up into the sprouts and form conidiophores on the outer surface of the growing shoot. The conidiophore production may be delayed until the plant has attained considerable size, when lesions may appear near the base of the stem and extend upward until a leaf-petiole is included. The conidia borne on the seed piece or the stem may be washed or splashed by rain to the leaves and there produce an infection similar to that

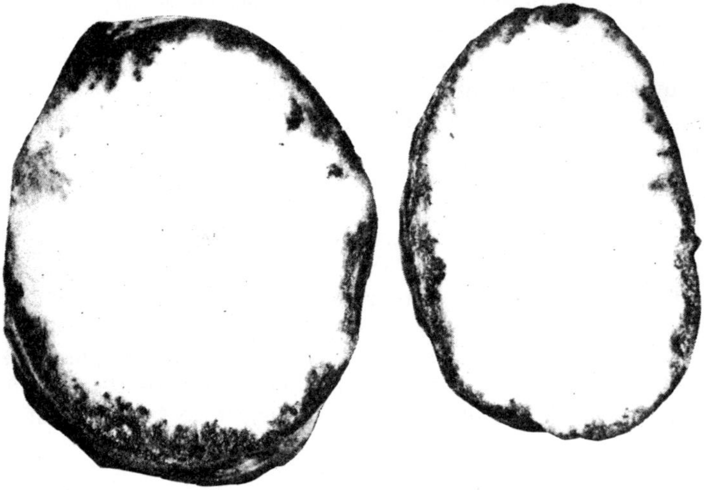

FIG. 95.—Cross-section of tuber showing brick-red discoloration caused by late-blight.

described for the tuber. The conidia are comparatively large, lemon-shaped, and have five to thirty nuclei. On germinating under favorable conditions each conidium gives birth to as many bi-ciliate swarm-spores as there are nuclei. Under high temperatures or other unfavorable conditions the conidium may germinate by means of a germ-tube. After the swarm-spores lose their cilia they send out a germ-tube which gains entrance into the host. The whole process may not take more than fifteen to twenty hours, and in five days or even less a second crop of conidia be produced. The white mold on the under side of the leaf is made up of conidiophores. These are rather long, thick-walled, slightly branched stalks. The ends of the branches swell where the conidia are borne. After the conidia drop the branches may grow a little longer and another crop of conidia be produced. The swelling and proliferation may take place four or five times before the stalk finally dies.

Aside from the splashing of the rain, it has been shown fairly well that Colorado potato-beetles or flea-beetles may help in disseminating the fungus. The wind also aids in scattering the spores. Many of the conidia drop directly to the ground where they are washed into the soil until finally they reach the growing tubers. Infection takes place readily when sufficient moisture is present.

It has been shown that the conidia germinate best at 41° to 57° F., but that the mycelium in the potato grows much more rapidly at a temperature of 73°-81° F. in the presence of a high humidity. The range of temperature explains why the late-blight occurs only where the nights are cool and days warm. Usually there is much more tuber-rot in clay soils than in sand, which acts as a filterer of the zoospores. A heavy frost that kills the potato vines also kills the conidia. In storage 40° F. or less permits very little tuber-rotting.

Control of late-blight.

Wherever late-blight occurs spraying the potato vines with bordeaux mixture 4-4-50 or 5-5-50 has proved very profitable. From 50 to 150 gallons, depending on the size of the plants, are applied to an acre at each application, and it is recommended that the spraying be done with as high pressure as can be generated. An arrangement of nozzles like that found on the so-called Nixon boom has proved most successful. The discs in the nozzles should be changed often enough to keep the holes from wearing larger. The spraying ordinarily is begun when the vines are six inches tall and from four to eleven applications are made depending on weather conditions. Copper-lime dust may be used instead of the bordeaux mixture if the dusting is done when there is no wind, and the vines are wet with dew or rain. Care must be taken to apply enough dust so that the amount of metallic copper is the same as is recommended for spraying.

It has often been suggested that hilling the rows will cover the growing tubers more deeply and protect them from rot. Unless the hills come to a peak at the top, they often serve as a funnel to direct the spores to the tubers rather than away from them and thus increase the rot. Besides, hilling is often detrimental to potato plants, therefore is a questionable means of control.

When spraying has not been practiced earlier in the season and blight attacks the vines so that there is no hope of further growth, it has been found that some of the tubers may be saved from rot by spraying the affected vines and soil with a solution of copper sulfate in water (1 pound in 5 gallons). This kills the vines and tends to check the spread of the disease.

When the vines are blighted they should be permitted to die and become dry before digging is begun; otherwise the inoculum is spread during the harvesting. The tubers should be picked up soon after being dug, and when in piles should never be covered with the dead vines.

If all the visibly affected tubers are sorted out at storage time and the healthy tubers stored in a dry place at a temperature of 38° F., there should be a minimum of loss after digging. It has been suggested that when sulfur or lime are mixed with the stored potatoes, the disease will not spread in the pile. This has not proved true when put into actual practice.

Several varieties of potatoes, especially some of those in Europe, have been found at least partially resistant to the Phytophthora. So far, however, no popular commercial type has this much-sought quality, and the term "blight-proof" applied to certain strains on the market is decidedly a misnomer. A resistant potato, nevertheless, may finally become an accomplished fact, for breeding work is now in progress.

REFERENCES

Bary de A. Researches into the nature of the potato-fungus. *Phytophthora infestans.* Jour. Roy. Agr. Society, Series 2. (ff.) 12: 239-269. 1876.

Brooks, F. F. An account of some field observations on the development of potato blight. New Phytologist 18: 187-200. 1919.

Clinton, G. P. Oospores of potato blight, *Phytophthora infestans.* Conn. Agr. Exp. Sta. Biennial Rept. Part. X 1909-1910: 753-774. pls. 38-40. 1911.

Clinton G. P. Downy mildew, or blight, *Phytophthora infestans.* (Mont) De By., of potatoes. Conn. Agr. Sta. Part IV. Ann. Report 1904: 363-384. Pls. 32-37. 1905.

Giddings, N. J. et Anthony Berg. A comparison of the late blights of tomato and potato. Phytopath. 9: 209, 210, 1919.

Haskell, R. J. *Phytophthora infestans* on eggplant in the United States. Phytopath. 11: 504-505. 1921.

Kern, F. D. and C. R. Orton. *Phytophthora infestans* on tomato. Phytopath. 6: 284-287. 1916.

Jones, L. R. Disease resistance of potatoes. U. S. Dept. Agr. Bur. Plant Ind. Bull. 87: 1-39. 1905.

Jones, L. R., N. J. Giddings, and B. F. Lutman. Investigations of the potato fungus, *Phytophthora infestans.* U. S. Dept. Agr. Bur. Plant Ind. Bull. 245: 1-100. 1912.

Melhus, I. E. Germination and infection with the fungus of the late blight of potato. Wis. Agr. Exp. Sta. Res. Bull. 37: 1-64. 1915.

Melhus, I. E. Hibernation of *Phytophthora infestans* in the Irish potato. Jour. Agr. Research 5: 71-102. 1915.

Morse, W. J. The transference of potato late blight by insects. Phytopath. 11: 94-96. 1921.

Martin, Wm. H. Late-blight of potatoes and the weather. N. J. Agr. Exp. Sta. Bull. 384: 1-23. 1923.

Pachano, A. A fruit spot of cucumber (*Phytophthora infestans*). Bol. Agr. Quinta Normal (Ambato, Ecuador) 1: 201-206. 1920.

Pethybridge, G. H. Considerations and experiments on the supposed infection of the potato crop with the blight fungus (*Phytophthora infestans*) by means of mycelium derived directly from the planted tubers. Scient. Proc. Roy. Dublin Soc. 13: 12-27. 1911.

Pethybridge, G. H. and Paul A. Murphy. On pure cultures of *Phytophthora infestans* DeBary and the development of oospores. Scient. Proc. Roy. Dublin Soc. 13: 566-588. 1913.

Reddick, Donald. Ito's potato variety Ekishirazu in New York. Phytopath. 13: 55, 56. 1923.

Stewart, F. C., G. T. French, and F. A. Sirrine. Potato spraying experiments, 1902-1911. N. Y. (Geneva) Agr. Sta. 349: 99-139. 1912.

Stewart, F. C. The persistence of the potato late blight fungus in the soil. N. Y. (Geneva) Agr. Exp. Sta. Bull. 367: 357-361. 1913.
Stuart, W. Disease resistance of potatoes. Vt. Agr. Exp. Sta. Bull. 179: 147-183. 1914. *Also* Bull. 122: 107-136. 1906.

Tip-Burn and Hopper-Burn of Potatoes

Physiological or caused by *Empoasca mali* Le B.

There is probably no other disease of potatoes which causes such reduction in yield as do tip-burn and hopper-burn. The potato is a native of high cool altitudes, and in consequence suffers when grown where the temperatures are high and the winds correspondingly dry. This condition, augmented by the injury induced by apple leaf-hoppers, causes a loss of fully twenty million bushels a year in the United States.

It has been observed that certain varieties withstand both tip-burn and hopper-burn better than do others. The Russet type of Rurals are quite resistant. The smooth Rurals are ordinarily not so badly affected as are the green Mountain types. The Cobbler and Early Ohio are very susceptible, but even these do not have such sensitive leaves as does the Bliss Triumph, which because of this tendency cannot be grown successfully in many places.

Symptoms.

The first indication of tip-burn or hopper-burn is a light wilting and rolling inward of the edges of the leaves. The affected tissue soon dries and turns brown. The dead area may extend only a short distance along the margin of the leaf, or may include the entire periphery. The dying may extend inward no more than an eighth of an inch, or may include the whole leaf surface. Normally the tip-burn is in the form of more or less uniform scallops of brown tissues, while the hopper-burn manifests itself more often by V-shaped dead areas, the lower point of the area being the place where the leaf-hopper pierced the vein. Early blight or arsenical injury may be mistaken for tip-burn, since all these types of injury appear much alike. Early blight is usually differentiated by the concentric rings in the lesions.

Cause.

There has been some controversy between the physiologists and entomologists regarding the cause of tip-burn, but it is

FIG. 96.—Potato tuber with very large lenticels. This is not due to disease.

now quite definitely proved that there are at least two types of the injury. The first, and more generally recognized one, is the result of hot sunshine with warm dry winds, producing excessive transpiration, especially when the leaf is held at such an angle that the sunbeams strike directly the long sides of the palisade cells. Plasmolysis and wilting of the tissues follow. The edge of the leaf may not die at once, but the first plasmolysis is followed more easily by a second one, and when the process is repeated several times the cells are no longer able to recover from the heavy drainage of water from the pores.

Hopper-burn is induced by the puncturing of the leaf-vein by the apple leaf-hoppers (*Empoasca mali*) which may often be seen rising in swarms as one walks through an infested field. The adult leaf-hopper is a pale yellowish-green insect, about an eighth of an inch long, which migrates from the apple-tree to the potato where it deposits its eggs in the tender vine. After a few days the eggs hatch, and the young nymphs feed with the adults on the under side of the leaves. During the summer several generations are born on the potato, some of the adults of which find their way back in the fall to the apple-tree and there deposit their winter eggs.

Not only does the puncture of the insect cause injury, but it is also stated that when some of the nymphs or adults are macerated and the extract injected into the leaf-vein, typical hopper-burn is produced. This is suggested as supporting the argument that the leaf-hopper alone does not cause the injury, but that the insect is the carrier of some contagium which produces the dying of the tissue.

The true tip-burn depends altogether on the weather conditions. A quick drying wind and hot sunlight evaporate water from the leaf faster than it can be received from the roots. There are, however, other conditions which accelerate or retard the dying of the foliage. An old leaf is more susceptible to the injury than a young actively growing one. A plant affected with leaf-roll or any one of the many other diseases suffers more severely from tip-burn than does one that is healthy. Certain types of soil retain the moisture less efficiently and, therefore, are an aid to the trouble. Fertilizers, also, may have undesirable effects. It has been found that when only a very small amount of borax is mixed with the potash, tip-burn is much aggravated.

Control.

When selecting the field for potatoes, soil should be chosen which is heavy enough to hold the moisture. Plenty of humus in lighter soils will help to retain the necessary water supply. The soil-moisture may also be conserved by means of careful surface cultivation. The most important control measure, however, is that of spraying with bordeaux mixture as given for the control of late-blight. It has a direct effect on true tip-burn as well as acting as a repellant for the leaf-hoppers. If a pint of nicotine sulfate is added to each hundred gallons of the bordeaux mixture, many of the insects can also be killed. The spraying should be done about every ten days, beginning as soon as there is any danger of tip-burn or hopper-burn injury, and continued throughout the remainder of the season.

References

Ball, E. D. The potato leaf hopper and the hopperburn that it causes. Wis. State Dept. of Agr. Bull. **23**: 76-102. 1919.

Dudley, J. E. Control of the potato leaf hopper (*Empoasca mali* Leb.) and prevention of "hopperburn." Jour. Econ. Entom. **13**: 408-415. 1920.

Eyer, John R. Notes on the etiology and specificity of the potato tip burn produced by *Empoasca mali* LeBaron. Phytopath. **12**: 181-184. 1922.

Lutman, B. F. Tipburn of the potato and other plants. Vt. Agr. Sta. Bull. 214: 1-28. 1919.

Lutman, B. F. The relation of the water pores and stomata of the potato leaf to the early stages and advance of tipburn. Phytopath. **12**: 305-333. 1922.

Early-Blight of Potatoes

Caused by *Macrosporium solani* E. and M.

Early-blight or leaf-spot of potatoes was first observed in Missouri in 1885. After its description in 1891, it was found rather generally present in nearly all parts of the world where potatoes are grown. It does not usually occur in epidemic form, although occasionally it may be almost as severe as late-blight. In Bermuda early-blight has been known to infect the whole leaf, so that in its later stage it was difficult to distinguish from an attack by Phytophthora. In Wisconsin it has been reported that the annual loss, between 1900 and

Fig. 97.—Early-blight on potato leaf.

1907 and due to disease, varied from fourteen to twenty-five million bushels. Most of this loss was attributed to the leaf-spot.

The fungus is able to infect a number of hosts. Those which have been reported are: potato, tomato, eggplant, *Solanum aviculare, S. carolinensis, S. giganteum, S. nigrum, S. nigrum guineense, Hyoscyamus albus, Solanum rostratum, Nicandra physaloides, Solanum Warscewiczii, Hyoscyamus niger* and *Solanum Commersonii.* The spot on Jimson weed was reported as the same, but later this statement was disproved.

Symptoms.

The disease may affect any part of the host above ground. It was never known to injure tubers. The lesions are dark brown circular or oval spots on the foliage (Figs. 97, 98), or rarely on the stem. The first spots commonly occur on the lower, shaded and consequently weaker foliage. Under favorable conditions the lesions enlarge rapidly until in extreme cases a single spot involves half the leaf, although this is not so often true as it is to have the leaf covered with many small spots. As the brown tissue dies the mesophyll cells, near the lower side of the leaf, collapse, and throw the sturdier pallisade cells into folds or ridges. This gives the spot a target effect, since the folds are arranged in concentric rings. Such rings form one of the symptoms which distinguish the early-blight from other possible spots on the foliage.

When the leaf-spot occurs on the edge of the leaf, it may often be mistaken for tip-burn, although the latter is usually not marked with concentric rings. Arsenical burning of the foliage is such a close imitation of the leaf-spot that the two cannot be distinguished from each other if the history of the spray application is not known beforehand. The arsenical injury causes a spot somewhat lighter in color than does the early-blight but the difference in shade is not conspicuous enough to be a sure guide in a diagnosis.

Cause.

The organism is one of the imperfect fungi, known as *Macrosporium solani*, or (with equally good reason) *Alternaria solani.* The spores, carried over winter on old potato leaves or on the susceptible weeds or possibly directly in the soil, are splashed by rains or thrown by the cultivator onto the growing foliage, where a spot begins to show eight or ten days after inoculation. The ends of the mycelial threads which have penetrated the leaf grow outward through the upper and lower epidermis where on the surface they bear large brown muriform spores. The fruiting continues as long as nutrient material and moisture are present. The spores are blown or carried by insects to neighboring plants or leaves.

FIG. 98.—Enlarged view of early-blight spots showing concentric rings on potato leaf.

The fungus is for the most part not an active parasite. In fact, it usually happens that the weakened foliage is first to be attacked, and seldom does it occur in epidemic form until after the actively growing period of the host, or the plant is enfeebled by the hot weather which is favorable for the pathogene. It has been found that the spores germinate best at about 85° F. Dew alone does not furnish enough moisture to cause infection, but frequent rains with heavy dews supply ideal conditions for the production of spores. Then, if the vines have been weakened, abundant infection takes place.

Control of early-blight.

In controlling the fungus at least three measures are necessary: first, the eradication of all susceptible solanaceous weeds; second, rotations of three to five years, and third, spraying as for the control of late-blight. The Macrosporium is rather difficult to combat, therefore every effort should be made to get the bordeaux mixture on every part of the plant and applied often enough to keep all the recently grown tips covered.

References

Galloway, B. T. The Macrosporium potato disease. Proc. Soc. Prom. Agr. Sci. 14: 46-58. 1893.

Jones, L. R. Studies upon Macrosporium solani. E. and M. Vt. Agr. Exp. Sta. Ann. Rept. 1895: 79-88. 1896.

Milward, J. G. Observations upon prevalence of early potato blight in Wisconsin. Wis. Agr. Exp. Sta. Bull. 168: 1-27. 1907.

Rands, R. D. Early blight of potato and related plants. Wis. Agr. Exp. Sta. Res. Bull. 42: 1-48. 1917.

Rands, R. D. Alternaria on Datura and potato. Phytopath. 7: 327-337. 1917.

Sorauer, Paul. Auftreten einer dem amerikanschen "early blight" entsprechenden krankheit an den deutschen kartoffeln. Ztschr. f. Pflanzenkr. 6: 1-9. 1896.

Cercospora Leaf-Blotch of Potato

Caused by *Cercospora concors* (Casp.) Sacc.

Cercospora leaf-blotch is one of the least important diseases of potato, although it has been known since 1854, and reported from nearly every country in Europe as well as from the United States.

Symptoms.

When spots are borne on the green leaf they consist of small blackened dead areas, but when infection causes the yellowing of the leaflet, the spots retain their green color longer than does the remainder of the tissue. The first appearance of the lesion is usually on the lower side of the leaf, followed later by spots on the upper surface. No other part of the plant is affected.

Cause.

The fungus, known as *Cercospora concors*, invades the weakened leaf tissue, and after a brief period of growth produces conidiophores which push up through the stomata, and bear thin, long, septate conidia. These are able to live while being blown from one plant to another or from one field to that adjoining, but do not remain viable over winter. Thick-walled chlamydospores are borne on the mycelium in the host tissue, and probably serve in keeping the pathogene alive from fall until spring.

The fungus thrives best in a moist atmosphere when the temperature is at 64° to 70° F.

Control.

Spraying as recommended for the control of late-blight wholly eliminates any ravages which the fungus might otherwise be capable of producing.

References

Jones, L. R. and C. S. Pomeroy. The leaf blotch disease of the potato caused by *Cercospora concors*. Vt. Agr. Exp. Sta. Ann. Rept. 19: 236-257. 1907.

Keissler, Karl von. Auftreten der Cercosporakrankheit der kartoffel in Nieder-Oesterreich. Zeitschr. f. Pflanzenkr. 27: 111-114. 1917.

Rust on Potatoes

Caused by *Puccinia pittieriana* Henn.

So many diseases of potatoes have been introduced into the United States that it is a relief to know that at least a few pathogenes that might become troublesome have not been able to invade the country. Among such diseases is the rust found on potatoes and tomatoes in the highlands of Costa Rica and in Ecuador. Like the rusts of cereals, the symptoms are pronounced brownish pustules on the leaves. In these pustules are numerous teliospores which act as the source of inoculum in the dissemination of the organism known as *Puccinia pittieriana.* Since the pathogene is not carried on the tubers, there is a possibility that it may never be introduced into the United States. It would be unfortunate if the rust did become established here not only on account of the damage it might do to potatoes, but also because when tested in South America it showed a special predilection for tomatoes grown from seed obtained in North America. Exclusion is the only known control measure. If any suspicious looking specimens are ever found on either potato or tomato, they should at once be forwarded to the state plant pathologist or to the Department of Agriculture at Washington, D. C., for diagnosis. Many organisms now injurious to crops might have been eradicated if reported when the resulting diseases were first observed.

References

Arthur, J. C. Two destructive rusts ready to invade the United States. Phytopath. 10: 65, 66. 1920.

Arthur, J. C. Origin of potato rust. Science 53: 228, 229. 1921.

Bacterial Leaf Disease of Potatoes

Caused by *Bacillus tubifex* Dale

The bacterial leaf disease of potatoes has been reported only once from England. Several plants were found with wrinkled yellowed leaves which were covered with small brown lesions. These leaves gradually died and dropped off, beginning with the lowest ones first. The pathogene, *Bacillus tubifex,* was of interest because the bacteria gathered into a mass or tube and in this form penetrated the host tissue. Successful inoculations were made both on the potato and the tomato.

Reference

Dale, Elizabeth. A bacterial disease of potato leaves. Ann. Bot. 26: 133-154. 1912.

Rhizoctoniose on Potatoes

Caused by *Corticium vagum* B. and C.

No disease of potatoes is more widespread than rhizoctoniose. It has been reported from every country and state where potatoes are grown commercially. It probably was present in Europe soon after potatoes became an established garden and field crop, but the disease was not described until 1858. Strangely, it was not found in America until more than thirty years later. Being so general the disease has received a large number of local names, such as scurf, black-speck, stem-rot, russet-scab, black-pox, foot-rot, brown-stem, rosette, pitting. canker, pock disease, collar-rot, and other terms. It is impossible to use a name which will describe the symptoms of all the different stages, so that the name of the fungus, Rhizoctonia, has been applied to the disease as well. The more stilted term, rhizoctoniose, is used here because no other term seems to fit so well.

The fungus attacks many plants besides the potato. Among the seventy-five or more species which have been cited as hosts of the parasite, may be mentioned a few of the more common ones: beets, carrots, alfalfa, red clover, onion, turnips, peas, celery, lettuce, beans, cabbage, carnations, parsnip, pigweed, raspberry, rhubarb, violet, and spinach.

The scurf on the tubers being inconspicuous, the grower often does not realize the damage which this fungus is able to bring about in reduced stands, weakened hills, and misshapen, small, cankered tubers. There are many reports of losses varying from 5 to 50 per cent of the crop. Judging from the beneficial results of control measures, the average loss for New York state, where the disease is moderately severe, is from fifteen to twenty bushels an acre. Other states, Colorado and Ohio for example, have reported even a larger percentage.

Symptoms.

On the mature tubers are found small black fungous bodies, which resemble particles of dirt (Figs. 99, 100). These are sclerotia, which vary in size from those just visible to the

FIG. 99.—Rhizoctonia sclerotia on potato tuber.

naked eye to those a third of an inch in diameter. There may be only one or two on a tuber, or the tuber may be almost completely covered with them. Their resemblance to drops of tar or pitch has given the disease the name of lac-scab in some countries.

When the tubers have sprouted the sprouts may be attacked with a sepia or brick-brown rot which kills the growing point. These sprouts on being dug up often have the appearance of being chewed off by insects. Similarly colored cankers, also, may form a girdle near the base of the sprout and kill it or cause a new shoot to grow out below the girdle (Fig. 101). This, in turn, may be girdled and followed by another adven-

FIG. 100.—Small sclerotia caused by Rhizoctonia.

titious shoot. The girdling and renewed growth may be repeated a half dozen times or more until the sprout finally is killed, resulting in a missing hill, or the plant feebly breaks through the soil a long time after the healthy hills have come up. These puny yellow hills never yield enough to be profitable.

Often the base of the stem is not infected until the plant has gained considerable size. The canker then does not kill the plant but causes a shortening of the new petioles at the tip

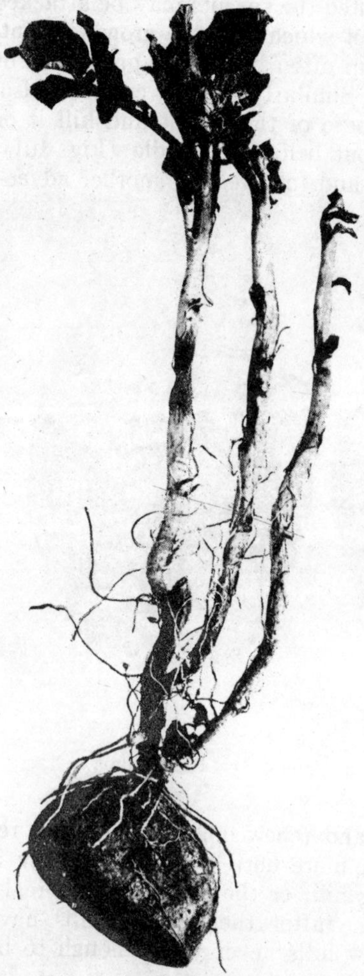

Fig. 101.—Cankers caused by Rhizoctonia on young potato shoots.

of the stem, so that the top leaves are massed closely together giving a rosette effect. The leaves of the rosette may be yellow or remain a normal green. Since the canker does not enter deep enough into the stem to shut off the supply of food and water from the soil, but does hinder the translocation of sugars from the leaves to the tubers, the excess food present often causes the vine to be abnormally large. These extremely large hills may set as many as fifty small knobby and russeted tubers, which never grow larger in size. Such plants are often referred to as "bastard hills" by the growers. Occasionally small aerial tubers are formed on the stem in the axils of the leaves (Fig. 102).

In wet weather and particularly on muck soil, the fungus forms a white cobwebby weft of mycelium about the base of the potato stem. It is not uncommon to find every plant in a field marked with these filmy fruit-bodies of the fungus.

FIG. 102.—Aerial tubers caused by Rhizoctonia.

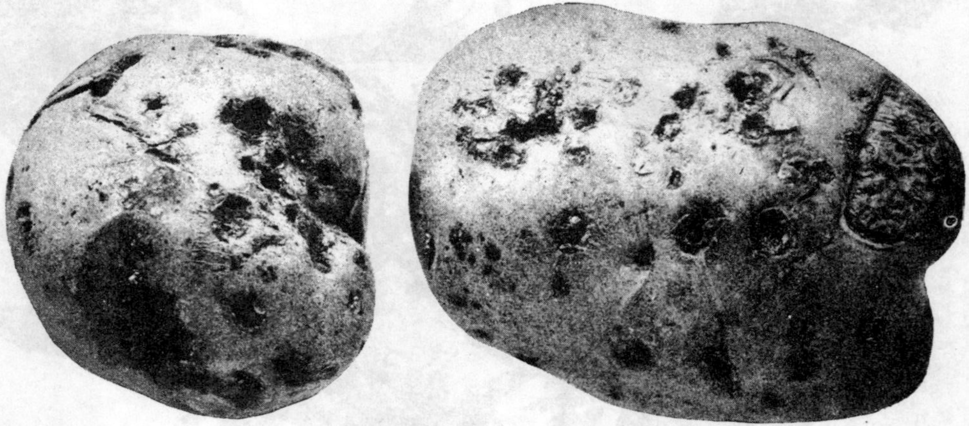

FIG. 103.—Cankers on potato tubers caused by Rhizoctonia.

A symptom which is rather rare is the pitting or tuber-canker (Fig. 103). A superficial examination of such infected potatoes would lead to the belief that the injury was produced by grubs or other insects. A close inspection shows that the decay begins before the skin is broken. A small thimble-shaped area, approximately a quarter of an inch across, is affected. The tissue disintegrates until the bursting epidermis reveals a cavity partly filled with decomposing tissue and mycelial threads. When the tubers are cooked the diseased area separates from that of the healthy so that the canker can be removed like a small core.

Still another symptom has lately been described on elongated pointed-end tubers of the Netted Gem and Burbank varieties. This is jelly-end rot. Previously it was believed that this black jelly-like decay of the stem-end of the tuber was due to Fusaria, but a certain percentage of this decay is now known to be caused by the scurf fungus. The rot penetrates only so far as the stem-end protrudes abnormally. The rot may also affect the knobs on the sides of the tubers. The affected tuber shrivels or succumbs to the attack of saprophytic organisms, following jelly-end rot.

Causes of rhizoctoniose.

The asexual stage of the fungus was described in 1858 and named *Rhizoctonia solani* Kühn. Forty-five years later the white weft at the base of the stem was proved to be the sexual stage of the organism and in consequence the name became *Corticium vagum* B. and C. The term Rhizoctonia is still used most commonly.

The fungus lives over winter in the soil and on the tuber in the form of mycelium and sclerotia. It is probable that few soils are free from it. In the spring when the seed pieces are planted the same warmth and moisture which starts the growth of the host also calls the parasite into activity. The sclerotia, which are nothing more than closely woven masses of hyphæ, send out threads which penetrate the tender sprouts and cause the cankers described above. Later in the season hyphæ grow over the developing tuber and gradually mass at certain points where the new sclerotia are formed. The fungus may be present on the tuber without forming sclerotia. In these cases the brown, septate, branched mycelium may be observed by the aid of the microscope.

In the asexual stage of the fungus no fruit-bodies or spores are produced. The sclerotia take the place of resting spores. During the summer, however, the white wefts mentioned under symptoms are covered with short club-shaped branches, which act as basidia and on the tips of which are borne two to four sporidia. They are not of great importance in the life history of the fungus.

The exact environmental conditions for the growth or suppression of Rhizoctonia have not been determined definitely but it is generally stated that heavy poorly drained soils increase the amount of fungous growth. It is also supposed to be favored by an acid soil. Rather limited experiments in Germany with ammonium sulfate applied as a soil fertilizer seemed to show that this chemical accelerated the growth of the organism. Several authors state that an increased humus-content of the soil retards the development of the disease.

Control of rhizoctoniose.

Tubers as free from sclerotia as possible should be chosen for planting, and as long rotations of crops should be practiced as can be carried out on the individual farm. In addition, the tubers should be treated before being cut for planting. In fact, the treating should be done as early in the spring as the weather will permit, and then the tubers spread out in thin layers to sprout in the light. The tubers with thick strong sprouts will come up quickly when planted, thus running less chance of becoming infected with Rhizoctonia. Furthermore, the tubers with sprindling sprouts can be eliminated before planting.

The treating usually recommended is that of using four

ounces of corrosive sublimate (mercuric chloride) in thirty gallons of water and soaking the seed in this solution for an hour and a half. Only wooden, glass or earthenware receptacles should be used. The poison is dissolved in a little hot water, then enough cold water is added to make the thirty gallons. The solution is used to treat three batches of potatoes and is then discarded. Since the poison reacts slowly or fast, depending on the temperature of the solution, the latter should

FIG. 104.—Arrangement of barrels for treating potatoes with cold corrosive sublimate.

not be less than 45° F., nor more than 70° F. unless the time of treating is reduced. The potatoes must be dried thoroughly after treating or the corrosive sublimate will continue to react and cause injury to the seed.

A simple home-made device for treating is that of having one barrel for preparing the solution and two or more additional barrels set on a platform raised twelve or fifteen inches above the ground (Fig. 104). A large hole is bored in the side of each elevated barrel close to the bottom from which, when the tubers have been treated, all of the solution can quickly be drained into a wooden pail. Many variations of this equipment have been used satisfactorily. Cement tanks are to be avoided and the potatoes should never be dipped when in cloth sacks for cement and sacks as well as loose dirt containing potato refuse quickly lessen the toxic effect of the disinfectant.

In order to avoid the discarding of the solution after the third treatment, some growers have added an ounce of the poison for each barrelful of tubers treated. This is merely guess work, and as a result the solution may become too weak to be effective or so concentrated that injury of the seed is pronounced. Therefore, a means of titrating the solution has been devised by Cross whereby its approximate strength may be determined at any time. Five grams of potassium iodide are dissolved in 1000 cubic centimeters of water, after which a gram of copper sulfate dissolved in a little of the water is added. The original corrosive sublimate solution is added drop by drop into a measuring cylinder containing ten cubic centimeters of the potassium iodide until an orange-yellow precipitate is formed. If both the disinfectant and the potassium solutions are of the correct strength, it will take twice as much of the former as of the latter to produce the precipitate. The amount of corrosive sublimate solution required for the titration is recorded. After using the disinfectant once or twice for treating potatoes, the solution is titrated again. If it has been weakened, it will take more of the disinfectant to cause a precipitation. If it takes a fifth or fourth more, the used solution is weaker in that same proportion, and from this ratio the amount of corrosive sublimate to be added can easily be figured. Before each successive titration enough water should be added to the disinfecting solution to bring it to its original volume.

Formaldehyde used in a cold solution or as a gas has been tried as a disinfectant, but has not proved successful against Rhizoctonia. In Iowa a method has been devised whereby a solution made by adding one pint of commercial formaldehyde to fifteen gallons of water is heated to 120° F. by means of a stove or live steam, and the potatoes dipped for two or three minutes then removed and placed in a covered pile for several hours. Later they are spread out to dry. Formaldehyde has

FIG. 105—Machinery for treating potatoes with hot corrosive sublimate. See long tank near the top through which tubers are drawn on a slowly moving belt. Dry warm air is fanned over the tubers while they are returning. They are then ready to be sacked.

some advantages in that it is not a rank poison, its disinfecting qualities are not easily destroyed, and it can be used in metal containers without fear of corroding. On the other hand gas is unpleasant to the eyes, nose, and throat, the tubers must be handled twice because covering is required, and even when heated it is not so effective against Rhizoctonia as corrosive sublimate.

Hot corrosive sublimate has been substituted in some localities for the hot formaldehyde and has proved the most effective

Fig. 106.—Tuber injured by seed treatment,—not dried after immersion in a corrosive sublimate solution.

disinfectant yet suggested (Fig. 105). A wooden tank is required. A pound of the poison is added to each one hundred and twenty gallons of water and the solution heated to 126° F. by injecting live steam through a rubber hose from an engine or heating plant. The potatoes in slat crates are dipped for two minutes after which they are dried. They may be left in the crates for drying if the weather is warm and the air not too humid. The solution should be titrated frequently as suggested for the cold corrosive sublimate method. A good type of tank to use is one just wide enough to admit the setting in of a bushel crate, and long enough for eight crates to sit in a row. It is well to have it twice as deep as a crate so that sufficient solution may be used to keep the temperature uniform. The crates may be set on a false bottom made of narrow strips of board. All metal parts should be covered with asphalt roof paint to prevent corroding. It is often expedient for some farmer with a traction engine to build such a tank on a trailer and go about the community treating the potatoes much as threshers move about to thresh grain. The apparatus can be built very cheaply.

References

Gloyer, W. O. The efficiency of formaldehyde in the treatment of seed potatoes for Rhizoctonia. N. Y. (Geneva) Agr. Exp. Sta. Bull. 370: 417-431. 1913.

Melhus, I. E. and J. C. Gilman. Measuring certain variable factors in potato seed treatment experiment. Phytopath. **11**: 6-17. 1921.

Morse, W. J. and M. Shapovalov. The Rhizoctonia disease of potatoes. Me. Agr. Exp. Sta. Bull. 230: 193-216. 1914.

Ramsey, Glen B. A form of potato disease produced by Rhizoctonia. Jour. Agr. Research **9**: 421-426. 1917.

Richards, B. L. Pathogenicity of Corticium vagum on the potato as affected by soil temperature. Jour. Agr. Research **21**: 459-482. 1921.

Rolfs, F. M. Corticium vagum, B. and C. var. solani. Burt. A fruiting stage of Rhizoctonia solani. Science n. s. **18**: 729. 1903.

Rolfs, F. M. Potato failures. Colo. Agr. Exp. Sta. Bull. 70: 1-20. 1902.

Rolfs, F. M. Potato failures. Colo. Agr. Exp. Sta. Bull. 91: 1-33. 1904.

Selby, A. D. A rosette disease of potatoes. Ohio Agr. Exp. Sta. Bull. 139: 53-66. 1903.

Selby, A. D. Studies in potato rosette. Ohio Agr. Exp. Sta. Bull. 145. 15-28. 1903.

Shapovalov, M. *Rhizoctonia solani* as a potato tuber rot fungus. Phytopath. **12**: 334-336. 1922.

Stewart, F. C. and W. O. Gloyer. The injurious effect of formaldehyde gas on potato tubers. N. Y. (Geneva) Agr. Exp. Sta. Bull. 369: 385-416. 1913.

Weimar, J. L. Reduction in the strength of the mercuric-chloric solution used for disinfecting sweet potatoes. Jour. Agr. Research **21**: 575-587. 1921.

Wollenweber, H. W. Der Kartoffelschorf: Arbeiten des Forschungsinstitutes für Kartoffelbau, pp. 1-102. Paul Parey. Berlin, 1920.

Common Scab of Potatoes

Caused by *Actinomyces scabies* (Thax.) Güssow

Common scab of potatoes, having been known for a century, is familiar to both the consuming public and to the grower. The pathogene being a common soil organism, the disease has probably been present since potatoes were introduced into Europe. The first description was published in 1825, but the cause of the disease was not discovered until 1890 when Thaxter's and Bolley's papers were printed. The investigations of the two men evidently caused considerable excitement among pathologists, and served as a milestone in the history of plant pathology, for from that date until the present an endless number of dissertations on the common scab have appeared. The literature, although plentiful, is not always enlightening, for often the investigator confuses this scab with other diseases, particularly with powdery scab. There are still many points regarding the disease which need to be investigated.

The disease is not only present on potatoes, but has also been observed on beets, turnips, mangels, rutabagas, parsnips, radishes, carrots, and possibly other fleshy roots.

In Europe a severe outbreak of scab is not supposed to cause such a commercial loss as would a similar epidemic in America, for the reason that on the former continent many potatoes are fed to cattle and the scabby tubers can readily be sorted for this purpose. Besides, the custom of cooking potatoes with their jackets on is not practiced extensively in Europe, and a peeled scabby potato is as good as a healthy one, excepting that there is a slightly greater loss in peeling. In fact, there are persons who insist that scabby potatoes are sweeter to the taste than healthy ones, and consequently more to be desired. In America where the market demands a smooth potato, the loss from scab is very great each year. It is difficult to make an exact estimate, but by summarizing all the available reports it would seem conservative to say that 5 per cent of the crop is lost because of the scab and its attendant insect injuries.

Symptoms.

The term, scab, signifies very aptly the type of lesion which this disease makes on the potato. It begins as a minute brown speck at a breathing pore of the young tuber. As the tuber enlarges the lesion also increases its width. There may be only one spot on the tuber or there may be so many that they coalesce, forming a more or less continuous scabbing of the epidermis (Fig. 107).

FIG. 107.—Common scab of potato.

Scab spots are often divided into three types, namely, warty scab, surface scab, and pitted scab. These types, as the names indicate, are lesions which are swollen, or depressed, or even with the surface of the tuber. The three kinds may occasionally be observed on the same tuber. In each case the corky ridges of the lesion are arranged more or less plainly in concentric ever-widening circles. This rifted condition is caused by the attempt of the host to shut out the parasite by a barrier of cork, each successive layer of which is killed, and a new one begun a little farther from the center of infection.

Cause.

Some growers still are of the opinion that scab is caused by insects or certain conditions in the soil, as the presence of manure, lime, or ashes. However, it has long been proved experimentally that an organism known as *Actinomyces scabies* which is sometimes classed as a higher bacterium and sometimes as one of the lowest forms of true fungi, produces the roughened lesions. Several of the investigators in Europe suggest that not only this species but possibly ten more of the same genus are capable of causing scab, and make the further statement that the three types of scab are the result of different species of Actinomyces attacking the potato.

The pathogene is able to live over winter on the humus in the soil as well as on the seed tuber. Since it attacks various hosts which have fleshy roots, it may be found occasionally where no potatoes have been grown before. The vegetative part of the organism is made up of aborted threads, which may be arranged in spirals, and which by constrictions are broken up into spores. When the inoculum of spores or threads come in touch with a newly-forming tuber, infection takes place, producing a lesion which becomes visible after one to seven weeks. The infected tuber when planted the next season serves as a source of inoculum for the new crop.

The scab pathogene is very susceptible to the influence of alkalinity or acidity in the soil. It often has been observed that when lime, ashes or barnyard manure are spread over a field, the amount of scab is increased at once. Alkalinity accelerates growth of the organism and acidity suppresses growth. It has been shown that a neutral soil with a pH value of 7 gave a crop almost 100 per cent scabby while an acid one with a pH value of 5.60 gave almost a clean crop, the pH values representing degrees of alkalinity and of acidity.

Temperature and moisture also have a strong bearing on the severity of the disease. In northern Europe where the mean summer temperature is relatively low there is not so much common scab as there is in many parts of the United States. Careful experiments have shown that the optimum temperature for the development of the parasite is approximately 72° F. According to Sanford, the same soil that when dry will produce scabby tubers will yield a clean crop if kept very moist during the season; that is, a combination of drought, medium temperature, and soil alkalinity is necessary for an epidemic of scab.

Control of scab.

When a farm is thoroughly infested with the Actinomyces, it is difficult to grow a smooth clean crop. There are certain recommendations, however, which when put into effect will produce commercially profitable results. As is true of nearly every disease, as long rotations as possible should be practiced. Healthy seed potatoes are required. If they must be bought, it is well to get certified stock, a record of which may be obtained at the time the purchase is made. No matter how smooth the tubers appear, they should be treated as is suggested for the control of Rhizoctonia.

If it is necessary to apply manure, lime or ashes to the soil, it is best to use such alkaline-producing material at some other time in the crop rotation, preferably on the new grass seeding. The turning under of a heavy cover-crop aids in reducing the amount of the disease and also builds up the land. Where the potatoes are grown each year in the same field it is quite desirable to sow rye in the fall and then plow it under in the spring. It may even be possible where early potatoes are grown to sow oats for the fall plowing followed by rye to be turned under in the spring. This furnishes a double amount of humus for the crop.

Commercial fertilizers differ in their effect on the prevalence of the disease. Chemicals containing chlorine, such as magnesium chloride and potassium chloride, have been shown to cause a reduction in the amount of scab, but they also reduce the yield when used in sufficient amounts. A fertilizer which can be employed profitably is ammonium sulfate instead of nitrate of soda, as the source of nitrogen. The former substance aids in acidifying the soil and thus in lessening the amount of scab.

A large amount of work has been done in New Jersey on the effects of sulfur in controlling the disease. Not only have flowers of sulfur and finely ground sulfur been used, but also sulfur to which has been added a sulfur oxidizing bacterium, the preparation being known as inoculated sulfur. There are certain soils in which the sulfofying bacteria are rare, so that when sulfur alone is applied it does not change into sulfurous or sulfuric acid. The inoculation is made to obviate this difficulty. It has been found that the infested soil needs an application of 300 to 450 pounds of the inoculated sulfur broadcast on the field and harrowed in before the potatoes are planted. The use of sulfur is to be recommended especially in soils like those of New Jersey and Long Island, where rotations are not usually practiced. In localities in which rotations extend over five to seven years and one of the crops is alfalfa or clover, much greater caution is required in the application of sulfur. In the latter case only the spots in the field where scab is severe should be treated. Generally the other measures enumerated above will suffice for the more lightly infested areas.

References

Arthur, J. C. Potato scab and its prevention. Ind. Agr. Exp. Sta. Bull. 56: 69-80. 1895.

Arthur, J. C. Formalin for prevention of potato scab. Ind. Agr. Exp. Sta. Bull. 65: 20-36. 1897.

Bolley, H. L. Potato Scab and possibilities of prevention. N. D. Agr. Exp. Sta. Bull. 4: 1-14, 21-31. 1891.

Bolley, H. L. The effectiveness of corrosive sublimate as a preventive of potato scab. Proc. Soc. Promotion Agr. Sci. **1896**: 75-81. 1896.

Chester, F. D. Treatment of seed potatoes with sulfur as a preventive of scab. Del. Agr. Exp. Sta. Bull. 34: 19-20. 1896.

Güssow, H. T. The systematic position of the organism of the common potato scab. Science n. s. **39**: 431-433. 1914.

Jones, L. R. McKinney, H. H. and Fellows, H. The Influence of Soil Temperature on Potato Scab. Wis. Agr. Exp. Sta. Res. Bull. 53: 1-35. 1922.

Lutman, B. F. and G. C. Cunningham. Potato scab. Vt. Agr. Exp. Sta. Bull. 184: 1-64. 1914.

Martin, Wm. H. Potato fertilizers bear relation to scab. *In* Hints to Potato Growers. N. J. State Potato Assoc. **3**: No. 7. 1922.

Martin, W. H. The influence of sulfur on soil acidity and the control of potato scab. N. J. Agr. Exp. Sta. Ann. Rept. **1919-1920**: 590-598. 1921.

Millard, W. A. Common scab of potatoes. Jour. Min. Agr. (London) **28**: 49-53. 1921.

Sanford, G. B. The relation of soil moisture to the development of common scab of potato. Phytopath. **13**: 231-236. 1923.

Sherbakoff, C. D. Potato scab and sulfur disinfection. N. Y. (Cornell) Agr. Sta. Bull. 350: 706-743. 1914.

Stewart, F. C. and W. O. Gloyer. The injurious effect of formaldehyde gas on potato tubers. N. Y. (Geneva) Agr. Exp. Sta. Bull. 369: 385-416. 1913.

Stuart, William. Disease resistance of potatoes. Vt. Agr. Exp. Sta. Bull. 179: 147-183. 1914.

Thaxter, Roland. The potato "scab." Conn. Agr. Exp. Sta. Ann. Rept. **14**: 3-17. 1890.

Wollenweber, H. W. Kartoffel Schorf. Arbeiten des Forschungsinstitutes für Kartoffelbau, pp. 102. Paul Parey, Berlin, Publisher. 1920.

Black-Leg of Potatoes

Caused by *Bacillus atrosepticus* van Hall

Black-leg, found in Vermont in 1906, was probably imported from Europe, where the disease has been known for many years. It has now spread to all the important potato districts in the United States. It may do considerable damage in some fields or in some areas one season and then disappear for several years. In the Maritime Provinces of Canada in 1915 10 per cent of the crop was lost. In Maine the estimated loss is 1 to 2 per cent of the yield from 50,000 acres. A 5 to 10 per cent loss is not uncommon in Germany, and much greater losses have been reported. In Great Britain large losses have been recorded of tubers stored in pits. The disease apparently spreads from diseased to healthy tubers, when conditions are favorable.

Symptoms.

The first symptom which attracts the eye when looking over a field where black-leg occurs is the sprangled yellow tops of the affected plants. The leaves near the tip of the vine are small, reddish-yellow, and curl upward, or the vine may be rosetted as is true when Rhizoctonia attacks the host. On examining the base of the diseased plant it is found that the blackening and rotting of the tap-root begins at the mother seed piece and extends upward until the stem for three or four inches above the ground is included (Fig. 108). Some of the vines will show a silvery coating of bacterial slime over the black lesion. This charcoal-black slimy rot can hardly be mistaken for any other disease. The tops are usually large enough to break over the weakened stem, so that the prostrate plants are a common symptom. If the plants are infected

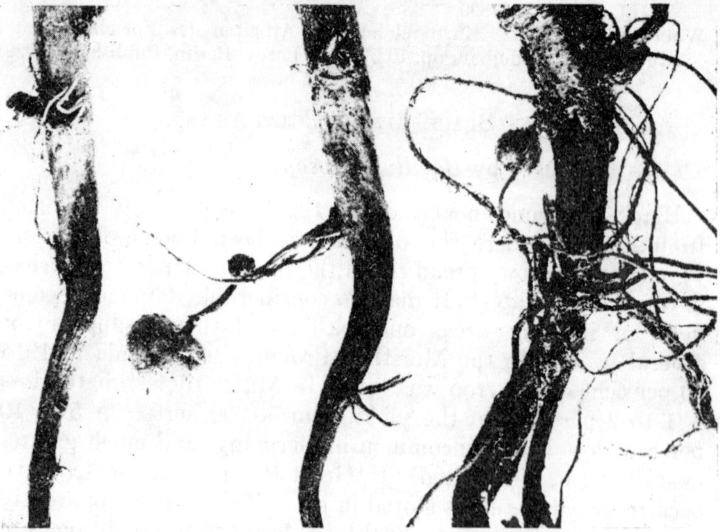

FIG. 108.—Black-leg on potato stems and roots. White areas oozing bacterial slime.

while still young, they may die and wither so that later in the season a missing hill is the only evidence that any disease has been present. Such seedlings, however, may continue to live if the environment becomes unfavorable for the pathogene, and then the plants are dwarfed and feeble.

The tuber is infected at the stem end through the stolon or may show a lesion on the side (Fig. 109). Usually the tissue softens, but remains a normal color until the epidermis, which peels easily from the dead tissue, is removed when the soft pulp turns reddish, then almost completely black. Bounding this diseased area and adjacent to the healthy cells is a black line of demarcation. This line advances as the decay progresses. When the infection is through the stolon it frequently happens that the tissue is not softened at once, but the vascular bundles are discolored, forming the darkened ring which is such a characteristic symptom of wilt. Unfortunately, in such cases, there are no means of telling the two diseases apart except by culturing the organisms. In some of the western irrigated sections, the affected tissue of the tuber may not

FIG. 109.—Black-leg rot on potato tubers.

soften, but lose its moisture and become leathery in texture. When the ends of the tubers are wilted and shrunken they may have the appearance of jellyend rot, or may give the impression that they were frozen.

Cause of black-leg.

Many names have been applied to the bacterium which causes black-leg. For some time it was considered that there were at least four or five distinct strains which were known as *Bacillus phytophthorus* (Frank) Appel, *Bacillus solanicola* Del., *Bacillus solanisaprus* Har., *Bacillus melanogenes* Peth. and Mur. and *Bacillus atrosepticus* van Hall. Later work has shown that these are probably identical, but there still is some controversy as to the name to be used. Following the rule of priority of description the pathogene should be known as *Bacillus atrosepticus* van Hall, although there are investigators who still prefer to use the name suggested by Appel. The index number of the organism is 5312-32120-2111.

So far as is known the bacillus does not live over winter in the soil or in diseased tubers and vines left in the field. The usual method of dissemination is with the seed tubers. Millions of bacteria invade the intercellular spaces of the host tissue. When the potatoes are cut, some of the organisms cling to the knife, from which they are wiped when succeeding cuts are made. In this manner one diseased tuber may be the source of inoculum for a dozen healthy ones. The young plant becomes infected through the mother seed piece.

The black-leg is more severe in low wet soils and in wet seasons. A high humidity with a temperature of about 80° F. is favorable for the bacillus. Evidently these conditions are not common in the South for the disease is most often found in the northern tier of states.

Control of black-leg.

As black-leg is relatively easy to control, the disease never need be feared. The first recommendation is that of procuring disease-free seed. If such potatoes are not available on the farm where they are to be grown, they may be bought as certified seed. Even though certified stock is obtained, the record of the seed should be acquired, and should show that no black-leg or other obnoxious diseases were present. The seed should in all cases be treated either with formaldehyde or corrosive sublimate according to the directions given for the control of Rhizoctonia. While cutting the tubers for planting, the operator should have two knives, so that one could be left in formalin to disinfect and the other used for cutting. As soon as the operator cuts into decayed tissue of any kind, he should change knives, placing the one he had been using into the disinfectant. This will eliminate the possibility of disseminating the pathogene from diseased to healthy tubers. As a fourth precaution there should be a rotation of crops, even though it is generally considered that the parasite does not live over in potato débris.

So far no resistant varieties or strains have been found. Although Green Mountains are less susceptible than Cobblers, they are not nearly enough immune to form a parent plant from which to breed desirable types.

References

Appel, Otto. Untersuchungen über die Schwarzbeinigkeit und die durch Bakterien hervorgerufene Knollerfäule der kartoffel. Arb. Biol. Abt. Land und Forstwirtschaft am Kaiserlichen Gesundheitsamte **3**: 364-432. 1903.

Artschwagger, Ernest F. Pathological anatomy of potato black leg. Jour. Agr. Research **20**: 325-330. 1920.

Cotton, A. D. and H. V. Taylor. The causes of decay in potato clamps with special reference to the season of 1918. Jour. Bd. Agr. Great Brit. Suppl. **18**: 48-60. 1919.

Delacroix, G. A new bacterial disease of the potato. Bull. Min. Agr. (France) **20**: 1013-1033. 1901.

Harrison, F. C. A bacterial rot of the potato caused by *Bacillus solanisaprus*. Centralbl. f. Bakt., etc., Ab. 2. **17**: 34-39, 120-128, 166-174, 384-398. 1906.

Jennison, H. M. Potato black leg with special reference to the etiological agent. Phytopath. **12**: 444. 1922; Ann. Mo. Bot. Gard. **10**: 1-72. 1923.

Morse, W. J. Black leg of potatoes. Me. Agr. Exp. Sta. Bull. **149**: 323-329. 1907.

Morse, W. J. Control of the black leg or black stem disease of the potato. Me. Agr. Exp. Sta. Bull. **194**: 201-228. 1911.

Morse, W. J. Studies upon the black leg disease of the potato with special reference to the relationship of the causal organisms. Jour. Agr. Research **8**: 79-126. 1917.

Murphy, P. A. The black leg disease of potatoes caused by *Bacillus solanisaprus* Har. Canada Dept. Agr. Exp. Farms. Div. of Bot. Circ. **11**: 1-8. 1916.

Pethybridge, Geo. H. and Paul A. Murphy. A bacterial disease of the potato plant in Ireland and the organism causing it. Proc. Roy. Irish Acadamy. **29**: 1-37. 1911.

Ramsey, Glenn B. Studies on the viability of the potato black leg organism. Phytopath. **9**: 285-288. 1919.

Rosenbaum, J. and G. B. Ramsey. Influence of temperature and precipitation on the black leg of potato. Jour. Agr. Research 13: 507-513. 1918.

Shapovalov, M. and H. A. Edson. Black leg tuber rot under irrigation. Jour. Agr. Research 22: 81-92. 1921.

Smith, Erwin, F. The bacterial black rot of the potato. *In* Bacterial diseases of plants, pp. 253-279. 1920.

Powdery-Scab of Potatoes

Caused by *Spongospora subterranea* (Wallr.) Johnson

Powdery-scab of potatoes evidently is indigenous to Ecuador and Peru from which place the disease was introduced into Europe and there discovered in 1841. In 1913 it was reported from Canada and from Maine and in 1914 it was found in New York. Since then it has been observed in Florida, South Carolina, Minnesota, Oregon, Washington, and Pennsylvania. It is present to some extent in nearly all of the other potato-producing countries.

The disease usually is not of great economic importance. In cold wet soil it may become severe enough to cause considerable damage both by scabbing and by means of the canker form. Furthermore, it has been shown that when such scabby tubers are placed in storage, they run a much greater chance of becoming infected by different storage-rot organisms.

Powdery-scab is the most commonly applied name of the disease, but it is also known as canker, corky-scab, spongy-scab, spongosporous-scab and corky-end. The organism ordinarily attacks only potatoes, but has also been made to infect the tomato, *Solanum ciliatum*, *S. marginatum*, *S. mamosum*, *S. hæmotocladum*, and *S. Warscewiczii*.

Symptoms.

The lesions first appear on the young tubers when the latter are less than an inch in diameter, and are made up of small slightly raised pimples, which are of varying shades of brown on the surface and faintly purplish below the epidermis (Fig. 110). Some of these lesions may remain as closed pimples until the tuber is grown, or infection may take place later in the season, so that when the tubers are dug they are covered with the small raised places about the margin of which are ring-like depressions suggesting the name skin-spot. Shapovalov believes that the skin-spot is one stage of powdery-

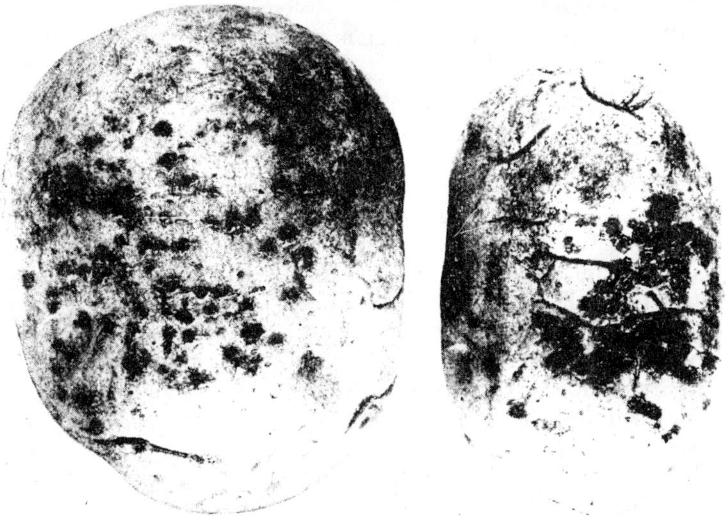

FIG. 110.—Powdery-scab on potato tubers.

scab, but European investigators deny this emphatically. According to them the organisms which are supposed to produce the pimples are *Oospora pustulans* and possibly *Phoma eupyrena*, and *Spicaria solani*.

The epidermis does not keep up in its growth with the increasing size of the pimple, thus finally breaking and curling backwards around the spot. The ruptured pustules, which occur either singly or in clumps on the tuber, are almost circular or oval in shape unless several pustules have coalesced when the lesion is elongated. The pustule is filled with a brown powder made up of the fruit-bodies of the pathogene, and the decomposed parts of the host tissue.

Under favorable conditions for the disease, the parasite may bar the laying down of a protective cork layer by the potato, and invade the deeper tissue. This causes the canker form of the disease, which may include a large part of the tuber in its course of decay.

The organism also attacks the stolons and roots producing small galls. An extremely large number of such galls are often found on a single plant.

Cause.

The cause of powdery-scab is a slime mold or Myxomycete, known as *Spongospora subterranea.* It lives over winter on the stored tubers and will remain alive in the soil for many years. Large masses of the spores are united into a hollow somewhat spherical spore-ball. Each spore in the presence of a potato and sufficient moisture germinates producing a swarm-spore. As many of the swarm-spores from one spore-ball are grouped closely together in the soil, they unite forming a plasmodium. The plasmodium dissolves the cuticles of the potato, and passes into the tissue between and through the cells. It cannot pass directly through the healthy cell-wall, but is able to kill the cell by partly surrounding it. By eating away the tissue it forms the pustules. When the conditions are favorable the plasmodia unite forming much larger bodies, which are virulent enough to produce the deep canker. When the slime mold has reached maturity, or when the food supply diminishes the plasmodia change into the hollow spore-balls, which remain in the tuber or drop out into the soil where they are ready to attack another crop of potatoes.

The organism grows fastest in wet cool soil. When infested tubers were planted even in southern New York or in Pennsylvania, the summer was warm enough entirely to prevent the infection of the new crop. Lime also favors the disease, and mechanical injury tends to increase infection.

Control.

When powdery-scab was first found in the United States, a quarantine was established in an attempt to check the dissemination of the slime-mold. It was determined later by experimentation that the disease thrives only in low soggy ground when the temperature is near 60° F., so that a quarantine is not necessary. Careful drainage and cultivation are sufficient to hold the disease in check except in certain types of soil. Where such soils exist an application of inoculated sulfur, as recommended for common scab, may be made. Additional precautions are seed treatment as for Rhizoctonia, care in not using contaminated manure, and long crop rotations.

References

Horne, A. S. On tumor and canker in potato. Jour. Roy. Hort. Soc. **37**: 362-389. 1911.

Kunkel, L. O. A contribution to the life history of *Spongospora subterranea*. Jour. Agr. Research **4**: 265-278. 1915.

Johnson, T. *Spongospora solani* Brunch. (Corky scab). Economic Proc. Roy. Dublin Soc. **1**: 453-464. 1908.

Johnson T. Further observations on powdery potato scab. *Spongospora subterranea* (Wallr.). Scientific Proc. Roy. Dublin Soc. n. s. **12**: 165-174. 1909.

Melhus, I. E. Powdery Scab (*Spongospora subterranea*) of potatoes. U. S. Dept. Agr. Bull. 82: 1-16. 1914.

Melhus, I. E. and J. Rosenbaum. Spongospora on the roots of the potato and on seven other new hosts. Phytopath. **6**: 108. 1916.

Melhus, I. E. J. Rosenbaum, and E. S. Shultz. *Spongospora subterranea* and *Phoma tuberosa* on the Irish potato. Jour. Agr. Research **7**: 213-254. 1916.

Millard, W. A. and Sydney Burr. The causative organism of skin-spot of potatoes. Roy. Bot. Gard. Kew. Misc. Inf. Bull. 8: 273-287. 1923.

Morse, W. J. Powdery scab of potatoes. Me. Agr. Exp. Sta. Bull. 227: 89-104. 1914.

Osborn, T. G. B. *Spongospora subterranea* (Wallroth) Johnson. Annals Botany **25**: 327, 341. 1911.

Pethybridge G. H. On the nomenclature of the organism causing "corky" or "powdery" scab in the potato tuber. *Spongospora sub-*

terranea (Wallr.) Johnson. Jour. Roy. Hort. Soc. (London) **38**: 524-530. 1913.

Ramsey, G. B. Influence of moisture and temperature upon infection by *Spongospora subterranea*. Phytopath. **8**: 29-31. 1918.

Shapovalov, Michael. Relation of potato skin spot to powdery scab. Jour. Agr. Research **23**: 285-294. 1923.

Black-Wart of Potatoes

Caused by *Synchytrium endobioticum* (Schilb.) Perc.

Black-wart of potatoes, although not described until 1896 in Hungary and six years later in England, probably was present much before that time. It, however, could not have been of great importance before its discovery because the symptoms on the tubers are so striking that it seems impossible to be long overlooked. The disease is now prevalent in parts of Hungary, Germany, Holland, Scandinavia, Italy, Great Britain, South Africa, Canada, and the United States. It is most generally distributed and causes most loss in Great Britain. In September, 1918, the disease was discovered in gardens of the mining town, Highland, Pennsylvania. An intensive survey has since shown that it is present in other mining communities of Cambria, Clearfield, Center, Huntingdon, Armstrong, and Luzerne counties in Pennsylvania; Randolph and Tucker counties in West Virginia; and Alleghany County in Maryland. All these places represent mining towns, the back yard gardens of which are infested. Evidently the people, many of whom are of foreign birth, either planted diseased tubers which were shipped to this country for table stock before the Plant Quarantine Law went into effect in 1912, or smuggled into the country when they immigrated some of the European varieties of potatoes. Fortunately no fields where potatoes are grown commercially have become infested.

Although in Europe infection is reported on *Solanum nigrum* and *Solanum Dulcamara,* the disease has been observed in America only on the potato and the tomato.

The disease is known locally by many names, most of which are more or less descriptive of the symptoms, such as black-scab, potato rosette, cauliflower disease, canker, tumor, cancer, and yellow-wart. Wart and black-wart are quite generally used.

Symptoms.

When looking over a field of potatoes the disease ordinarily cannot be detected. It has been remarked that vines with diseased tubers may be larger and slightly darker green than the normal, but the difference is not great enough to be of value when inspections are made. Unlike in tomatoes where the warts are larger on the stem than they are on the roots, the potato stem above the surface of the soil seldom shows hypertrophy unless it lies flat on moist earth. The only indications of the presence of the trouble are warty tubers that may stick up out of the ground.

The disease affects all underground parts of the potato, including the stolons and the roots (Fig. 111). On the tuber the first symptoms appear at the eye, where instead of the normal shoot are nodule-like protuberances which may be single or arranged into a mass resembling the head of a cauliflower. The size of the tumor may vary from those that are almost microscopic to a swelling larger than the original potato. One eye only may be included or infection may take place at any number of them. In an advanced stage the tubers are wholly covered by the excrescences, having lost every semblance of potatoes. The protuberance is rusty brown with a lighter tint at its base. When exposed to the light it may take on a dirty green color.

Cause of black-wart.

The parasite belongs to one of the lower groups of fungi, and is known as *Synchytrium endobioticum.* The generic term, Chrysophlyctis, has also been used. Because of its power in totally destroying a potato crop when it once becomes established, it has been much advertised in news articles and bulletins. Luckily it does not spread so rapidly nor so easily as does the late-blight fungus. When it once has infested a soil, however, it is much more to be dreaded.

Motile spores from zoosporangia in the soil serve as inoculum for the infection of the underground plant part. They are able

FIG. 111.—Black-wart of potato.

to penetrate the host epidermis or get through the breathing pores. Starting as naked masses of protoplasm, these zoospores rapidly enlarge after they have entered the host cell, the nucleus of which they engulf. The enlarged mass divides into five or six thin-walled zoosporangia, all of which are inclosed in one sorus by a firm yellow outer and a thin colorless inner membrane. In the meanwhile the action of the parasite has stimulated the host cells into abnormal growth and division until finally the whole tuber is converted into an unsightly tumor. When the potato tissue disintegrates the large yellow fruit-bodies are set free in the soil, where if the occasion requires they may lie dormant for years. In the presence of the potato plant and enough moisture, the sorus absorbs water until it bursts, thereby freeing the large number of zoospores.

Infection does not take place in soil temperatures above 72° F.

Control of black-wart.

No practical methods have been devised by which the fungus can be eradicated, or the plant protected in the presence of the pathogene. The only recourse is that of having a strict quarantine over the infested area, growing there only varieties that are immune, and having every one cooperate fully in carrying out the spirit of the quarantine. This may work a hardship for some of the growers who are involved, but it is a slight sacrifice to make when it is considered what a calamity there would be if the disease became prevalent in the commercial potato districts.

In very recent years much emphasis has been placed on varieties that are immune to black-wart. In Great Britain and other European countries the government is constantly carrying on tests in an attempt to determine which variety may safely be planted on infested soil, and inspectors are trained to differentiate between the desirable and non-desirable varieties so that crops from the proper fields may be certified for seed stock. In America no black-wart certification is practiced but the known varieties, as well as a number of seedlings, are being tested for immunity. Some of those which have stood the test are Irish Cobbler, Early Petoskey, and others of the Cobbler group; Early Harvest and certain others of the Early Michigan group; Spaulding Rose, Northern King, and White Rose of the Rose group; Burbank of the Burbank group; Green Mountain, Norcross, Gold Coin, and others of the Green Mountain group; and Round Pink Eye and McCormack of the Peachblow group. In glancing over the above list it is disappointing not to find any of the popular Rural group, including Rural New York No. 2, Sir Walter Raleigh, Rural Russet, and others among those that are immune. This group together with the Early Rose, Netted Gem, American Giant, and other much valued varieties are susceptible. It remains to be seen whether their places can be supplied by strains bred for immunity.

Not only each grower but each consumer should be on the lookout for black-wart, and should send any suspicious looking specimens to his state pathologist or to the United States Department of Agriculture for identification. Any one desiring to become familiar with the quarantine law on the subject may get the information by addressing the Federal Horticultural Board at Washington, D. C.

References

Curtis, K. M. The life history and cytology of *Synchytrium endobioticum* (Schilb.) Perc., the cause of wart disease in potatoes. Phil. Trans. Roy. Soc. London, Ser. B **210**: 409-478. 1921.

Güssow, H. T. Outbreak of a serious potato disease in Newfoundland, potato canker (*Chrysophlyctis endobiotica* Schilb.). Canada Dept. of Agr. Central Exp. Farms, Ottawa, Bull. 63: 1-8. 1909.

Horne, A. S. On tumour and canker in potato. Jour. Roy. Hort. Soc. **37**: 362-389. 1911.

Johnson, T. *Chrysophlyctis endobiotica* Schilb. (Potato wart or black scab), and other Chytridiaceæ. Scient. Proc. Roy. Dublin Soc. n. s. **12**: 131-144. 1909.

Lyman, G. R., L. O. Kunkel, and C. R. Orton. Potato wart. U. S. Dept. Agr. Circ. 111: 1-19. 1920.

Orton, C. R. and F. D. Kern. The potato wart disease. Penn. Agr. Exp. Sta. Bull. 156: 1-16. 1919.

Orton, C. R. and Freeman Weiss. The regeneration of first generation hybrid potatoes to the wart disease. Phytopath. **11**: 306-310. 1921.

Percival, John. Potato "Wart" disease: the life history and cytology of

Synchytrium endobioticum (Schilb.) Percl. Centralbl. f. Bakt., etc., Abt. 2. **25**: 440-446. 1910.

Potter, M. C. A new potato disease. Jour. Bd. Agr. (London) **9**: 320-323. 1902.

Schilberszky, K. Ein neuer Schorf-parasit der Kartoffelknollen. Ber. Deut. Bot. Ges. **14**: 36-37. 1896.

Weiss, Freeman, C. R. Orton and R. E. Hartman. Investigations of potato wart. U. S. Dept. Agr. Bull. 1156: 1-22. 1923.

Silvery-Scurf of Potatoes

Caused by *Spondylocladium atrovirens* Harz

Silvery-scurf of potatoes, also known as scab, dry-rot or dry-scab, is probably present in all potato-growing districts but has never proved of much importance. The fungus causing the trouble was first described in 1871 in Europe, but was not observed in the United States until 1907. Since then it has been reported from nearly every state.

Symptoms.

The first symptoms are varying sized areas of a light brown discoloration on the surface of the tuber. Occasionally the lesions have the appearance of watery blisters. Later the skin is lifted somewhat so that an air space is formed, giving the surface a marked silvery appearance. The discoloration may cover only a small spot or include the whole surface of the tuber. When the disease is very severe the grower often confuses it with late blight-rot, although there is not much resemblance from the viewpoint of the trained observer. As silvery-scurf is only skin deep the disease may easily be diagnosed by cutting into the affected potato. If diseased tubers are placed in a warm moist chamber the surface of the spot becomes olive-colored, due to the fruit-bodies of the fungus. If badly affected tubers are placed in a dry atmosphere, they lose their moisture through the injured epidermis and become mummified.

In connection with the silvery appearance and shriveling there are often minute black sclerotia, not much larger than a pin point and partly embedded in the diseased epidermis, much like those described for anthracnose.

Cause.

The fungus is now known as *Spondylocladium atrovirens*, although for a short time it bore the equally long name of *Phellomyces sclerotiophorus* Frank. The life history of this pathogene is very simple. It is made up of dark septate mycelium, which in the presence of warmth and moisture produces conidiophores on the surface of the tuber. A large number of five- to six-septate conidia are borne in whorls on the conidiophores. The conidia germinate readily, sending out a germ-tube which enters the host. The fungus can live over in the soil or on the tubers in storage.

It was long considered that the small sclerotia usually present on the lesion were a stage of the fungus. According to Taubenhaus, however, they belong to an entirely different organism, which follows Spondylocladium and which he has named *Colletotrichum atramentarium.* (Berk. and Br.) Taub. The latter fungus though found on the living stems and tubers and persisting in the soil, has never proved to be of economic importance. When the diseased epidermis of the potato sloughs off, as it frequently does, the sclerotia are disseminated with it in the field or the storage-house.

Control.

No control methods are known for silvery-scurf. Treating the potatoes with cold and hot formaldehyde and corrosive sublimate have been tried, but the death point of the potato is lower than that of the fungus. Long rotations may be of some benefit.

References

Melhus, I. E. Silver scurf, a disease of the potato. U. S. Dept. Agr. Bur. Plant Ind. Circ. 127: 15-24. 1913.

Pethybridge, J. H. "Silver Scurf" and tuber disinfection. Jour. Dept. Agr. and Tech. Inst. Ireland 19: 282-286. 1919.

Schultz, Eugene S. Silver scurf of the Irish potato caused by *Spondylocladium atrovirens*. Jour. Agr. Research **6**: 339-350. 1916.

Taubenhaus, J. J. A contribution to our knowledge of silver scurf (*Spondylocladium atrovirens* Harz) of the white potato. Memoir N. Y. Bot. Gard. **6**: 549-560. 1916.

Bacterial Brown-Rot of Potatoes

Caused by *Bacterium solanacearum* EFS.

Although there were many previous reports of a bacterial disease of potatoes, there were no statements until Smith isolated the causal organism in 1895, which would prove that the trouble was the one now known as brown-rot. Since then it has been found in all the countries where the temperature is not too low, extending northward in the United States to the southern part of New England, and the lower Hudson Valley of New York. The bacterium produces not only a destructive potato disease, but also causes great losses by a wilt or blight of such important crops as tobacco, tomato, peanut, eggplant, pepper, and soybean, as well as appearing on geranium, bean, and many other cultivated and wild plants widely separated botanically. In America the disease is generally recognized as brown-rot, but in India where it also does much damage, the trouble is known as bacterial ring disease.

Symptoms.

The first apparent symptom is the slight wilting of the vine, which may again become turgid during the night, followed on the second day by a more pronounced collapsing, so that the plant can never recover its erectness. When such a vine is examined it will be found water-soaked and flabby. The vascular bundles are colored brown and may be seen through the tissue, especially in the petioles and leaves, and sometimes also in the stem. The affected plant soon dies. The tubers are infected through the stolon, and at first show a brown discoloration of the vascular ring, followed by the spreading of the disease outward until the surface is reached. The invaded water-soaked tissue while still whole is not shrunken, but gradually becomes soft and mushy. In extreme cases, large cavities lined by softened ill-smelling tissue are found in the tubers, the cavities sometimes involving almost the whole potato. Between the cavity and the healthy tissue there is a water-soaked zone. Unlike in black-leg, a disease that might be confused with brown-rot, the stem is not thoroughly blackened, and the roots of a badly diseased plant may be left with their normal healthy appearance.

Cause.

The pathogene is a bacterium, motile by means of a single polar flagellum, dirty white in culture, liquefies gelatine, and known as *Bacterium solanacearum.* The organism lives over winter in slightly affected tubers, in potato refuse left in the field, and on the other crop and weed hosts. Infection takes place through wounds in the roots, or through parts of the plant above ground that have been injured by insects or otherwise. The bacteria invade the intercellular spaces and soften the tissue by dissolving the middle lamellæ.

The organism is partial to wet weather and high temperatures, thriving best at about 95° F. It seems to be most prevalent in the eastern United States on washed sand, which contains coarse gravel.

Control.

Potatoes grown far enough in the North to avoid infection should be used for seed. These should be planted on soil that has been cropped for at least several years with immune hosts such as cereals and grasses, and in fields where care has been taken to eliminate the susceptible weeds. If local seed must be used it should be treated as suggested for the control of Rhizoctonia. If the farm offers a variation in the types of soil, it is well to avoid coarse sand or gravel and select for the potato field a loam with plenty of humus.

REFERENCES

Fulton, H. R. and E. E. Stanford. Two wild hosts of Bacterium solanacearum. Phytopath. 6: 108, 109. 1916.

Smith, E. F. Brown rot of Solanaceæ. *In* Bacteria in relation to plant diseases 3: 174-219. 1914.

Smith, E. F. A bacterial disease of the tomato, eggplant, and Irish potato. U. S. Dept. Agr. Div. of Veg. Phys. and Pathology Bull. 12: 1-26. 1896.

Smith, E. F. The brown rot of Solanaceæ. *In* Bacterial diseases of plants, pp. 177-291. 1920.

Stanford, E. E. and F. A. Wolf. Studies on Bacterium solanacearum. Phytopath. 7: 155-165. 1917.

ROOT-KNOT OF POTATO

Caused by *Heterodera radicicola* (Greef) Mueller

(See Root-Knot of Tomato, page 550, for a detailed discussion)

The root-knot of potato caused by eel-worm or nematodes is not present in the most important potato states, but is found in the South and West.

The tuber which is affected may or may not show signs on the epidermis of the trouble. Usually a large number of small pimples are present on the surface of the potato, each pimple being surrounded by a slight depression. When the tuber is cut open there is a darkened ring in the flesh just below the pimple, and frequently the dark ring may be present even when no pimple has been produced. In the center of the discolored area, a pearly white body the size of a pin point may sometimes be observed. This is the female parasite. The outside of the tuber changes very little if any, in color, but when severely infected the potato withers until it becomes dry and hard.

Since many fields still remain uninfested, care should be taken in keeping the nematodes from contaminating the soil. Where nematodes are present, long rotations with immune crops are desirable. Nearly all of the cereals and grasses may be used for this purpose. Corn and sorghum are especially desirable because they permit clean tillage. In rare cases it may be possible to keep every vestige of plant life out of an infested field for two years, and in this manner starve the nematodes. Holding the land fallow so long, however, proves expensive when other crops may be grown. Sometimes it may prove practicable to flood the land for a few months and in this manner kill the eel-worms. In Europe fair success was obtained by alternating potatoes with barley and two catch-crops, then killing the catch-crops by spraying them heavily with a corrosive sublimate solution. Any plants which are very susceptible and can be grown cheaply may be used for the catch-crop.

Seed potatoes should always be obtained from a source known to be free from the trouble, and so far as possible planted on clean soil. When a farm once becomes heavily infested it probably is better to discontinue the growing of potatoes for a number of years, even though it means a decrease in the current income from the farm.

References

Bessey, E. A. Root-knot and its control. U. S. Dept. Agr. Bur. Plant Ind. Bull. 217: 1-82. 1911.

Doten, S B. and P. Frandsen. The potato eelworm. Nev. Agr. Exp. Sta. Bull. 76: 1-7. 1911.

Holmes, E. S. Scab and eelworm in potatoes. Jour. Dept. Agr. Victoria 8: 570-582. 1910.

Müller, H. C. and E. Molz. Catch plant methods for combating nematodes. Zentbl. Agr. Chem. 50: 262-266. 1921.

Scofield, C. S. The nematode gall worms on potatoes and other crop plants in Nevada. U. S. Dept. Agr. Bur. Plant Ind. Circ. 91: 1-15. 1912.

Zimmerman, Hans. Nematoden (Heterodera) an kartoffeln. Zeitschr. f. Pflanzenkr. 30: 139-145. 1920.

Mosaic of Potatoes

Cause not determined

The history of mosaic is like that of leaf-roll. No doubt it has been present many years but, being confused with other troubles some of which were known under the term leaf-curl or later curly-dwarf, it was not considered as a distinct disease until about 1911 in Germany. In 1912 it was first recognized in Maine. Since then it has been reported from every potato country in the world, and is so common in the United States that it is difficult to find a field in which it does not occur.

The losses due to mosaic have never been calculated accurately, but an idea can be gained of the reduction in yield, when it is known that computations made on data from many sources show that the average plant with mosaic yields two-thirds as much as does a healthy one. If there were 30 per cent of mosaic, it would mean that the actual loss was approximately 10 per cent. On Long Island where strain tests were conducted on a large scale it was found that potatoes with 2/10 per cent mosaic averaged 301 bushels an acre while those with 48 per cent yielded only 189 bushels. In another series when the strains were divided into healthy, slightly diseased, and badly diseased groups, the average yield for each group was 442, 328, and 157 bushels an acre respectively.

The literature is of such a confusing nature that it is impossible at present to give the host range of potato mosaic. Crosses have been made not only on various solanaceous plants, but it appears that cucurbits and other hosts differing widely in morphology are also susceptible. Judging from the trend of recent investigations mosaics are rather cosmospolitan in their nature. The tomato mosaic will infect the potato but the symptoms on the latter host are not the same as those associated with the disease when the inoculum comes from a diseased potato plant.

Symptoms.

Mosaic has lately been divided into several different types, each of which is supposed to be distinct, although occasionally occurring together on the same plant. The mosaic or mottling of the foliage (Figs. 112, 113) is the most characteristic symptom, and the type which exhibits this mottling most conspicuously is named mild mosaic. Lighter green or yellowish-green spots with no regular or definite outline appear promiscuously over the upper surface of the leaf. The lighter color gradually shades off into the normal green of the tissue adjacent to the spots. When the leaf is held up between the eye and the sun, the varying degrees of color density may be

FIG. 112.—Mottled potato leaves affected by mosaic.

observed. Often this mottling is not evident on plants grown in the bright sunlight or in high temperatures, but is very striking on those which have been partly shaded. For the same reason, the mosaic may not be visible on the upper leaves but may show distinctly on the shaded leaves of the same plant. In some western sections of the country, especially in some of the elevated mountain districts where the days seldom are cloudy, mottling is totally lacking even in plants which when transferred to more favorable places show unmistakable signs of the disease. Even when the mottling is present, it may be seen more clearly by shading the plant when making the examination. Inspectors have found that mosaic can be detected more easily in the early morning before there is brilliant sunlight, or after the sun has set in the evening.

FIG. 113.—Corrugated leaves produced by mosaic on potato plant at the left compared with healthy foliage.

Some plants with undoubted symptoms of mosaic have in addition to the mottling or the corrugation of the leaf a rolling of the upper leaves somewhat similar to that displayed by leaf-roll. It, however, is not related to the latter disease, although it has received the name leaf-rolling mosaic.

The plants may be dwarfed (Fig. 114), yet this is not necessarily a concomitant symptom of mottling. When the plant is dwarfed, the internodes and petioles are shortened, so that it is somewhat stockier and more bushy. When the plant is much dwarfed and has an exaggerated case of rugose leaves, it is usually known as curly-dwarf. There may be several types of this dwarfing, also, but for the grower who is concerned only with control measures they may be considered together.

In certain cases of mosaic the leaves may not be mottled, but they are much roughened or corrugated, due to the faster

FIG. 114.—Comparative size of potato plant with severe mosaic and that is healthy.

development of some spots in the leaf than in others. Often the leaves are smaller and may be cupped downward. Such a form of the disease which is named rugose mosaic is more effective in reducing the yield than is the type where mottling alone is present. In a field where the Colorado potato-beetles are plentiful, infected hills may often be over-run by the beetles while the healthy vines are scarcely molested.

There are no internal symptoms of the stem, leaves, or tubers which would differentiate diseased plants from those which are healthy. Although the yield may not be so great, the individual tubers from a diseased plant may be as large and appear as perfect as those from a normal stalk. For this reason it is impossible by bin selection alone to pick out seed-tubers that will produce a healthy crop.

Cause of mosaic.

Such factors as plant and animal pathogenes, physiological disturbances, inheritance, fertilizer, soil, weather and so forth have been suggested as the cause of mosaic. Proof is still lacking for each of them, but with the efforts of so many pathologists massed on this one subject it will not be surprising to have the problem solved reasonably soon.

The disease increases rapidly in a field. The tubers from a diseased vine are usually affected. In this manner, the inoculum is carried over from one year to the next. Furthermore, it is possible that the mosaic may live over on certain perennial plants as does the tomato mosaic. Aphids are known to be carriers of the infectious inoculum. The pink and green aphid, one of those present on potatoes, has for its winter host the wild and cultivated rose. In the spring the aphid migrates from the rose to potatoes and many other hosts. However, it has been observed that even when no aphids are present, mosaic may spread in a given field where healthy plants are grown adjacent to diseased ones. It is possible that additional research may show that the leaf-hopper or one of the many other insects on the potato may act as a carrier. The virus may be transmitted by bruising the diseased leaf, then rubbing the sap into the healthy leaf, or by grafting, or in any other way by which sap is transferred.

Environmental conditions do not have much effect on the amount or severity of mosaic. As has been shown, bright sunlight and high temperatures often mask the symptoms so that the disease cannot be detected readily, but it does not seem to lessen the virulence of the virus.

Control of mosaic.

The control measures for mosaic are the same as for leaf-roll, although the latter is somewhat easier to hold in check. It is possible by careful roguing of potato stock with 10 to 15 per cent of the disease materially to reduce the percentage. It is doubtful, however, whether stock with many more diseased hills would be worth the attempt to rogue. The trouble spreads so rapidly that only a slight percentage can be dealt with satisfactorily by the eradication in the field of the affected plants.

In addition to the isolated seed-plot, indexing the seed-tubers, roguing, and spraying, it would be well to remove all wild rose bushes that might harbor aphids, and treat the cultivated roses with nicotine in order to kill these green lice before they have the opportunity to migrate to neighboring potato fields.

An attempt has been made to kill the contagium of mosaic and leaf-roll in the tubers by treating with hot water or with dry heat. This has been unsuccessful because the tubers withstand less heat than does the virus.

REFERENCES

Blodgett, F. M. and Karl Fernow. Testing seed potatoes for mosaic and leaf roll. Phytopath. 11: 58, 59. 1921.

Murphy, P. A. Investigation of potato diseases. Canada Dept. Agr. Dom. Exp. Farms. Bull. 44: 1-86. 1921.

Patch, Edith M. Rose bushes in relation to potato culture. Me. Agr. Exp. Sta. Bull. 303: 321-344. 1921.

Quanjer, H. M. The mosaic disease of the Solanaceæ, its relation to the phloem-necrosis and its effect upon potato culture. Phytopath. 10: 35-47. 1920.

Schultz, E. S. Donald Folsom, Frank Merrill Hildebrandt, and Lon A. Hawkins. Investigations on the mosaic disease of the Irish potato. Jour. Agr. Research 17: 247-273. 1919.

Schultz, E. S. and Donald Folsom. Transmission of the mosaic disease of Irish potatoes. Jour. Agr. Research 19: 315-337. 1920.

Schultz, E. S. and Donald Folsom. Transmission, variation, and control of certain degeneration diseases of Irish potatoes. Jour. Agr. Research 25: 43-118. 1923.

Leaf-Roll of Potatoes

Cause undetermined

Even though the name for this trouble has been suggested comparatively recently, leaf-roll is evidently one of the oldest diseases with which potato-growers have had to contend. As far back as the history of potato diseases can be traced in literature, mention is made of certain curly leaf. Those early horticulturists confused the term with two or three possible troubles, so that the present day pathologist is not certain just what diseases were prevalent so soon after the potato was introduced into Europe. It is probable, however, that leaf-roll was one of the maladies. It is now present wherever potatoes are grown, and of such importance that the recent years have been marked by an avalanche of literature on leaf-roll and other related degeneration diseases. Even with this tremendous amount of writing, there are still many points which need further investigation.

The disease, so far as is known, attacks only potatoes. The loss on this host is equal to that caused by any of the other more important diseases. In fields where no attempt has been made to obtain healthy seed, it is not surprising to find from 50 to 90 per cent of the hills affected. A badly diseased hill may produce no marketable potatoes, but taking all stages of severity into consideration it has been calculated that the average diseased hill yields about one-third of that of a normal hill. Thus 30 per cent of infection would mean a loss of approximately 20 per cent of the crop.

Symptoms.

The first symptom to attract attention is the rolling of the leaves upward from the edges and parallel to the midrib (Fig. 115), the amount of rolling depending on the variety of potato and also on the position of the leaf on the plant. The lower leaves show this symptom in a more pronounced manner than do those nearer the top. The rolling may be only an almost imperceptible upward turn of the leaf edge, or in an aggravated form the leaves may be rolled tightly. The tips of the leaves often are tinted with shades of red or purple, the colored areas later dying leaving the plant marked with tip-burn. Plants affected with leaf-roll may stand stiffly erect, hold their leaves close to the stem, and often are more or less dwarfed (Fig. 116). In some varieties of potatoes and in certain environments, or possibly as different types of the disease, the plants are conspicuously procumbent, with more or less rolling of the foliage. In either the erect (Fig. 117) or the reclining type of leaf-roll, the plants are inclined to age prematurely and die.

FIG. 115.—Severe leaf-roll on young potato plant.

There are other influences which tend to imitate the symptoms of this disease. The lowest leaves of a healthy plant when next to the overheated soil in midsummer will tend to curl upward. The wilt disease or drought will also cause a curling of the foliage as well as yellowing of the plant. The leaves, however, in these cases are soft and flabby, while in leaf-roll they are stiff and crisp, so that they crackle when pressed between the thumb and finger.

There are one or two fairly diagnostic symptoms under ground. When the plant is dug up carefully, leaving the tubers attached to the stolons, it is evident that the number of tubers is much less than that of a normal hill, and that those tubers which are present are often almost if not entirely sessile, being borne tightly against the tap-root. It is very common to find just two tubers, the larger one being near the surface of the ground on one side of the root, and the smaller one being

FIG. 116.—Relative size of a potato plant with a severe case of leaf-roll as compared with a healthy plant.

somewhat deeper in the soil, and on the opposite side. The reason for the number, size and arrangement of these tubers is that the sugar which is manufactured in the leaves cannot be translocated for tuber formation. When the sessile tubers are cut they may also show marked crispness or glassiness. A familiar sight when digging up the diseased hill is the persistent seed piece, which may remain unrotted until harvest time. This seed piece on being cut will be found hard and glassy.

The crispness of the tuber is not always due to leaf-roll. Chilling, yellow dwarf, and other injuries or diseases may bring about similar symptoms.

If the seed-tubers are placed in the light early in the spring, giving the healthy potatoes a chance to grow short thick normal sprouts, tubers with leaf-roll will often put out thin spindling shoots, presenting a means by which the healthy and diseased tubers may be separated. As there are many other causes for weakened sprouts, this symptom alone cannot be used to determine the presence of leaf-roll.

Fig. 117.—Leaf-roll of potato showing the upright growth of the branches and the rolling of the leaves.

In Maine and Vermont net necrosis of the tuber has occasionally been observed as an associated symptom of the disease. Thousands of affected potatoes in New York have been examined for any internal discoloration, but none was ever found. It is possible that certain environmental conditions govern this necrosis, permitting it to appear in certain localities and not in others. Generally there is no internal discoloration either in the plant or in the tuber to distinguish it from one which is healthy. Under the microscope cross-sections of the stem, particularly of the upper nodes or of the petiole, will show thickening of some of the cell-walls by cutinization, and a yellow crystalline deposit filling some of the cell cavities. It is mostly the outer phloem which is affected, although other tissues may exhibit pathological changes. Because of these internal symptoms, the term phloem necrosis is sometimes used as an appellation of the disease.

Cause of leaf-roll.

Nearly every causal factor imaginable from that of divine displeasure to trypanosomes has been suggested as the one producing leaf-roll. The cause has not yet been determined, although investigators have known for some time that the virus is infectious and that the disease can be transmitted by the transferring of sap from affected to healthy plants. The disease does not live over in the soil, unless tubers that might produce volunteer plants are left in the field at digging time. So far as is now known, aphids are the carriers of the virus, although it is probable that they are not the only carriers. When a tuber from a diseased vine is planted the following spring, it will in nearly every case produce affected progeny. During the growing season aphids, and possibly other insects, disseminate the inoculum to plants from healthy seed, so that the percentage of leaf-roll increases very rapidly. The first year that the plant is infected the symptoms do not show at all, or if they do it is merely by a curling and a slight reduction in size of the leaves at the top of the stem. This is known as primary infection. As it is so difficult to recognize, it is impossible to select healthy hills from a field in which there is much leaf-roll. If tubers from a vine with primary infection are planted, the progeny have secondary infection or an advanced stage of the disease. Each succeeding year the plants are less vigorous and lower yielding until in three or four years they are eliminated entirely.

The climatic conditions favorable for the production of the disease have not been determined, yet it is known that leaf-roll is more prevalent in some areas than in others. For example, on Long Island and Bermuda the Bliss Triumphs are very susceptible to the trouble but in northeastern Maine they are relatively free from the disease. The Rural type of potato when grown in southern Ontario or in the section of

New York adjoining Lake Ontario has more leaf-roll than when planted farther away from the lake. It may be that in each of the above cases the presence of the disease depends only on the number of aphids present, and that the climate has no direct influence on the disease apart from its carriers. No doubt when the cause of leaf-roll is known, the ecological factors can be determined with a greater degree of assurance.

Control of leaf-roll.

The virus being constantly within the plant, seed treatment is of no direct benefit, but is of value in helping to eliminate other diseases that might be confused with leaf-roll. Spraying with bordeaux mixture is advised for two reasons; first, as protection against late-blight, early-blight, flea-beetles and tipburn, which often mask the symptoms of the disease, and secondly, with nicotine for the eradication of aphids, thereby checking the dissemination of the virus. Sprouting the seed before it is planted and discarding all tubers that show symptoms of spindling sprout, will aid in reducing the number of infected parent tubers.

The all-important means of control is the use of healthy seed. The man who is growing potatoes for table stock can eliminate most of his loss from leaf-roll by purchasing each year enough certified potatoes to plant his entire field, or at least enough for a propagating plot, from which he can secure his seed for the following year. The one important precaution is that such healthy seed must not be planted near other seed that is contaminated, for the disease is very infectious.

The average seed-grower can follow the same plan, but the one who is improving a potato strain must use more exact methods which involve an isolated seed-plot and indexing of the mother seed-tubers.

Large true to type tubers are selected for indexing from as healthy potato stock as can be obtained. As many small clay flower-pots as there are tubers to be indexed are filled with good potato soil and numbered consecutively. A month or more before planting time the selected tubers are numbered with india ink in the same order as are the flower-pots, after which one eye of each tuber is cut off, care being taken to take a liberal slice, and planted in the pot having the corresponding numeral. The pots are then placed in a suitable greenhouse or in a well-lighted adequately warmed room, while the cut surfaces of the remaining parts of the tubers are sprinkled with sulfur to keep them from shriveling and the tubers stored properly. When the eyes which were planted in the pots are up four to six inches, leaf-roll or any of the other degeneration diseases will begin to show. The tubers with the numerals corresponding to those on the pots in which are found the diseased plants are discarded. As the moisture, temperature, and light can be controlled indoors, the symptoms of leaf-roll and mosaic will show even more strikingly in potted plants than they would in the field, thus permitting the recognition of nearly every tuber which is diseased. These potatoes, the eyes of which produce healthy plants in the pots, are cut into small seed pieces so that they will cover a large area, and planted in a seed-plot isolated more than a quarter of a mile from any potato field. The plot is cultivated carefully, sprayed and rogued. The best hills are used for the seed-plot of next season, while those remaining may be planted in the field.

Indexing is a rather complicated process, but has proved its worth for starting a clean strain from which good seed can be propagated. An attempt was made to simplify it by testing only one tuber in a hill but it was easily proved that some tubers of a plant might be healthy and others diseased, thus vitiating the results of any hill indexing. If a dozen intelligent growers in each state practice indexing in connection with isolated rogued propagating plats, they can supply healthy high yielding strains for nearly every seed-grower in that state.

It must be borne in mind that no matter how good the potatoes may be, they need constant attention or they will begin to deteriorate, due to traces of disease that are found even in the most elaborately bred strains.

References

Appel, Otto and Otto Schlumberger. Die Blattrollkrankheit und unsere Kartoffelernten. Arb. d. Deut. Landw. Gesell. **190**: 1-102. 1911.

Artschwager, Ernst F. Histological studies on potato leaf roll. Jour. Agr. Research **15**: 559--570. 1918.

Botjes, J. G. Oortwijn. De bladrolziekte van de aardappelplant, pp. 1-136. Printed by H. Veenman, Wägeningen, 1920.

Murphy, P. A. and E. J. Wortley. Relation of climate to the development and control of leaf-roll of potato. Phytopath. **10**: 407-414. 1920.

Orton, W. A. Potato wilt, leaf-roll, and related diseases. U. S. Dept. Agr. Bull. 64: 1-48. 1914.

Perry, F. R. Methods in the improvement of seed potatoes. Proc. Potato Assoc. Amer. 9: 64-68. 1922.

Quanjer, H. M., J. C. Dorst, M. D. Dijt, et A. W. v. d. Haar. De mozaiekziekte van de solanaceeën hare verwantschap met de phloeemnecrose en hare beteekenis voor de aardappelcultuur. Mededeelingen van de Landbouwhoogeschool deel XVII. H. Veenman, Wageningen, pp. 1-74. 1919.

Schultz, E. S. and Donald Folsom. Leaf-roll, net necrosis, and spindling sprout of the Irish potato. Jour. Agr. Research **21**: 47-80. 1921.

Yellow-Dwarf of Potatoes

Cause not determined

Yellow-dwarf of potatoes was first observed in Clinton County, New York, in 1917. The grower stated that it had been present on his farm fully ten years before that date. In 1922 it was observed in nearly every county of the state and was also reported from New Jersey, Pennsylvania, and Vermont. In most cases the amount of disease present in a field is only from a trace to 2 or 3 per cent, but on a few farms from 20 to 50 per cent of infection has been found, showing that the disease can become very destructive. Judging from the limited available knowledge of the disease, it is potentially the most malignant trouble known on potatoes, and more to be feared even than black-wart. Of course, a further study of the subject may reveal an easy means of control.

Symptoms.

The most marked characters as seen in the field are the dwarfed condition of the plant and yellow color of the foliage. The stalks are shorter than are those of healthy plants, but have about the same diameter except toward the top where the stem thickens, giving affected plants a stockier appearance. The growing apex dies early and later the axillary buds of the upper part are also killed. Meanwhile the stalk becomes yellowish-green throughout. A longitudinal section of the entire length of a stalk reveals at the upper nodes the presence of rusty colored specks in the pith and cortex. Occasionally the specks may even be found in the tissue of the internodes. The amount of discoloration depends on the severity of the disease or the lateness of the stage in which it is found. The leaves may have a tendency to roll, or show an extreme case of rugosity. This condition might be mistaken for a type known as curly-dwarf were it not for the specks in the nodes of the stem and the symptoms present in the tubers, which the plant bears

FIG. 118.—Knobbiness of potatoes caused by yellow-dwarf.

prematurely before its death. When infection does not take place until the plants are fully grown, the first symptom is such an imperceptible yellowing and rosetting of one of the terminal

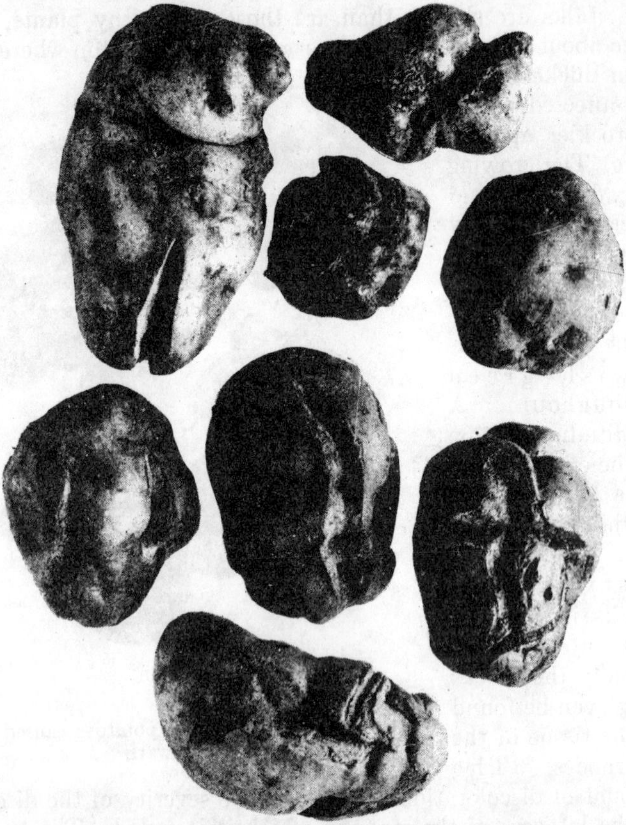

FIG. 119.—Cracking of potato tubers caused by yellow-dwarf.

shoots that only the practiced eye can detect it. As the yellowing becomes more pronounced, the disease spreads downward until within a few weeks the entire hill is dead.

The tuber, in common with those of leaf-roll, are small, few in number, often sessile, and quite brittle when cut. Further than this they frequently are irregular in shape, may be knobby (Fig. 118), and often have prominent lenticels. The most important symptoms are the cracked surface, the cracks running parallel with the long axis of the tuber, and the brown specks on the inside of the potato (Figs. 119, 120). The specks, which seldom occur in the vascular bundles, are few and scattered in some tubers and numerous or massed in others, but

FIG. 120.—Internal brown spotting of a potato tuber caused by yellow-dwarf.

never appearing in streaks or rings. The larger part of the discoloration is usually near the eye end of the tuber. Not all the potatoes in the same hill will show the disease, and occasionally a plant may be found which has no apparent symptoms above ground, but the tubers begin to show gray or light brown spots. It is not uncommon to find at harvest time the affected potatoes beginning to rot because of some saprophyte following the yellow-dwarf. Rarely will a badly diseased tuber escape decay during the winter in storage. The infected tubers which have no apparent lesions may be abnormally large in size, and when planted do not decay but persist as hard brittle seed pieces until fall.

Cause of yellow-dwarf.

The cause is yet undetermined. There are certain evidences that an organism will be found associated with it, and that it does not belong to the degeneration types of trouble. It has been shown conclusively that the inoculum is carried in the seed, and judging from observation the inoculum remains in the soil for at least four or five years, if not indefinitely. At first it was thought that the disease by self-elimination would never become serious, but it is now known that tubers from plants infected late in the season will apparently be healthy, but when planted will result in diseased progeny. The symptoms will continue to appear throughout the entire summer, so that roguing, as for leaf-roll, will not eliminate all the affected hills. The manner of transmission, the point of infection, and many other important phases of this threatening disease have not been made clear. Investigational work is now in progress. So far as observations have been made, all varieties are about equally susceptible.

Control of yellow-dwarf.

Control measures for an infested field are not known. Many farms are still uninfested, and in these cases the greatest caution should be practiced in buying seed stock. No potatoes should be purchased until they have been proved free from the disease, not even a trace of the trouble being tolerated. When it is once introduced on a farm it evidently increases slowly, until the growing of potatoes becomes unprofitable. There are already a certain number of seed-growers whose potatoes can no longer qualify for certification, and this means a great loss to them.

If inadvertently a few diseased tubers are planted, the hills, as soon as they are recognized, should be removed, and the spot disinfected with formaldehyde (diluted 1-50) as a precaution, even though the success of this disinfection has not yet been demonstrated by carefully planned experiments. The field should not be used again for potatoes for five or ten years.

Reference

Barrus, M. F. and Charles Chupp. Yellow dwarf of potatoes. Phytopath. **12:** 123-132. 1922.

Other Degeneration Diseases of Potatoes

Aside from the various known stages of leaf-roll and mosaic, certain other degeneration troubles of potatoes have been described, which may or may not be connected in some manner with the above two diseases. The exact relationship will probably not be worked out until the cause of these types of diseases is determined definitely. Until then it seems proper to treat each one separately.

Streak.

The first symptoms of streak appear on the upper side of the potato leaves, and bear a close resemblance to the beginning spots of early-blight. The spots elongate by following the veins and invade slightly the adjoining tissue. The large lesions appear as dark narrow streaks, which may be seen on both sides of the leaf, and also extending up and down on the petioles and stems. Such infected stems are brittle, so that they may be snapped off easily with the thumb and finger. A badly diseased leaflet may show a network of the dark brown lines. Later the leaf dies but remains hanging to the plant. The death of the entire vine results prematurely. The nature of this disease is so severe that self elimination of the affected hills keep the percentage of disease very low in any one field.

Marginal leaf-roll or spindling-tuber.

This disease was first differentiated from other troubles in Holland within the last five years. The association of the abnormally long tubers with the symptoms was not observed until it was pointed out by later workers. In the reduction of yield, in its transmission, and in the erectness of the affected stalk, it resembles true leaf-roll, but differs in not having shortened stolons nor extreme rolling of the leaves nor phloem necrosis. On plants affected with marginal leaf-roll, the leaves frequently are rolled upward at the margins which are often crinkled, giving the leaf such a characteristic appearance that the disease has been named spoon-shaped leaf-roll. The terminal leaflet of each leaf may be twisted and is then generally covered by the two next leaflets, which in turn are distorted in such a way as to give the whole leaf the appearance of being more pointed than normal. The leaves are often smaller and darker green than are those of the healthy plant. The tubers are elongated, with more numerous and more conspicuous eyes and a smoother skin than are found on an unaffected potato. The disease probably is present in all varieties of potatoes in the United States. It is transmitted through the tuber or by aphids. When this disease and leaf-rolling mosaic attack a plant at the same time, mottled curly-dwarf results.

Unmottled curly-dwarf.

This disease is distinct from the above mentioned curly-dwarf. The unmottled curly-dwarf does not seem to be connected with any other disease. The pronounced dwarfing of the affected plant is the most noticeable symptom. The leaves are corrugated and may be slightly rolled but are not mottled. The tubers are few in number and often cracked. The plant dies prematurely.

Crinkle.

Although crinkle of potatoes has been definitely reported only from Holland, England, France, and Canada, the same type of injury is present in the United States. It is supposed to affect potatoes, tomatoes, and other solanaceous plants. The most marked symptom is the curling downward of the margins and tips of the leaves. The plants resemble mosaic somewhat in being dwarfed and having corrugated leaves. In some varieties there are rusty-brown spots beginning near the tips of the leaves, which become very brittle; they yellow prematurely and the lower leaves which sometimes resemble true leaf-roll may drop off easily when touched. The whole plant dies before normal maturity. The disease can readily be transmitted from a diseased to a healthy plant by means of grafting. No report has been made on aphid transmission. It probably is one phase of mosaic.

Leaf-drop.

This disease has been described by Murphy in Canada who considered it as distinct from mottled curly-dwarf. The affected plants grow stiffly erect usually with only one shoot, and often with extreme dwarfing. The stem is poorly covered with foliage, which may be puckered and curved downward or normal in appearance. A constant characteristic, which separates it from other types of injury, is that the leaves die and fall off one after another from the ground upwards, leaving in the latter part of the season a small bare stalk topped by a cluster of yellow leaves. Streaks may be present in the leaf-petioles and stems, and occasionally an almost imperceptible browning of the vascular ring of the tuber may be observed. When the plant dies the stem may become so limp that it falls over. Leaf-drop can probably be classed as a symptom of rugose mosaic.

Russet-dwarf.

Hungerford of Idaho describes still another degeneration disease of potatoes, which he calls russet-dwarf. It was first brought to his attention in 1919, when it caused considerable damage in a number of fields of Idaho Rural potatoes. The affected plants are dwarfed and have a sickly yellow color. The most conspicuous symptom, however, is the rusty appearance of the leaves, especially the lower ones. On close examination it will be found that the veins of the lower leaf surface are water-soaked, later turning brown. Leaves so affected die prematurely and fall off from the ground upward as in the case of leaf-drop. Similarly also dark brown streaks may be present on the diseased petioles and stems of the lower part of the plant. In many respects this disease is comparable to crinkle and leaf-drop, the slight variation of symptoms possibly being the result of differences in the environmental conditions under which the specimens were grown. It was shown experimentally that the disease was transmitted by means of tubers, and that it spread from one plant to another in the field.

Giant hills.

Nixon of Pennsylvania has called attention to the large number of hills in some strains of seed potatoes, which stay green late in the fall, even withstanding the first lighter frosts. Such plants have extremely large coarse tops with small upper leaves, the edges of the leaflets of which are rolled upward slightly. He has named them "giant hills." They can be recognized most easily when the normal plants have died or are frosted. When the tubers from such a hill are dug, they also are found to be large and coarse, often with very irregular shape. When such tubers are planted in the northern states it has been proved that the progeny will also be giant hills. It has been suggested that farther south, owing to the higher temperature, giant-hill tubers will produce leaf-roll plants in a large percentage of cases. This has not yet been proved by using a strain of known origin, and growing them for several generations while isolated from true leaf-roll plants. Investigational work probably will make this point clear. It has been demonstrated that roguing out the diseased hills will quickly reduce the percentage of giant hills, not more than two seasons being necessary to eliminate almost perfectly these abnormally large vines.

References

Fernow, Karl H. Spindling tuber or marginal leaf-roll. Phytopath. **13:** 40. 1923.

Folsom, D. Potato spindle-tuber. Me. Agr. Exp. Sta. Bull. 312: 21-44. 1923.

Hungerford, Chas. W. Leaf-roll, mosaic, and certain other related diseases in Idaho. Phytopath. 12: 133-139. 1922.

Martin, W. H. "Spindle Tuber," A new potato trouble, N. J. St. Potato Assoc. Hints to Potato Growers. 3: No. 8. 1922.

Murphy, P. A. New or little known diseases of potatoes which cause "running out" of seed. Phytopath. 10: 316-317. 1920.

Murphy, P. A. Investigation of potato diseases. Canada Dept. Agr. Dominion Exp. Farms Bull. ser. 2. 44: 1-68. 1921.

Orton, W. A. Streak disease of potato. Phytopath. 10: 97-100. 1920.

Quanjer, H. M. New work on leaf-curl and allied diseases in Holland. Report of the International Potato Conference (London), pp. 127-145. 1921.

Schultz, E. S. and Donald Folsom. Spindling tuber and other degeneration diseases of Irish potatoes. Phytopath. 13: 40. 1923.

Schultz, E. S. and Donald Folsom. Transmission, variation, and control of certain degeneration diseases of Irish potatoes. Jour. Agr. Research 25: 43-118. 1923.

Wilt of Potato

Caused by *Fusarium oxysporum* Schlecht

The Fusarium wilt of potatoes, though not a serious disease in Europe and Australia, is generally present in the United States where the loss incurred is sometimes very large. It was first observed by Stewart of New York in the early nineties. After the careful description by Smith and Swingle in 1904, it was recognized in many states as well as abroad. In Europe a trouble of the potato known as leaf-curl had appeared in literature even as early as the last quarter of the eighteenth century. After 1904 it was possible to show that the term leaf-curl included several distinct diseases among which were wilt, leaf-roll, curly-dwarf, and probably others. This disentangling of the disease symptoms and causes was of prime importance in the potato industry, for it really was the beginning of intensive study of maladies which affect the crop.

The fungus is able to infect potato, tomato, cowpea, pea, sweet-potato, and possibly other hosts.

The term, Fusarium wilt, is as suggestive of the disease as any of the numerous terms which have been applied, such as Fusarium blight, stem-end-browning, vascular mycosis, and brown-ring. The injury to the tuber is also known as dry-rot, Fusarium necrosis, vascular browning, and other similar terms.

The loss resulting from wilt has never been determined accurately, but the estimated reports sent to the Plant Disease Survey Office average from 1.6 per cent to 2.3 per cent annually. These figures represent a loss in bushels of seven to eleven millions. Individual states have reported losses as high as 10 per cent, and single fields have been observed where infection showed on nearly every hill.

Symptoms.

Unless the disease is of extreme severity, the affected plants do not show in the earlier part of the season any marked

Fig. 121.—Wilt on potato vine in the field.

changes from those that are healthy. When the vines are about a foot tall dwarfing may begin, but more often after a period of high temperature in midsummer the vines seemingly in full vigor will take on a yellowish cast. The lower leaves may die, and partly break from the stem, hanging only by a thread of the epidermis (Fig. 121). When the roots and base of the stem have become thoroughly invaded, the plant wilts, as if water in the soil were lacking. Soon after wilting, the host dies. It is not unusual to find first one stalk in a hill dying, and this followed later by the succumbing of each one of the remaining stalks in succession.

FIG. 122.—Stem-end browning of tubers caused by wilt.

One of the well-known symptoms is the browning of the vascular tissues. If, with the thumb-nail, the peeling is removed from part of the stem below the surface of the soil, the exposed tissue of the healthy vine is pearly white. The diseased vine, on the other hand, will show a marked discoloration ranging from light brown to almost black. Similarly when the affected tubers are cut open very near the stem end the discolored vascular bundle will show as a brown ring (Fig. 122), or the discoloration may appear occasionally as net-necrosis. The presence of such a brown ring or necrosis, however, cannot be taken as proof of the presence of wilt unless other symptoms support the assumption. Black-leg may cause a similar discoloration, as may also Verticillium wilt, chilling, or subjection to abnormally high temperature, while the tuber is either in the soil or in the storage-house. Affected tubers are often elongated, and instead of the normal depression where the stem is attached to the healthy potato, the tuber will end in a drawn-out sharpened point. When diseased tubers are stored for the winter, they sometimes will contract a stem-end rot or dry rot (Fig. 123). This may be the result of the wilt

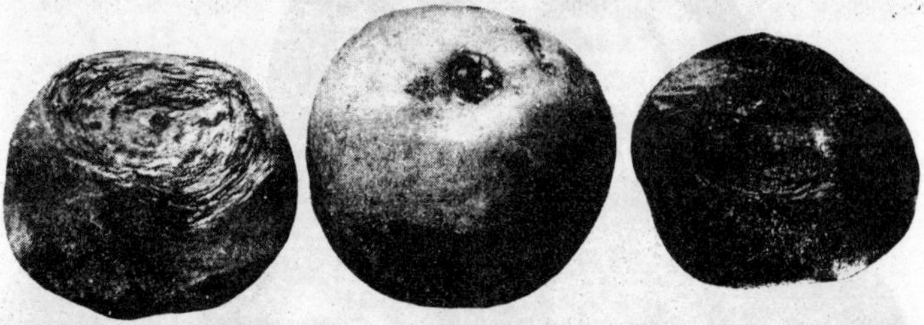

FIG. 123.—Stem-end decay on tubers caused by the wilt organism.

pathogene but may also be caused by other species of Fusaria following wilt, or entering through bruises. Frequently when tubers affected with wilt are sprouted in the light, they exhibit marked symptoms of spindling-sprout (Fig. 124).

Cause of wilt.

The fungus, which is known as *Fusarium oxysporum,* lives over winter in the soil and in diseased tubers. It is made up of long mycelial threads that invade the plant tissues, especially the water-ducts. When the stalk is killed, and the air or soil wet enough, the mycelium grows to the surface of the host where it fruits. The spores are long and sickle-shaped with varying numbers of septa, and when massed are pale salmon colored. The infection may take place from the seed-tuber into the new plant, from the soil into the seed piece, and then into the sprout, or directly from the soil into the roots. In most cases when the mother plant becomes diseased, the fungus grows downward through the stolons into the young tubers. A discoloration may or may not be formed. When the infected tubers are planted they produce a diseased hill if weather conditions are favorable for the fungus, but it is possible to plant

Fig. 124.—Spindling sprouts of potatoes compared with thick short healthy sprouts. The spindling sprouts in this case are caused by wilt.

nothing but affected seed pieces and get vines every one of which apparently are healthy. The yield, however, will be much reduced when such diseased stock is used. In a carefully conducted demonstration where healthy and diseased tubers of six potato varieties were planted in parallel adjacent rows, the yield from the diseased tubers was only half that of the healthy stock, although the progeny in no case was infected. In most cases the soil is the source of inoculum.

A similar wilt is caused by *Fusarium eumartii* (Fig. 125),

which is much more virulent than the common wilt parasite. It starts from a central point of infection and gradually spreads

FIG. 125.—Rot of potato tubers caused by *Fusarium eumartii*.

outward in an ever widening circle, killing the plants as it progresses.

High temperatures are favorable for the growth of the fungus. In 1918 when the summer was much warmer than the average, there was almost twice as much wilt reported as there was during the season before or the season following the above date. Experimental evidence shows that the optimum temperature for the Fusarium is between 77° and 86° F. Apparently the amount of soil-moisture is not so important, although it has been stated that a wet period at the time of infection followed later by a drought will increase the amount of disease. Due probably to the temperatures involved the disease is more prevalent on a southern slope than on one facing the north, and is more common in the valleys than on the high hillsides.

Control of wilt.

Wilt of potatoes will probably never be controlled fully until immune varieties are available. At present no strains are distinctly resistant; therefore, the recommendations made for the control of other soil organisms are to be followed. First in order are always long rotations in which are planted cereals and grasses. Second, clean seed is needed and may be grown successfully in some localities. The seed-plat should be isolated, and three or four times during the season inspection should be made in which all diseased hills are carefully rogued out. In pulling up the affected plants it is necessary to dig up all the tubers, which, with the tops, should be removed from the field. The best hills from the plat are saved for planting. If the seed is to be purchased it should be guaranteed by being certified, and the record of the stock obtained with the purchase. When the seed is cut it is possible to discard any tubers which show internal browning. It has often been stated that merely clipping the stem-end of the seed-tuber is sufficient. Such clipping has not proved reliable under all conditions. Unfortunately, the chief source of infection is from the soil.

REFERENCES

Goss, R. W. Relation of environment and other factors to potato wilt caused by *Fusarium oxysporum*. Neb. Agr. Exp. Sta. Res. Bull. 23: 1-84. 1923.

Edson, H. A. Vascular discoloration of Irish potato tubers. Jour. Agr. Research **20**: 277-294. 1920.

Edson, H. A. and M. Shapovalov. Temperature relations of certain potato-rot and wilt-producing fungi. Jour. Agr. Research **18**: 511-524. 1920.

Haskell, R. J. Fusarium wilt of potatoes in the Hudson River Valley, New York. Phytopath. **9**: 223-260. 1919.

Link, Geo. K. A physiological study of two strains of Fusarium in their causal relation to tuber rot and wilt of potato. Neb. Agr. Exp. Sta. Res. Bull. 9: 1-45. 1917.

McKay, M. B. Transmission of some wilt diseases in seed potatoes. Jour. Agr. Research **21**: 821-848. 1921.

Manns, Thos. F. The Fusarium blight (wilt) and dry-rot of the potato. Ohio Agr. Exp. Sta. Bull. 229: 299-337. 1911.

Smith, E. F. and Deane B. Swingle. Dry rot of potatoes due to *Fusarium oxysporum*. U. S. Dept. Agr. Bur. Plant Ind. Bull. 55: 1-64. 1904.

Wollenweber, H. W. Studies on the Fusarium problem. Phytopath. **3**: 24-50. 1913.

Verticillium Wilt of Potatoes

Caused by *Verticillium alboatrum* Reinke and Berth

(See also Verticillium Wilts of Eggplant and Okra, pages 234 and 315, and Fusarium Wilt of Potatoes, page 427)

A Verticillium wilt of potatoes has been recognized for many years. Since the symptoms are identical with those caused by Fusarium, the only means of separating the two is by culturing the fungus. It has been suggested that the work of the two parasites in the potato results in a slightly different discoloration, but this is not evident enough to be depended on for a diagnosis. As the Verticillium will grow at slightly lower temperatures than does Fusarium, the statement has been made that the farther north in latitude the larger is the percentage of wilt caused by Verticillium; yet in New York where the summer temperatures are relatively low there are only occasional farms where the latter fungus is destructive. The disease appears to be regional in its distribution. For example, it is severe in western Oregon while Fusarium is unimportant there. It is also common in England and Germany.

The life history of the fungus is similar to that of Fusarium, except that Verticillium never causes a rot of the stems or tubers. It is wholly a vascular parasite.

The suggestions made for the control of Fusarium wilt apply equally well to the Verticillium wilt.

References

Dale, Elizabeth. On the cause of "blindness" in potato tubers. Ann. of Bot. **26**: 129-131. 1912.

Edson, H. A. and M. Shapovalov. Temperature relations of certain potato-rot and wilt-producing fungi. Jour. Agr. Research **18**: 511-524. 1920.

McKay, M. B. Transmission of some wilt diseases in seed potatoes. Jour. Agr. Research **21**: 821-848. 1921.

Pethybridge, G. H. The Verticillium disease of the potato. Scient. Proc. Roy. Dublin Soc. n. s. **15**: 63-92. 1916.

Stalk Disease of Potatoes

Caused by *Sclerotinia sclerotiorum* (Lib.) Mass.

The stalk disease of potatoes was not described until 1880, although supposed to have been present for many years. It is by far the most serious in the western part of Ireland, but has also been found in Scotland, England, Germany, and Norway. A similar trouble has been reported from South Africa, Hawaii, Canada, and the southern part of the United States. Where it is at its worst the disease is second only to late-blight. Aside from the name stalk disease, it is also known as yellow-blight, sclerotium disease, white-spot, wilt, stem-rot, haughing, and hocking.

Symptoms.

The tubers are not affected. The stem is attacked near the surface of the soil, and soon becomes so weakened that the plants fall over. There is little or no yellowing of the foliage before the stem collapses. In wet weather a white weft of mycelium covers the diseased part of the stem and large thick black sclerotia may be formed among the threads of the mycelium. If the weather is dry, the fungus is not visible on the outside, but the pith of the stem disintegrates, and sclerotia are formed in this central cavity. Occasionally the trouble might be mistaken for the Botrytis disease, but in the latter case the sclerotia are smaller and flat.

Cause.

In Ireland the causal fungus is known as *Sclerotinia sclerotiorum* (Lib.) which in the United States is used as a synonym of *Sclerotinia libertiana*. In Hawaii and in the southern states a similar disease is attributed to *Sclerotium rolfsii*. Lachaine, who found the disease in New Brunswick, has given no definite name to the organism, but is of the opinion that it is *Sclerotinia libertiana* Fuckl. Botrytis often accompanies the Sclerotinia and for a long while was supposed to be the conidial stage of the latter fungus. It is now known that *Sclerotinia sclerotiorum* does not have a conidial stage. The sclerotia live over winter, or for several years, in the soil. When conditions are favorable, the sclerotium produces one to numerous thin stalks on the tips of which are borne the apothecia. The length of the stalk depends on the depth in the soil at which the sclerotia are buried. The apothecia are borne at the surface of the soil, and may be sessile or attached to a stalk two inches long. The inner surface of the apothecium is lined with closely standing asci, each with eight spores. Although the spores are far too small to be seen with the unaided eye, they are shot into the air in such masses that they may be seen as a mist-like cloud. Some of these spores fall on the old dying leaves near the base of the stem, where infection occurs. The fungus passes down the leaf-petiole into the stalk where most of the injury is caused. The fungus is disseminated by wind-blown spores, splashing of rain, and with potato debris and infested soil.

Control.

Spraying has not proved effective. Deep plowing to bury the sclerotia and long rotations with grass and cereal crops have been suggested. It was found in Ireland that when the potatoes are planted late the number of diseased plants is much reduced. In Norway the disease is not so prevalent as it was once, due apparently to the use of resistant strains. In Ireland there are some varieties as Champion II, Clifden Seedling, and Summit that are free from the disease, while others are very susceptible.

References

Cotton, A. D. Potato diseases. Jour. Bd. Agr. (Gt. Brit.) Supplement **18**: 28-48. 1919.

Edson, H. A. and M. Shapovalov. Parasitism of *Sclerotium rolfsii* on Irish potatoes. Jour. Agr. Research **23**: 41-46. 1923.

Johnson, T. Diseases of the potato and other plants in Ireland. Jour. Dept. Agr. and Tech. Inst. Ireland **3**: 8-21. 1902.

LaChaine, O. W. Sclerotial disease of the potato. Quebec Soc. for Protection of plants. Ann. Rept. **14**: 105-109. 1922.

Pethybridge, Geo. H. The "stalk" or "sclerotium" disease (*Sclerotinia sclerotiorum* Mass.). Jour. Dept. Agr. and Tech. Inst. Ireland **10**: 248-251. 1910; **11**: 429-435. 1911; **12**: 345-348. 1912; **13**: 454-457. 1913; **14**: 447-450. 1914; **15**: 506-509. 1915; **16**: 574-579. 1916.

Gray Mold-Rot of Potatoes

Caused by *Botrytis cinerea* Pers.

The fungus, *Botrytis cinerea,* is a weak parasite which attacks many species of plants that have been weakened by some cause or another. Very few complaints have been made regarding this fungus by potato-growers in the United States, but it is serious in occasional fields in Ireland, England, Holland, and Germany.

Symptoms.

The symptoms are similar to those described for the stalk diseases, excepting that instead of the white covering of mycelium at the base of the stalk, the fungus grows more sparingly and produces an ashen-gray covering of conidia and conidiophores over the lesion. The affected part is well supplied with small flat black sclerotia that do not drop easily, but cling to the dead stem through the winter.

Cause.

The fungus, *Botrytis cinerea,* can easily be recognized by its masses of ashen-gray spores at the base of the potato stem and even on the affected leaves. The spores are blown or splashed to neighboring plants, where high temperature and plenty of moisture cause them to germinate. It is probable that they can enter the host only when the latter is weakened by hot weather or poorly drained soil. The mycelium masses at different places on the stem and forms the sclerotia. These flat sclerotia do not produce apothecia as does the Sclerotinia, but in the spring grew out into numerous short hyphæ on which are borne conidia.

Control.

As the organism is able to attack so many different plants, especially vegetables, long rotations with cereals and grasses are advised. Well drained soil is necessary, and where the disease is likely to be injurious, thin planting will help the vines to dry more quickly after a rain or dew. Burning the old diseased vines will also decrease the amount of inoculum and thereby lessen the chance for infection of the new crop.

References

Cotton, A. D. Potato disease. Jour. Bd. Agr. (Gt. Brit.) Suppl. **18:** 28-48. 1919.

Kirchner, O. Die Stengel-faüle, eine neu auftretende krankheit der Kartoffeln. Württembergisches Wochenblatt für Landwirtschaft. **20:** No. 34. 1893.

McWeeney, E. J. Two sclerotia diseases of potatoes. Trans. British Mycol. Soc. **1** (1897-1901): 67, 1898.

Pethybridge, Geo. H. The Botrytis disease. Jour. Dept. Agr. and Tech. Inst. Ireland **16:** 579-584. 1916.

Violet Root-Rot of Potato

Caused by *Rhizoctonia crocorum* (Pers.) DC.

The names for this disease, violet root-rot, red root-rot, red-root, and root-felt disease, indicate the striking appearance of the disease on the host. As such bright colors could not remain undiscovered long, this trouble has been known in Europe for two hundred years. It was not observed in the United States, however, until 1890 when it was found on alfalfa, and in 1915 on potato. It is not confined to these two hosts but has been reported on forty-seven genera belonging to twenty-one families of plants, most of which are dicotyledons, and among which are such well known plants as asparagus, pine, lily, tulip, narcissus, beet, hops, dock, apple, sweet clover, bean, turnip, artichoke, sea-kale, carrot, and dandelion. It is seldom of economic importance on the potato, having been found only a few times in some of the western states, and once in New York.

Symptoms.

The disease is confined to the underground parts of the plant unless the weather is wet continuously, which may aid the fungus in growing up the stem and even out on some of the leaves. The affected part of the host is covered with a loose weft of mycelium having a beautiful reddish-purple color when young and gradually changing to almost a chocolate-brown when old (Fig. 126). The mycelium is attached to the epidermis by means of haustoria through which it feeds. Embedded in the purple mat of mycelium are numerous small sclerotia much darker than the fungous threads. If enough of the roots are attacked, the parts above ground may show yellowing or wilting.

The lesions on the tuber may begin at any point on its surface, but seem to be more often on the stem-end. They begin as very minute purplish or violet colored spots that enlarge gradually until half or even the entire tuber may be included. In addition to the violet color which at once attracts attention, the diagnosis is made more certain by the presence of many minute sclerotia, the largest of which are not more than one-sixteenth of an inch in diameter, while most of them are just visible to the unaided eye. By the aid of a hand lens, purple strands of mycelium may be seen radiating

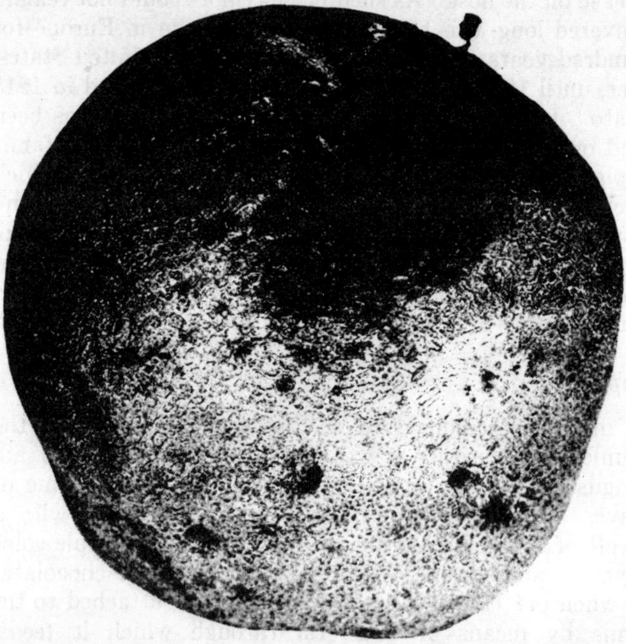

FIG. 126.—Violet rot of potato tuber. The affected part is covered with violet mycelium.

in all directions from each sclerotium and connecting it with all the neighboring sclerotia.

The dead tissue shrinks only slightly, if at all. Very soon after infection starts, the epidermis bursts irregularly with the broken corners turning upward exposing the tissue beneath, which becomes dry and powdery. The exposed surface remains ashen-gray, but immediately below the surface the tissue is purple, the depth of discoloration depending on the length and severity of infection. Beneath the purple colored tissue is a water-soaked zone which gradually merges with normal appearing cells. There is never any wet rot unless saprophytes follow the lesion caused by the Rhizoctonia.

Cause.

The pathogene is a non-spore-bearing fungus known as *Rhizoctonia crocorum* or as *Rhizoctonia violacea*. It over-winters in the form of sclerotia, which by clinging to parts of the plant or masses with loose particles of soil, serve as a means of dissemination. The mycelium grows from the sclerotia, and is formed in such profusion that it not only covers the host part but grows from one plant to the other.

Control.

The often repeated recommendations of obtaining seed from an uninfested area, seed treatment, long rotations with cereals and grasses, and destruction of susceptible weeds are also suggested for the control of violet root-rot.

References

Anonymous. Rhizoctonia diseases. Jour. Bd. Agr. London **20**: 416-419. 1913.

Duggar, B. M. *Rhizoctonia crocorum* (Pers.) DC. and *R. solani* Kühn (*Corticium vagum* B. and C.) with notes on other species. Annals Mo. Bot. Gard. **2**: 403-458. 1915.

Faris, James A. Violet root rot (Rhizoctonia crocorum DC.) in the United States. Phytopath **11**: 412-423. 1921.

Peltier, G. L. Parasitic Rhizoctonias in America. Ill. Agr. Exp. Sta. Bull. 189: 283-390. 1916.

Anthracnose of Potatoes

Caused by *Colletotrichum atramentarium* (B. and B.) Taub.

In 1909 a disease named dartrose was described in France. A similar disease was investigated later in Ireland, South

Africa, and Australia, where it was known as black-dot. Apparently the same disease was observed in Utah and then in many other states where it received the name anthracnose.

Symptoms.

The lesions on the stem are not unlike those caused by Rhizoctonia. Infection takes place below the surface of the soil, the brown dead areas enlarging both up and down the stem as well as around it. Stolons and roots may also become affected. The partial girdling injures the vines above ground so that the leaves lose their fresh green color, and the lower ones may die and drop off. In very severe cases the entire plant may be killed. On the diseased tissue numerous minute black dots are present. These are sclerotia embedded in the epidermal cells. They are also present in the inner side of the affected hollow stem. They may often be found on the tubers, and almost invariably accompany silvery-scurf lesions.

Cause.

The impression one receives when reviewing the literature is that Ducomet mistook sclerotia with setæ for a closed pycnidium with bristles and named the fungus *Vermicularia varians;* writers following him made the same error. Appel states that the sclerotia develop bodies that appear like setæ of Colletotrichum McAlpine and Miss Doidge, working separately, each described the formation of conidia on the sclerotia (if it were Vermicularia they would be borne within the body), but used the nomenclature of Ducomet. Lately, Foex of France has done the same thing, although he adds that he never saw typical fruiting bodies of the Vermicularia. Taubenhaus and O'Gara, on the other hand, have described a Colletotrichum which has long setæ on the minute sclerotia, and the latter author, after inoculation work and studying type material, has named the fungus parasitic on potatoes *Colletotrichum atramentarium.*

The fungus lives over winter in the form of sclerotia which

may remain attached to the tubers or the old stems, or drop to the ground with the host tissue which has sloughed off. In the spring the sclerotia produce conidia sparingly and the crop is again affected. The fungus is probably only a wound parasite.

The injury is of such a slight nature ordinarily that control measures need not be practiced. Recommendations which have been made are the burning of old diseased vines, planting clean seed, and practicing long rotations in which tomatoes and potatoes do not follow each other in the same soil.

References

Foex, Et. La dartrose de la pomme de terre. Comp. Rend. d. Séances Acad. D'Agr. de France **8**: 844-848. 1922.

McAlpine, D. Black Dot disease. *In* Potato diseases in Australia, pp 92, 93. 1911.

O'Gara, P. J. New species of Colletotrichum and Phoma. Mycologia **7**: 38-41. 1915.

Pethybridge, Geo. H. Notes on some saprophytic species of fungi, associated with diseased potato plants and tubers. Trans. British Mycol. Soc. for 1918. **6**: 104-120. 1919.

Plant Disease Survey Office. Anthracnose caused by *Colletotrichum atramentarium*. Plant Disease Bull. Suppl. **10**: 203. 1920.

Taubenhaus, J. J. A contribution to our knowledge of silver scurf (*Spondylocladium atrovirens* Harz) of the white potato. N. Y. Bot Gard. Mem. **6**: 549-560. 1916.

Bacterial Ring Disease of Potatoes

Probably caused by *Bacterium sepedonicum* (Spieck and Kot.)

Since 1905, when the group of potato diseases known formerly under the term leaf-curl was divided, one of the divisions has been much discussed in Germany where it is known as bacterial ring disease. As the description of this disease does not agree with any potato troubles found in other countries, it is considered as distinct, even though the reports regarding it are incomplete or conflicting. The bacterium associated with the lesions has been inoculated successfully into tomato and other solanaceous plants. A bacterial ring disease has been described in India, but is caused by another organism.

Symptoms.

In many respects the symptoms on the vines resemble leaf-roll, or later in the season the brittleness of the stem and flecks on the leaf-veins are similar to those described as streak. When affected tubers are planted, the sprouts may not grow at all, or if they do sickly looking plants with rolled leaves are produced. Unlike leaf-roll, one stalk of the plant may be badly diseased while another is barely affected, or some of the petioles may turn yellow and die while the remainder apparently are healthy. In dry weather when the symptoms are aggravated, much tip-burn follows, and the plant dies early. If the season is wet the plants are often able to survive and produce at least a fair yield.

The stem may be marked with dark streaks which extend outward into the petiole. If the stem is cut open white specks in the pith may be observed under the hand lens. The tissue in these areas, as well as the xylem ducts, are filled with bacteria. There are no definite symptoms on the outside of the tuber, except an occasional blackening about the eyes, but when the tuber is cut open the vascular ring is found to be darkened and the tissue on each side of the bundles softened. In severe cases the softening extends outward almost to the epidermis and inward almost to the center, so that an apparently healthy tuber will consist of a mere outer shell and a central solid core with the intervening space filled with a slimy rot. Saprophytic organisms may cause the decay of the whole tuber.

Cause.

An organism, resembling very closely the one causing bacterial brown-rot of potatoes, has been isolated by pathologists in Germany, and named *Bacterium sepedonicum.* It is supposed to be a soil organism which penetrates wounded roots and stems, and from there passes into the tubers. According to the results of inoculations the bacterium can enter only through exposed vascular bundles. Under field conditions such exposures are made by insect bites or mechanical injury. It has never been observed that insects transmit the pathogene from diseased to healthy plants. After inoculation takes place six to eight weeks elapse before infection is visible. At no time does the infection cause a local rot except within the tuber. If diseased tubers are planted, the progeny will also be affected. In extreme cases the sprout is killed before it breaks through the ground.

The optimum temperature for the growth of the organism is between 68° and 77° F.

Control.

Since this disease is not known to occur in America, it is hoped that the quarantine laws will be strict enough to exclude it. Where it does occur, the control measures recommended are the procuring of disease-free seed, and long rotations with crops which are not susceptible to the ring disease.

References

Appel, O. Neuere Untersuchungen über Kartoffel und tomaten-erkrankungen. Jahresbericht der Vereinigung f. Angew. Bot. 3: 122-136. 1906.

Appel, O. Leaf roll diseases of the potato. Phytopath. 5: 139-148. 1915.

Spieckermann, A. and P. Kotthoff. Untersuchungen über die Kartoffelpflanze und ihre krankheiten. 1. Die Bakterienringfäule der Kartoffelpflanze. Landwirtschaftliche Jahrbücher 46: 659-732. 1914.

Speck-Rot of Potatoes

Caused by *Stysanus stemonites* (Pers.) Corda

Speck-rot of potatoes has not been reported often, possibly being overlooked because of its unimportance. It was observed in Nebraska in 1901, in Oregon in 1913, and in Germany in 1906.

The lesions are never large. Warren in Nebraska associated the organism with a browning of the vascular ring in the tuber.

Appel in Germany did not find the brown ring, but local brown decayed spots or specks on the surface of the tuber. Bailey in Oregon found the disease also on the stem.

The fungus, *Stysanus stemonites*, evidently enters the host only through wounds. A very small amount of mycelium is present in the tissue, but on the surface the dark coremium-form of fruit-body is rather prominent. The fungus lives over winter in the tuber, and probably also in the soil.

The disease has not been severe enough to demand control measures.

References

Appel, O. and W. F. Bruck. *Stysanus stemonitis* (Persoon) Corda und seine rolle als parasit der kartoffel. Arb. Kais. Biol. Anst. f. L. u. F. —wirtschaft 5: 441-448. 1907.

Bailey, F. D. Notes on potato diseases from the North. Phytopath. 4: 321-322. 1914.

Bessey, Charles E. The "brown disease" of potatoes. Science 15: 274. 1902.

Tuber-Rots

As the tuber is filled with starch, which can be broken down into sugars, it serves as an ideal pabulum for all kinds of fungi which are able either by direct penetration of the epidermis or by taking advantage of wounds to enter the heavily stored cells. Certain fungi produce a characteristic decay, which may be diagnosed with a fair degree of certainty while others cause a general rot that can easily be mistaken for others of a similar nature. Tuber-rot is the all-embracing term used for such troubles, but frequently it is possible to subdivide this general term into some that are more specific such as, jelly end-rot, powdery dry-rot, leak, black-leg, mahogany-rot, dry-rot and pink-rot. Some of the organisms which cause tuber decay are well known pathogenes of the vines and consequently are treated separately.

In order to gain an idea of the immense loss resulting from tuber-rots (Fig. 127) it is only necessary to drive through the potato sections in the spring when the storage-houses are being cleaned. On nearly every farm may be seen a dump heap where the rotten tubers have been hauled, sometimes by the wagon load. More than an eighth of the potatoes which are dug are discarded before the consumers can use the remainder. This means a loss of forty to fifty million bushels a year, and

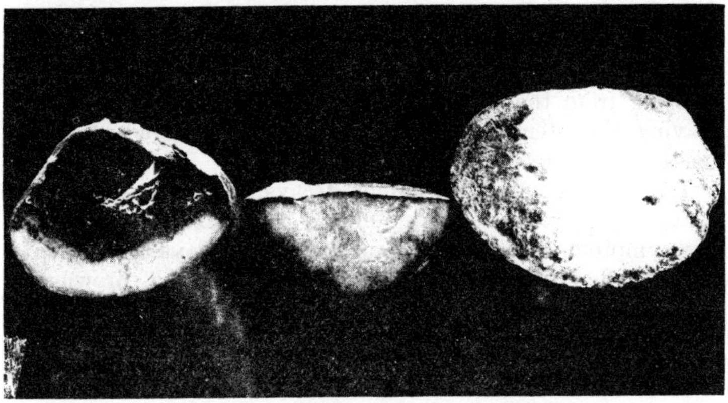

FIG. 127.—Storage rot of potato tubers.

a corresponding increase in the cost of production of the healthy stock.

Symptoms.

Leak, which is especially important in the delta region of the San Joaquin River, California, usually first shows as a brown discoloration around some wound and spreading rapidly soon includes the whole tuber. The diseased tissue is soft and easily crushed, and so watery that with slight pressure a brown liquid drips out through the bursted skin. When a pile of potatoes contains some of these melters or leakers, even the healthy tubers are smeared with the mushy dirty-colored tissue and become very disagreeable to handle.

The powdery dry-rot is quite the opposite of leak. It also starts from bruises but the sepia-brown tissue never becomes watery. The diseased portion of the tuber is shrunken and wrinkled. In an advanced stage pinkish growths of mycelium may be present. It is not unusual to find internal cavities caused by the decay and lined with mycelium. In some stages the tissue is almost black, and is then difficult to differentiate from what is termed black-rot and caused by another fusarial fungus.

Jelly end-rot, as the name implies, is a fairly soft decay and includes the stem-end of the tuber, or occasionally the knobs growing from the side. The rot gradually passes downward leaving the stem-end of the potato a discolored shriveled mass.

There are many other minor decays, but none distinctive enough to be easily separated merely by an external diagnosis. For symptoms of some of the more serious decays the reader is referred to late-blight, Fusarium wilt, black-leg, brown-rot, and pink-rot.

Causes.

Fusaria are by far the most numerous of any fungi found on potato tubers. A long list of Fusarium species which cause decay when inoculated into the potato is recorded. Some of these species seldom cause any appreciable injury while others are very common. One of the latter type is *Fusarium cœruleum* which causes much loss in storage. *Fusarium trichothecioides,* the cause of powdery dry-rot, is also important. Jelly end-rot is caused by *Fusarium radicicola, Fusarium oxysporum,* and *Rhizoctonia solani. Fusarium discolor* var. *sulphureum* is distinctive because of its ocherous-yellow mycelium. Other species are *F. hyperoxysporum, F. solani,* and *F. eumartii.*

All the Fusaria are characterized by abundant mycelium, which in some species is beautifully colored, and by sickle-shaped septate spores. They are able to live over winter in the soil or on tubers, some of them seemingly being omnipresent. Very few are able to enter the host unless the epidermis has been injured in some way. They vary in their reaction to high and low temperatures, but none of them grows well below 34° or 40° F. if the humidity of the air is low.

Leak is caused by fungi different from the Fusaria. Much of it is attributed to *Pythium debaryanum* Hesse, but some is caused by *Rhizopus nigricans* Ehrenb. These two fungi are present on a number of other hosts. The species of Pythium is the same which produces the damping-off of seedlings and is described under the damping-off of tomatoes (page 546), while the life history of *Rhizopus nigricans* is given briefly in the discussion of soft-rot of sweet-potatoes (page 519). Both enter the host through wounds and soften the tissue. They are present nearly everywhere, and when the temperature and humidity are high are very destructive. Infection may take place in the soil, during harvesting, or when the potatoes are in the bin. When infection has once occurred the disease can spread to neighboring injured tubers if environmental conditions are favorable for the parasites.

Control.

There are four requirements in controlling tuber-rots; namely, healthy vines in the field, extreme care in handling the tubers, a disinfected or clean storage-house, and low temperatures and humidity with plenty of space for air movement. The healthy vines may be obtained by following the control measures suggested for late-blight, rhizoctonia scab and leaf-roll. These include seed selection or the buying of certified seed, seed treatment, and spraying. The matter of handling the tubers carefully cannot be over-emphasized. They should be dug at a time when they can be dried quickly, and then stored. Certain types of diggers cause more injury to the tubers than others, therefore should be avoided. When the tubers are picked up from the ground they should not be tossed eight or ten feet to a crate or a wagon-bed, but the container should be near at hand so that the potatoes are never dropped farther than the depth of a bushel crate. It is a common practice to dump a wagon-load of potatoes through a trap-door into the basement, or shovel them through a window onto the rough cement floor. This treatment cannot help but cause severe bruising and skinning, through which storage decay organisms can readily pass. If the potatoes must be stored in a basement they should be lowered by means of a chute which has been padded with an old piece of carpet, or other soft material.

Too little attention is given to the storage-house. Often it is a large cellar with cement floors and stone walls below a home or a barn. The entire space from floor to ceiling is filled with potatoes, regardless of the depth or width of the pile. No attempt is made to furnish air for these living breathing tubers. In the spring when the crop is sold, the room is littered with dirt and refuse which is left to contaminate the crop in the fall. After emptying the storage-house, it should be cleaned scrupulously and the walls and floors washed with some disinfectant, such as blue vitriol (1 pound dissolved in 5 gallons of water). The floor should be made of boards laid so that the air can circulate below them and upward among the tubers. If the bin is deeper than six feet, partitions should be built by placing two-by-fours upright and nailing slats on each side, thus providing plenty of ventilation. The storage at all times should be kept dry and as near 38° F. as can be obtained. In early fall when the weather is still warm the windows can be opened at night to let in the cool air, and closed during the day.

The ideal storage-house needs no extra heating during the winter, for if a stove is used it is almost impossible to get a uniform temperature in all parts of the house; consequently the potatoes near the stove are over-heated while those next to the outer wall are chilled. It is possible to build a storage-bin above ground that will require no additional heat. In building such a place, the outer and inner walls must be lined heavily, and the cavity between the walls made as nearly air-tight as is possible. The cellar storage is probably more desirable for the average farm, since the amount of shrinkage of the tubers in the underground bins is less. The most natural condition for potatoes is storage in pits. Great care must be taken, however, in providing ventilation, and in the northern states where the soil freezes to a depth of several feet, too heavy a covering of straw and dirt are required to make this type of storage successful. Farther south it may be used advantageously. The pit should be placed on ground high enough to afford good drainage. A ventilator of boards or tile should be placed through the center of the pile and extended to the outside of the covered pit. The ventilator should never be placed in a vertical position or it will afford an entrance for rain or snow into the pit. Alternate layers of clean straw and dirt should be placed over the potatoes, increasing the thickness as the cold weather approaches. When there is danger of freezing the opening of the ventilator can be stuffed with straw.

References

Carpenter, C. W. Some potato tuber-rots caused by species of Fusarium. Jour. Agr. Research 5: 183-210. 1915.

Cotton, A. D. and H. V. Taylor. The cause of decay in potato clamps, with special reference to the season 1918. Jour. Bd. Agr. Gt. Britain. Supplement 18: 48-60. 1919.

Goss, R. W. Temperature and humidity studies of some Fusaria rots of the Irish potato. Jour. Agr. Research 22: 65-80. 1921.

Hawkins, Lon A. The disease of potatoes known as "leak" Jour. Agr. Research 6: 627-640. 1916.

Link, Geo. K. A physiological study of two strains of Fusarium in their causal relation to tuber-rot and wilt of potatoes. Neb. Agr. Exp. Sta. Res. Bull. 9: 1-45. 1917.

Pethybridge, Geo. H. and E. H. Bowers. Dry rot of the potato tuber. Econ. Proc. Roy. Dublin Soc. 1: 547-558. 1908.

Pratt, O. A. Control of the powdery dry rot of western potatoes caused by *Fusarium trichotheciodes*. Jour. Agr. Research 6: 817-832. 1916.

Pratt, O. A. A western field rot of the Irish potato tuber caused by *Fusarium radicicola*. Jour. Agr. Research 6: 297-310. 1916.

Shapovalov, M. Some potential parasites of the potato tuber. Phytopath. 9: 36-42. 1919.

Sherbakoff, C. D. Fusaria of potatoes. Cornell Agr. Exp. Sta. Memoir 6: 87-270. 1915.

Wilcox, E. Mead, George K. K. Link and Venus W. Pool. A dry rot of the Irish potato tuber. Neb. Agr. Sta. Res. Bull. 1: 1-88. 1913.

Black-Heart of Potatoes

Cause physiological

Black-heart of potatoes is one of the few plant diseases known not to be caused by an organism. The producer of the crop seldom sees this disease, therefore is not impressed with its seriousness as is the consumer, who may find a large percentage of a given shipment made worthless for food purposes by this trouble. Although it is found wherever potatoes are stored, it is more severe in the United States and Canada where potatoes are shipped long distances in heated cars during the winter.

Symptoms.

Affected tubers usually appear perfectly normal on the out-side. When they are cut open the tissue of the potato may at first be white, but when exposed to the air will change to pink, then to red or brown, and finally become inky-black (Fig. 128). Other tubers will show the blackening when they are first examined. The size of the darkened area depends on the severity of the disease. It usually starts near the center and spreads outward until, in the worst cases, the whole inner tissue may be blackened. Sometimes the discoloration is not massed in one area but is made up of several large blotches. The part toward the eye-end of the tuber seems to be more susceptible than that near the stem. Occasionally the disease is accompanied by the formation of cavities of varying sizes, each lined with blackened tissue. If the tuber is heated over a certain temperature, or until it is almost cooked, the tissue will not discolor but will remain even whiter than that of the normal potato.

Cause.

Black-heart was first observed in shipped potatoes and was considered to be connected in some way with frost-injury.

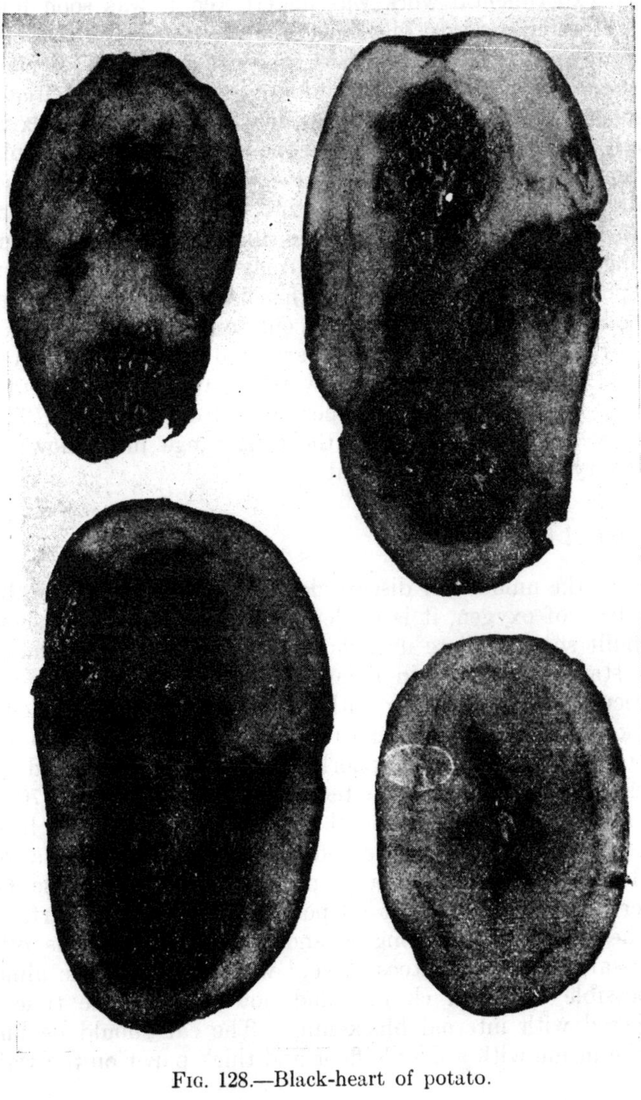

FIG. 128.—Black-heart of potato.

When experimental work was undertaken it was soon found that freezing retards black-heart formation. On the other hand, it was proved that heating tubers to about 100° F. causes blackening even when plenty of air is permitted to circulate about the potatoes. It was also shown that when the oxygen supply is shut off from the tubers, the blackening will take place at temperatures as low as 60° F. Thus the severity of the trouble depends on the temperature, the oxygen supply, and the length of exposure. When the sun is shining very warmly during the summer, black-heart may be induced by laying the potatoes in the sunlight. This has been known to take place when the tubers were laid out to dry after treating, or when the early crop was being harvested and left lying on the ground too long. In some hot-water treatment experiments it was observed that in all temperatures from 104° to 212° F. the disease was produced in the time range just below that necessary for killing the potatoes.

Control of black-heart.

Since the amount of disease depends on a high temperature and lack of oxygen, it is evident that a storage-house should be built so that these unfavorable conditions are eliminated. The statements made in regard to the control of tuber-rots, particularly those dealing with storage, apply equally well to black-heart. The two facts to be emphasized most are that no matter how well or how poorly the house is ventilated, the disease will be present if the temperature is much above 70° F. for any length of time, and that if the temperature is below 50° F. the tubers will withstand lack of ventilation for a long while. This means that a properly built storage-house (see tuber-rots) should protect all potatoes from black-heart.

The matter of shipping is another question. It is often necessary to ship potatoes in cold weather when it is almost impossible to avoid chilling and not at the same time be bothered with internal blackening. The car should be lined on the inside with a double floor and thick paper on the walls

If a stove is necessary proper care should be taken in keeping the fire as low as the temperature permits, and so arranging the middle partitions that the tubers nearest the stove are protected while the warm air is deflected to the farther side of the car.

References

Bartholomew, E. T. Black heart of potatoes. Phytopath. **3: 180-182.** 1913.

Bartholomew, E. T. A pathological and physiological study of the black heart of potato tubers. Centralblatt f. Bakt., etc., Abt. 2. 43: 609-639. 1915.

Stewart, F. C. and A. J. Mix. Blackheart and the aeration of potatoes in storage. N. Y. (Geneva) Agr. Exp. Sta. Bull. 436: 321-362. 1917.

Frost-Injury and Freezing of Potatoes

It is difficult to determine the amount of loss of potatoes due to freezing, but it is often estimated that in the northern states the loss is as heavy from this one cause as from all the known diseases combined. The injury may begin in the field if low temperatures occur early in the autumn or if the crop cannot be dug until too late in the season. After harvest many tubers must be placed in temporary storage so that when the market is dull they are not removed soon enough to avoid freezing. Poorly made pits, insufficiently protected storage-houses, and open-air shafts or windows result in added losses, although the greatest amount of freezing takes place in the cars during shipment.

There is no marked resistance or susceptibility shown by any variety of potato, which could be used commercially as a means of reducing the losses from freezing. It is interesting to note, however, that tubers from certain potato groups withstand almost a degree lower temperature than do those from other groups. The Peachblow type is the most resistant, with Green Mountain, Rurals and Up-to-date only slightly less so, while the American Giants are the most susceptible, followed by the Rose, Cobbler, Michigan and Triumph, respectively.

There is a relatively large difference in susceptibility of individual tubers of the same variety in any given bin. Furthermore the stem-end of the tuber is somewhat more sensitive than is the eye-end.

Symptoms.

MacMillan showed that a very common leaf injury results from the sudden dropping of the temperature during summer nights. The leaflets, particularly their basal parts along the midrib and near the petiole, are checkered with small irregular brown dead areas. The spots rarely appear on mature leaves, the injury being confined almost wholly to the tender tip foliage. Probably no particular weakening of the plant follows such types of lesions, although occasionally one may observe cases in which Cercospora evidently followed frost-injury.

The symptoms manifested by the chilled or frozen tuber vary from mere spots of grayish drier-appearing flesh cells slightly below the epidermis to a complete collapse of the inner tissues (Fig. 129). In fact, chilling of the tuber can be so slight that aside from a characteristic brittleness they cannot be recognized from the healthy potato. When such potatoes are planted the seed piece is usually persistent as in the case of leaf-roll or yellow-dwarf. If the tuber is once frozen solid it becomes soft and watery on thawing. The sap oozes out through the skin, making it wet and the inner tissue is so mushy that the potato crushes easily. Later decomposition sets in causing a disagreeable odor.

When the tuber is not frozen hard but chilled to a certain stage, its outer appearance does not change from that of the normal, but when the tuber is cut open very typical net-necrosis is present, or darkened blotches appear in the tissue adjacent to the vascular bundles. Occasionally the vascular ring may also be darkened as in the case of the Fusarial wilt. Still another type of symptom is the shriveling of the tuber until it becomes leathery in texture. When it is cut open, the color of the flesh is almost normal, except that it appears more nearly dry. After a brief exposure to the air the tissue becomes gray, followed soon by a faint tinge of pink, which gradually becomes more intense and darker until it is red, then purple and finally coal-black. This symptom compares very closely with the pink-rot described by Pethybridge.

Cause of frost-injury.

The injury on the tender leaves is produced by warm evenings in which the stomatal cavities are filled with water that freezes when the temperature drops suddenly in the night.

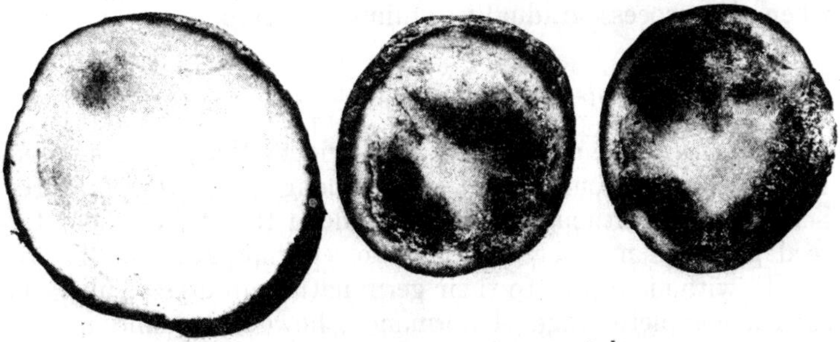

FIG. 129.—Frost-injury on potato tubers.

A few cells on each side of the stoma are killed, resulting in the small irregular lesions found so abundantly in many potato fields.

The freezing of the tuber takes place at different temperatures depending on the length of time of exposure, the variety of potato, and the treatment during exposure. It has long been known that concentrated sap freezes at a lower temperature than 32° F., but workers generally disagree as to the exact point in the refrigeration at which tuber injury begins. The general conclusions are that normally the tuber is injured when subjected to a temperature of 28° to 29° F. If the potato is unmolested during this sub-cooling process, the temperature may go considerably lower than 28° without serious results, but as soon as the potato is jarred or handled, ice particles begin to form and injury soon follows. The slight jarring is sufficient to start the process of congealing which otherwise would not take place.

When temperatures are just above freezing, the potato tissue is not injured, but the sugar-content is increased sufficiently to cause a sweetish taste of the tuber, thus destroying its salability. The increase in sugar-content is produced up to 45° F., although it is very slight above 38° F. When the potatoes are subjected to temperatures between 30° and 38° F. no material change takes place at once but the sugar production is gradually accelerated until four to six weeks have passed when the process gradually diminishes again.

Control of frost-injury.

From the above explanation it is evident that potatoes used for table stock should not be exposed long to any temperatures below 38° F., although it may be added that tubers held for seed purposes may be subjected to a temperature of 33° or 34° F. without injury to their germinating ability. Tubers in such a complete stage of dormancy, however, should not be planted until they have been permitted to be in the light and warmth long enough to allow sprout development.

The means for eliminating tuber injury from freezing is so simple that it needs no elaboration, yet it is fitting that at least a few words of caution be added. It seems such a sheer waste of capital and labor to grow a crop successfully, then permit that crop to spoil when it can be prevented in so many cases. Suitable storage-houses can be built at a reasonable cost, especially if placed partly underground. The description of the storage-house given in the recommendations for the control of tuber-rots applies equally well when building for protection from frost (page 449). The storage-room should always be provided with ample ventilation, and the temperature so regulated that it will remain as near 38° F. as possible. It is much more difficult to avoid losses during shipment, but even these can be decreased by giving proper attention to the car before it is loaded and while it is being shipped. The inside of the car should be paper-lined, the door spaces well protected, and if necessary heat supplied, care being taken that conditions favorable for the development of black-heart are not furnished by over-heating.

References

Appleman, Charles O. Changes in potatoes during storage. Md. Agr. Exp. Sta. Bull. 167: 327-334. 1912.

Jones, L. R., M. Miller and E. Bailey. Frost necrosis of potato tubers. Wis. Agr. Exp. Sta. Res. Bull. 46: 1-46. 1919.

Kotila, J. E. Frost injury of potato tubers. Mich. Acad. Sci. Ann. Rept. **20**: 451-460. 1918.

MacMillan, H. G. A frost injury of potatoes. Phytopath. **10**: 423-424. 1920.

Wright, R. C. and R. B. Harvey. The freezing point of potatoes as determined by the thermoelectric method. U. S. Dept. Agr. Bull. 895: 1-7. 1921.

Wright, R. C. and Geo. F. Taylor. Freezing injury to potatoes when undercooled. U. S. Dept. Agr. Bull. 916: 1-15. 1921.

Pink-Rot of Potatoes

Caused by *Phytophthora erythroseptica* Pethyb.

Pink-rot and wilt of potatoes has been a serious disease in Ireland for a number of years, and recently has been reported from England, Scotland, and Holland. It has not been observed in the United States. In well drained soil with proper cultural methods it is not serious, but when conditions are favorable for the organism the disease may become as destructive as late-blight.

Symptoms.

The rot of the tuber begins at the stem and gradually spreads toward the eye-end. The affected skin is slightly browned, with a distinct black band marking the division between the healthy and diseased areas. When the entire tuber is affected, it is somewhat softer and more pliable than is the healthy one, although there are no signs of extreme softness or mushiness. When the tuber is cut open the tissue at first has the normal color, but when exposed to the air for a short time the flesh becomes pink, gradually darkening to red and finally turning black. In this country tubers that are flexible and with the same color changes when cut may occasionally be observed. They must not be confused with pink-rot, however, since the trouble in this case is due to chilling.

The rot affects the tubers while they are still in the soil. If wounds are present the exposed tissue is black, as are also the lenticels. Green mold may cover them, or a liquid having exuded from the diseased lenticels causes dirt to be caked on the outside of the potato giving it the appearance of a clod.

The fungus also causes a disease of the vines. The symptoms are similar to those described for black-leg. The base of the stem rots, the foliage becomes pale green or yellow, and frequently small aerial tubers are formed close to the surface of the soil.

Cause.

The fungus is closely related to the one which causes late-blight, and is known as *Phytophthora erythroseptica.* It is able to exist in the soil for at least four years and probably much longer. Infection evidently takes place through the stolon when the plant is still young. The fungus grows downward into the tuber and upward into the base of the stem. The fungus fruits in the soil on decaying tubers, stalks, stolons or roots. Besides the conidia which are produced, oospores are formed in pure cultures of the organism. They may play some part in helping the parasite to live in the soil for a number of years. Infected seed-tubers are not known to serve as a source of infection for the new crop.

Control.

It is hoped that the Federal Plant Quarantine can be enforced strictly enough so that pink-rot will never be introduced.

If the grower finds specimens which look suspicious, he should send them to the state pathologist or to the United States Department of Agriculture at Washington, D. C. Where it does occur it is suggested that healthy seed be obtained, long rotations practiced with grasses and cereals as the intervening crops, using well drained soil, and roguing and destroying affected hills.

References

Cotton, A. D. Potato pink rot: A disease new to England. Jour. Min. Agr. Gt. Brit. **28**: 1126-1130. 1920.

Pethybridge, Geo. H. On the rotting of potato tubers by a new species of Phytophthora having a method of sexual reproduction hitherto undescribed. Scientific Proc. Roy. Dublin Soc. **13**: 529-565. 1913.

Pethybridge, Geo. H. Pink rot and wilt (*Phytophthora erythroseptica* Pethyb.) Jour. Dept. Agr. and Tech. Instr. Ireland **13**: 461-465. 1913. **14**: 452-455. 1914; **15**: 510-511. 1915; **16**: 596. 1916; **19**: 281-282. 1919.

Armillaria Rot of Potatoes

Caused by *Armillaria mellea* Vahl.

The appearance of the honey fungus on the potato is so rare that it has been reported only from Washington and Australia. The fungous threads combine to make chocolate-colored strands or rhizomorphs, which grow about the roots and tubers (Fig. 130). At certain points they are attached to the host tissue from which they draw their sustenance. The affected potatoes become shriveled and worthless for food. The fungus, *Armillaria mellea*, is found on the roots of fruit hosts, as well as around dead logs and stumps. It is not an active parasite, and attacks potatoes, rhubarb and possibly other vegetables only when they are planted in newly cleared soil where the humus has permitted the fungus to grow luxuriantly. The rhizomorphs grow long distances underground, and when food is plentiful produce the mushroom fruit-bodies. Because of the color of the mushroom, the parasite is known as the honey fungus. It produces on its gills a great mass of brownish

FIG. 130.—Armillaria rot of potato tuber showing the large mycelial strands.

spores which are disseminated by the wind, and serve in establishing the organism in other favorable places.

Armillaria rot will never become troublesome on well-tilled land. In certain localities it might be advisable not to use a newly cleared field for potatoes, but to plant corn or some other crop until the rhizomorphs in the soil have been killed.

References

Bailey, F. D. Notes on potato diseases from the Northwest. Phytopath. **4**: 321-322. 1914.

Herbert, D. A. Report on plant diseases. West Aust. Dept. Agr. and Ind. Ann. Rept. **1919**: 29. 1920.

Johnson, F. H. Potato diseases. Agr. Gaz. N. S. Wales **21**: 699-701. 1910.

Sprain of Potatoes

Cause not determined

In Great Britain, Ireland, and some of the European countries various types of internal discoloration of the potato have been mentioned. The writers were not always sure whether the symptoms should be classed under one disease or whether they were the result of several distinct causes. At least one phase of the trouble has lately been attributed to a bacterium, but there still seems to remain another form of the discoloration known as sprain. It is characterized by brown blotches or streaks in the flesh of the tuber, unaccompanied by any outward signs either upon the potato or the plant above ground. During certain seasons it is of enough importance so that buyers cut sample tubers for the appearance of the disease before they will accept the stock. It seems to be more prevalent on sandy soil and in dry seasons. It is doubtful whether a similar trouble has ever been observed in the United States.

No control measures are known on dry sandy soil. (It has been demonstrated, however, that affected tubers may be planted on medium or heavy soil and that the resulting crop will be free from the trouble.)

References

Cotton, A. D. Potato diseases. Jour. Bd. Agr. (Great Britain) Supplement 18: 46. 1919.

Horne, A. S. The symptoms of internal disease and sprain (streak-disease) in potato. Jour. Agr. Science 3: 322-332. 1910.

Pethybridge, Geo. H. Sprain. Jour. Dept. Agr. and Tech. Inst. Ireland. 12: 356-357. 1912.

CHAPTER XV

DISEASES OF RHUBARB AND SALSIFY

RHUBARB and salsify or vegetable-oyster are grown only in small acreages for commercial use, but are planted generally in gardens. Salsify is a hardy plant with few diseases, while rhubarb is susceptible to very destructive root- and crown-rots.

PHYTOPHTHORA FOOT-ROT OF RHUBARB

Caused by *Phytophthora parasitica* var. *rhei* Godfrey

In 1917 a foot-rot of rhubarb was observed at Brookland, D. C. Since then the same trouble has been found in Maryland, Virginia, Pennsylvania, Tennessee, and Illinois. It may be much more wide-spread than these few places indicate, but the similarity of this trouble to the symptoms caused by crown-rot makes a superficial diagnosis unreliable.

The parasite is extremely virulent, and when the environment is favorable the fungus may destroy an entire planting.

Symptoms.

The first spots appear on the base of the leaf-stalk. The lesions are sunken brown spots, not large at first but spreading so rapidly that in twenty-four hours the leaf will have wilted on its petiole and fallen to the ground. Soft-rot bacteria enter the wound and aid the parasite in quickly transforming the infected parts into a soft rotten mass. Stalks ranging from those whose leaves are not yet unfurled to those which are mature may be invaded. All the stalks of one plant are usually not attacked simultaneously but the fungus advancing through the upper part of the root kills one stalk after the other until, in severe cases, the whole plant is destroyed. The inoculum spreads rapidly so that in a brief time there may be many vacant areas in the field.

Cause.

After comparing the causal organism with various Phytophthoras and making successful inoculations, the parasite was named *Pytophthora parasitica* var. *rhei.* It is one of the destructive downy-mildew fungi, and reproduces by forming oospores and conidia, the latter giving rise to zoospores. The spores may be carried by splashing rain or on cultivating tools from plant to plant. The organism enters the host through wounds and probably also directly through the epidermis. When slightly infected roots are employed to start a new planting, the organism in the tissue may be introduced into uncontaminated soil. The oospores and possibly the hyphæ in the roots live over winter and furnish inoculum for the spring infection.

The disease is most prevalent in those seasons when there is a continued rainy period sometime during the summer months. The parasite apparently is susceptible to low temperatures as it has not been reported definitely from any of the far northern states. Its maximum temperature for growth is approximately 97°, its optimum 86°, and its minimum 55° F. These favorable conditions are met with in the latitude of Maryland and southern Illinois.

Control.

Although the disease may be very virulent if proper precautions are not taken by the grower, it is encouraging to know that control measures have been markedly successful. First, when a new planting is made, great care should be exercised in obtaining the roots from a field where the disease has never been present. This recommendation is fairly easy to follow since the disease as yet is not widespread. If it is impossible to make sure that the roots are healthy, just before planting and while in a dormant stage, they should be dipped in corrosive sublimate (1 ounce in 7.5 gallons of water) or in formaldehyde (1 pint in 30 gallons of water) for one-half hour, then thoroughly rinsed in clean water. If on breaking the roots, any internal parts show discoloration, the whole clump of roots should be discarded. Healthy or treated roots should never be set in a field where rhubarb has been grown in recent years.

As the seat of the trouble is within the tissue, spraying will not aid a plant in recovering when it once has become diseased, but it will protect the stalks from new infections. The plants from which stalks are being harvested cannot be sprayed, for the fungicide spots the surface and makes the product unsightly. Spraying with bordeaux mixture 4-4-50, however, may be practiced on young plants or on the old specimens after harvest. The spray is applied with high pressure directly into the head of the plant so that the bases of the stalks are covered thoroughly and the soil about the crown of the root is well saturated.

If an occasional plant shows infection, it should be dug up carefully so that no diseased parts or adjoining soil is scattered over the field. The affected plants can be destroyed by fire and the spots where they stood disinfected by wetting the soil with formaldehyde (1 gallon in 49 gallons of water).

Reference

Godfrey, G. H. A Phytophthora foot-rot of rhubarb. Jour. Agr. Research 23: 1-26. 1923.

Phytophthora Crown-Rot of Rhubarb

Caused by *Phytophthora cactorum* (Leb. and Cohn) Schroet.

(See also Phytophthora Foot-Rot of Rhubarb, page 465.)

A rot of rhubarb stems was observed in Pennsylvania, and as the symptoms were almost identical with that of foot-rot the two troubles were considered the same. Rather surprisingly when isolations were made, a different species of Phytophthora was found to be the cause. Consequently the term crown-rot was employed to distinguish this type of disease from the foot-rot.

The parasite is identical with *Phytophthora cactorum*, which has previously been reported as the cause of damping-off and foot-rot of seedlings of many plants. Cultures of the fungus isolated from a decaying apple caused the disease on rhubarb as quickly as did a culture from rhubarb itself.

The life history and control measures are the same for *Phytophthora cactorum* as for *Phytophthora parasitica* var. *rhei.* There is, however, a significant difference in their temperature requirements. The latter grows best at about 86° F. The maximum, optimum, and minimum temperatures for the former are approximately 92°, 77°, and 43° F. respectively. Judging from these temperatures, the organism discovered in Pennsylvania will be destructive much farther north than the one present near Washington, D. C. Therefore, any Phytophthora disease on rhubarb in the northern tier of states will probably be of the crown-rot type.

Reference

Beach, W. S. A crown-rot of rhubarb caused by *Phytophthora cactorum*. Penn. Agr. Exp. Sta. Bull. 174: 1-28. 1922.

Rhizoctonia Foot-Rot of Rhubarb

Caused by *Corticium vagum* B. and C.

(See also Rhizoctoniose of Potato, page 367.)

In several of the northern states a Rhizoctonia rot has been observed at the base of rhubarb stalks. Apparently the disease is not of great importance. The plants have a yellow unthrifty appearance, and gradually the leaves begin to wilt. There is a brown-rot at the base of the petiole, which by girdling the stalk finally causes it to fall prostrate. The leaves in the meanwhile die and shrivel. The whole plant rarely is affected, although from one-fourth to three-fourths of the stalks may be killed.

No control measures have been suggested.

References

Clinton, G. P. Rhubarb stem-rot. Rhizoctonia sp. Conn. Agr. Exp. Sta. Ann. Rept. 1904: 327. 1905.

Duggar, B. M. and F. C. Stewart. The sterile fungus Rhizoctonia as a cause of plant diseases in America. N. Y. (Cornell) Agr. Exp. Sta. Bull. 186: 51-76. 1901.

Phyllosticta Leaf-Spot of Rhubarb

Caused by *Phyllosticta straminella* Bres.

A Phyllosticta leaf-spot has long been recognized on *Rumex acetosa*, known to gardeners as schaff. The same disease has been observed rather generally on rhubarb, which frequently is injured seriously.

The first indications of the disease are numerous small greenish-yellow areas on the upper sides of the leaves (Fig. 131). When these spots coalesce, as they often do, the leaf has the appearance of excellent mosaic mottling. Within less than a week, the invaded tissue turns brown and dies, resulting in circular to angular spots, variable in size, and having a minute white spot in the center surrounded by a wide red zone and this in turn bordered by a grayish-green zone. In some of the smallest spots only the red color is present. When the affected tissue dies, it may drop out leaving a shot-hole effect on the foliage. The leaf-spot is sometimes mistaken for that caused by other fungi, since the black fruit-bodies frequently are not visible on the surface of the leaf. The opening of the pycnidium is then even with the leaf cuticle and may be seen only by the aid of a hand-lens. A cross-section of the spot usually is necessary to make a diagnosis of the trouble.

The causal fungus has been referred to *Phyllosticta straminella*, although the spores are long and large unlike those of most Phyllostictas. The globose pycnidium is nearly always buried in the host tissue where either it or the mycelium lives over the winter. In the spring new infections take place, first on the lower leaves and gradually spreading to all parts of the foliage.

FIG. 131.—Phyllosticta leaf-spot of rhubarb, showing the white spot in the center of the lesion surrounded by brown and red zones.

When the disease is severe enough to demand control measures, the stalks with affected leaves can be removed during the regular harvest. Late in the autumn after the tops have been frozen, all the dead leaves should be removed and burned. Early in the spring and also after the harvest is completed applications of bordeaux mixture may be given. Sprays cannot be applied at cutting time, because of the discoloration on the stalks caused by the fungicide.

REFERENCE

Stevens, F. L. Rhubarb leaf-spot due to *Phyllosticta straminella*. Ill. Agr. Exp. Sta. Bull. 213: 308-312. 1919.

ANTHRACNOSE OF RHUBARB

Caused by *Colletotrichum erumpens* Sacc.

In 1918 a leaf-petiole rot of rhubarb was observed in Illinois. The leaves on the affected stems wilt and die. The spots on the petioles begin as oval, soft, watery areas. These rapidly increase in number and size until the whole stalk is included. The disease is more prevalent on the older stems, although the young vigorously growing stalks may also be affected. The fungus is particularly abundant on stalks that have been killed by Phytophthora. After the tissue is softened, the fungus produces small dark fruit-bodies in great abundance. This is true especially of the stems which have dropped to the ground.

The disease is important not only in reducing the yield of the crop, but also in making the rhubarb bunches unsightly and therefore unsalable. Infection often takes place before the stalks are cut, and although apparently healthy when harvested, they may show much spotting by the time they reach the market.

The small black fruit-bodies of the fungus are small stromæ covered with dark setæ and hyaline conidiophores bearing small single-celled spores. The stromæ survive the winter and produce conidia in the spring. The mycelium probably also remains alive in the old decaying stems.

No control measures have been tried experimentally. If all the diseased petioles are removed as soon as lesions begin to show on them, the amount of disease will probably be much less the following year.

REFERENCE

Stevens, F. L. Rhubarb anthracnose due to *Colletotrichum erumpens*. Ill. Agr. Exp. Sta. Bull. 213: 299-306. 1919.

Downy-Mildew of Rhubarb

Caused by *Peronospora jaapiana* Magnus

A downy-mildew of rhubarb has been known in Europe for a number of years, but there seems to be no record of its occurrence in the United States. Apparently the disease is seldom or never destructive. Large brown lesions are formed on the leaves, the lower surfaces of which are covered with a violet to white fungus growth. In severe cases the leaf may die.

Three species of Peronospora have been reported on rhubarb and closely related plants: *Peronospora rumicis* Cda., *Peronospora polygoni* Thm. and *Peronospora jaapiana*. It is possible that the three species are distinct, or it may be found later that the same species has been given different names. The species *rumicis* is supposed to live over winter in the root-stalks and to grow up with the new leaves. This has not been proved for *jaapiana*. According to Magnus, the latter remains in the old diseased plant parts as dormant mycelium, for no oospores are found. Infection takes place through stomata, and when the mycelium has become well established the very long conidiophores grow out of the breathing pores in clumps, and like those of other Peronosporas bear oval conidia on the tips of small branches.

If the downy-mildew ever becomes serious enough to demand attention, spraying the leaves as suggested for the control of Phytophthora on rhubarb will probably protect the plants from the parasites. In addition to spraying, the diseased leaves should be removed and destroyed by fire as soon as infection is visible.

References

Magnus, P. Erkrankung des rhabarbers durch *Peronospora jaapiana.* Ber. Deut. Bot. Ges. 28: 250-253. 1910.

Osterwalder, A. Peronospora auf *Rheum undulatum* L. Centralbl. Bakt. u. Par. II 10: 775-777. 1903.

Rust of Rhubarb

Caused by *Puccinia phragmitis* (Schum.) Korn.

The rust produces large crimson beautiful spots on rhubarb leaves. The central part of the spot is crowded with the cluster-cups of the fungus, the edges of the cups being torn into a fringed border. The disease is relatively rare, having been observed only several times in Europe and in Nebraska.

The aecial stage, known as *Aecidium rubellum* DC., is present on rhubarb and the various species of dock. The other two stages, as is common for rusts, are found on another host, which in this case is one of the grasses named Phragmitis. Evidently the spores can blow long distances to infect the alternate hosts.

The disease is not of enough importance to demand control measures.

References

Bates, J. M. The finding of *Puccinia phragmitis* (Schum.) Korn. in Nebraska. Science **16**: 138. 1902; Jour. Mycol. **9**: 219-220. 1903.

Cooke, M. C. Pests of garden vegetables. Jour. Roy. Hort. Soc. **27**: 801-831. 1903.

Gray Mold-Rot of Rhubarb

Caused by *Botrytis* sp.

(See also Gray Mold-Rot of Lettuce, page 253.)

Occasionally a gray mold-rot of rhubarb is observed. Normally growing plants are seldom affected. The first indication of the disease is a wilting of some of the leaves, which become yellow in color or at least partly so. The yellow later changes to brown, while the leaves become limp upon the petiole and droop until they touch the ground. The characteristic ashen-gray fruit-bodies of the parasite are formed on the under side of the principal veins which traverse the dead part of the leaf. The gray mold also forms about the petiole where it touches the leaf-blade and so weakens the stalk that the leaf droops. The rot and concomitant gray mold continue to spread until the whole leaf and petiole are invaded.

The fungus lives over winter in the discarded stalks and leaves. In the spring infection occurs through wounds made at harvest time.

In combating the trouble, the diseased stalks should be destroyed·if possible, the plants kept in an actively growing condition, and spraying with bordeaux mixture 4-4-50 practiced after the harvest is completed. Both the leaf and petioles should be covered with the fungicide. Another important step in protecting the rhubarb from Botrytis is in placing the plants far enough apart so that foliage can dry quickly after dew or rain. The fungus requires the presence of water for infection, and if the plants dry within a few hours such infection cannot take place.

Reference

Leendertz, Czn. J. Eene Botrytis-ziekte op roode bessen en rabahber. Tijdsch. o. plantenz. **26**: 173-175. 1920.

White-Rust of Salsify

Caused by *Albugo tragopogonis* (DC.) Gray

The white-rust is common on the white and black salsify, probably wherever the hosts are grown. The same organism occurs on a long list of related plants. Although the disease is frequently very destructive, no investigational work has been conducted with the malady or with the parasite in its relation to the host.

The disease begins as light yellow areas on the leaves. The epidermis over the spots is soon forced upward into a dome-like swelling and finally bursts, revealing a white sorus of chalk-like appearance. These sori usually occur in great numbers on the leaves and finally cause the death of at least part of the foliage. The roots are not attacked directly, but in severe cases of the disease they are dwarfed in size due to the decreased leaf surface.

The fungus is *Albugo tragopogonis*, or as it is sometimes known, *Cystopus tragopogonis*. It is composed of rather large mycelial strands in the host tissue. A fruiting layer of closely packed short conidiophores is formed under the epidermis, and while the spores are being produced in chains, they push upward until finally they burst out. They are disseminated by the wind. When the tissue has become thoroughly invaded, oospores are formed among the dying cells. They are thick-walled spores that are able to survive long periods of inactivity, and aid the fungus in living through the winter. The fungus is never a soil organism and does not grow saprophytically.

Moist warm weather is said to favor the appearance of the white-rust.

No successful control demonstrations have been reported, yet judging from the slight work that has been done the plants are protected from infection if frequent applications of bordeaux mixture 4-4-50 are made, beginning before the fungus has an opportunity to inoculate the leaves. In small garden plots the spraying may well be accompanied by the removal of all diseased foliage as soon as pustules are evident.

References

Ritzema-Bos, J. Bestrijding van de zoogenaamde "witte roest" der schorzeneceren, veroorzaakt door *Cystopus tragopogonis* (Pers.) Schroter. Tijds. o. Plantenz. **26**: 216-220. 1920.

Wilson, G. W. Studies in North American Peronosporales—I. The genus Albugo. Bull. Torrey Bot. Club **34**: 61-84. 1907.

Bacterial Soft-Rot of Salsify

Caused by *Bacillus carotovorus* Jones

(See Bacterial Soft-Rot of Carrot, page 76.)

Sporidesmium Leaf-Spot of Salsify

Caused by *Sporidesmium scorzoneræ* Aderh.

This leaf-spot has evidently been reported only once on the black salsify. In one section of Germany where much salsify is grown there were no leaf diseases of the crop until 1901 when suddenly this trouble appeared in epidemic form. Frequently hundreds of circular spots varying from a mere point to those one-fourth inch in diameter are found on a single plant. The necrotic areas are brown with red borders. When the spots are numerous or coalesce, the entire leaf may be killed. When the whole top dies, the roots remain small and unsalable.

The fungus, *Sporidesmium scorzoneræ*, has been shown experimentally to cause the leaf-spot. The organism is characterized by brown elongated spores, which are muriform at the base where they are much wider than at the tip. In fact, the spore is very similar to that of Macrosporium. The fungus lives over winter both as mycelium and as spores. When infection takes place, the life cycle of the parasite is completed in about six weeks. No perfect stage is known.

Spraying with bordeaux mixture has been attempted, but with indifferent results. It is possible that the fungicide was applied too late, and that if the bordeaux mixture were forced under high pressure on the plants before there was any opportunity for infection, the development of the Sporidesmium could be checked.

Reference

Aderhold, R. Ueber eine bisher nicht beobachtete Krankheit der Schwarzwurzeln. Arb. Biol. Abt. f. Land- und Forstw. 3: 439-440. 1903.

CHAPTER XVI

DISEASES OF SPINACH

More than half of the spinach in this country is grown in four states: California, Maryland, Virginia, and Texas. The total acreage is slightly over ten thousand. The average percentage of loss due to diseases is not available, but undoubtedly it is very high.

Spinach Blight

Cause undetermined

The disease on spinach is known as blight, yellows, or mosaic. It was discovered in Norfolk County, Virginia, about 1905, and since then has become prevalent in most, if not all, of the spinach-growing sections in the country. It is not certain that it occurs on any other continent unless it has been described by Schoevers in Holland. The trouble is very destructive when it once becomes well established. In eastern Virginia the estimated loss to spinach-growers is over two hundred thousand dollars a year.

Symptoms.

The symptoms of blight are typically those of mosaic as described on other vegetables, except that blight is even more severe in its effect and soon kills the host. The disease begins as a mottling of the young inner leaves, later changing them to a yellow color and finally killing them. The disease gradually spreads to the outer leaves, which in turn are changed to yellow. As the disease progresses and before there is any killing of the tissue, the foliage curls and wrinkles. If the plant is affected when young, and it may contract the disease in the seedling stage, it remains much dwarfed. The dwarfing, the yellowing, the corrugation, and the killing are all such conspicuous symptoms that the disease could hardly be mistaken for any other trouble. It is true that malnutrition may cause somewhat similar malformations, but the latter is more localized in individual fields or areas, causes no mottling, produces yellowing of tips and edges of older leaves, and frequently results in a red color along the base of the leaf-blade and on the petiole.

Cause.

As is true of many virus diseases, the cause of blight was attributed first to unfavorable conditions. After considerable research the fact was established that the disease is due to a virus which is transmitted from diseased to healthy plants by means of potato and spinach aphids. The disease appears on the spinach within twelve to thirty days after the feeding of the contaminated insect. An infectious virus is present in the plant as soon as there is any indication of disease on the host. The infecting agent, however, is more virulent when taken from plants in the later stages of the disease for then the incubating period is not much more than half as long as when the inoculum is obtained from recently infected individuals. Some infections are obtained even though the aphids remain on the plants only five minutes. The insects are also able to inoculate five or more plants in succession after feeding once on diseased foliage. Molting of the aphid does not remove the contamination from its body. Furthermore, the virus may be transmitted from parent aphid to offspring for four or more generations.

In the South the spinach is grown during the winter months, while in the North the spinach is raised in the summer, except in greenhouses. It is not definitely known how the infectious virus is retained in the North during this period of inactivity. Both the potato and spinach aphid feed on many other plants. It is possible that some of these hosts are susceptible and perennial, permitting the virus to remain in the roots until the following spring. Such a bridging host, however, has not been discovered. In the southern states, the aphids themselves are capable of retaining the virus from one spinach crop until the next, and this may account for the fact that the disease is much more serious there than when the vegetable is grown as a summer crop.

Control of blight.

In localities in which blight is general, the only method of combating the trouble is that of procuring resistant varieties of the host. A large number of spinach varieties was tested for immunity in Virginia. The Manchuria was found to be the most resistant to the disease, but was not a desirable type for the market. It therefore was crossed with the Savoy spinach, which is one of the more favored varieties, and from the progeny of this cross selections were made of plants with the characters of size, savoying, color, closeness and shape of leaf of the Savoy parent, and the resistance of the Manchurian plant. The new variety was named the Virginia Savoy. It might be well for those growers who have suffered losses due to the blight, to request their state pathologist or county agricultural agent to procure a sample of this seed for them, and try it on their own farms.

REFERENCES

Jodidi, S. L. A mosaic disease of spinach as revealed by its nitrogen constituents. Jour. Amer. Chem. Soc. **42**: 1061-1070, 1883-1892. 1920.

McClintock, J. A. and L. B. Smith. True nature of spinach blight and the relation of insects to its transmission. Jour. Agr. Research **14**: 1-60. 1918.

Schoevers, T. A. C. Nieuwe ziekten waarop gelet moet worden. Tijdsch. o. Plantenz. **25**: 95-98. 1919.

Smith, L. B. Breeding mosaic resistant spinach and notes on malnutrition. Va. Truck Exp. Sta. Bull. 31-32: 137-160. 1920.

Smith, L. B. Notes on spinach breeding. Proc. Amer. Soc. Hort. Sci. 17: 146-155. 1921.
True, R. H. et al. Physiological studies of normal and blighted spinach. Jour. Agr. Research 15: 369-408. 1918.

Malnutrition of Spinach

Cause physiological

(See Malnutrition of Crucifers, page 167.)

The cause, effect, and control of malnutrition of spinach are the same as was described for a similar trouble on cabbage.

The plant affected with malnutrition is dwarfed, and has yellowed tips and edges of the foliage. The diseased leaves do not exhibit the normal savoy effect of healthy leaves, but are unusually thick and the tissue is very brittle. The petioles and basal portion of the leaf show a marked red discoloration. In severe cases the whole leaf turns red. The plant dies prematurely.

The symptoms of malnutrition are somewhat like those of blight or mosaic. Malnutrition, however, may be distinguished by it producing no true mottling, in its destroying the savoy effect of the leaf, in thickening the leaf, and causing the appearance of the abnormal red color.

References

Harter, L. L. The control of malnutrition diseases of truck crops. Va. Truck Exp. Sta. Bull. 1: 1-16. 1909.
Harter, L. L. Spinach troubles at Norfolk and improvement of trucking soils. Va. Truck Exp. Sta. Bull. 4: 61-80. 1910.
Smith, L. B. Breeding mosaic resistant spinach and notes on malnutrition. Va. Truck Exp. Sta. Bull. 31-32: 137-160. 1920.

Downy-Mildew of Spinach

Caused by *Peronospora effusa* (Grév.) Rabenh.

The downy-mildew or leaf-mold is prevalent wherever spinach is grown. It may be absent during certain seasons and then later destroy whole fields, or by appearing year after year, it constantly may menace the spinach crop. Plant disease survey reports show that in California where three hundred carloads of the vegetable were expected one season, the mildew cut the yield to twelve carloads.

According to the statements in literature, the parasite affects not only spinach but also swiss chard, lamb's quarter, and other plants belonging to the goosefoot family, although none of the investigators seems to have conducted careful cross inoculations.

Symptoms.

The disease on the leaves is manifested by large—or sometimes small—coalescing spots, which are differentiated from the healthy tissue by their pale yellow color. Any fractional part of the leaf up to the entire surface may be included in the spotting. The diseased leaves are scattered promiscuously on the plant, although the lower ones are first to be attacked. On the upper side of the leaf the spot is bare except in rare cases, while on the lower side the lesion is covered with a gray to violet-gray mold. The affected part finally decays or dries, and the whole leaf succumbs. In severe cases the entire plant is affected.

Cause.

The name *Peronospora effusa* is generally given to the spinach parasite, although most authors are agreed that there are two distinct varieties, which have been designated as *major* and *minor*. The variety *major* is distinct from the latter in that it has wider and taller conidiophores and the tips of the branches are not so narrowly or deeply cleft. Laubert, in order to simplify the matter, has suggested that the variety *minor* be named *Peronospora spinaciæ*. His suggestion, however, has not been accepted by later investigators.

The fungus in its life history and morphology is much like the other downy-mildew fungi on vegetables. The gray to violet mold is composed of tall branched conidiophores, the tips of the smaller branches being divided into two sterigmata arranged so that they resemble the claws of a bird. Oval or globose spores are borne on the sterigmata, and when mature germinate by means of a germ-tube. As the disease advances in the leaf, the fungus produces a large number of thick-walled oospores, which are able to remain alive during prolonged periods of inactivity. Newhall states that there is some basis for believing that the oospores are carried with the spinach seed, and serve as sources of inoculum for the new crop. In the field the conidia are scattered by rain, on tools, and on the clothing of workmen.

The parasite demands a large amount of moisture. Consequently it develops best when the crop is grown in undrained soil or irrigation water is used too freely, when the plants stand too closely together, or when weeds are permitted to grow between the rows and hinder drying of the foliage. Like other mildew fungi, the conidia need a relatively low temperature for germination. When such low temperatures are followed by the increased warmth necessary for infection, the fungus proves to be very destructive.

Control of downy-mildew.

Aside from the use of a fungicide, it is well to correct any unfavorable environmental conditions to which the spinach may be subjected such as supplying good drainage, practicing careful tillage, destruction of weeds, applications of suitable fertilizers, and not crowding the plants. The fungicide which has given some indication of success is composed of one-half pound of blue-vitriol dissolved in twenty-five gallons of water and five pounds of caustic-potash fish-oil soap dissolved also in twenty-five gallons of water. The two dilute solutions are then poured together, being stirred violently during the mixing. The first spray applications are made before there is any chance of infection, and repeated if necessary at three- to seven-day intervals. All parts of the plants should be covered with the liquid and the spray applied under high pressure.

References

Eriksson, J. Zur entwickelungsgeschichte des spinatschimmels. Arkiv. f. Bot. 15: 1-25. 1918.

Halsted, B. D. Some fungous diseases of the spinach. N. J. Agr. Exp. Sta. Bull. 70: 1-15. 1890.

Laubert, R. Der "falsche mehltau" (Peronospora) des spinats und des gänsfüsses. Gartenflora 55: 435-440, 461-464. 1906.

Smith, L. B. Control of spinach leaf-mold (downy mildew) by spraying. Va. State Crop and Pest Comm. Quart. Bull. 3: 1-4. 1921.

Heterosporium Leaf-Spot of Spinach

Caused by *Heterosporium variable* Cke.

The leaf-spot of spinach, or rust as growers name it, has been known for more than fifty years, and probably is present wherever the host is grown extensively. When weather conditions are favorable for the parasite or the plants are weakened for some reason, the leaf-spot may cause the loss of almost the entire crop.

Symptoms.

The lesions on the foliage begin as small brown spots, which quickly enlarge and multiply in number until so much of the leaf surface is included that the remainder turns yellow and later dies. While the spots are enlarging, they become much more conspicuous by the formation of a greenish-black mold on both the upper and lower surfaces of the leaf. The whole plant may die, although frequently there are just enough injured leaves to make the crop unsalable.

Cause.

The fungus was named *Heterosporium variable* in 1877. Before this date and since then, it has also been classed as a Helminthosporium and a Cladosporium. It is a weak parasite, gaining entrance into the host after the latter has been injured by too much rain, by cold weather, by insects, by other dis-cases, or in any other manner in which the vitality of the plant is reduced. It is especially prevalent in fields in which the crop has been affected by malnutrition. The coarse abundant mycelium in the leaf gives rise to large olive-colored one- to three-septate spores. It is this fruiting layer that gives the green moldy appearance to the older lesions. Small oval spores may also be formed by budding at the end of mycelial threads or on differentiated conidiophores. They may be borne singly or in chains.

It is not known exactly how the fungus survives the winter, as the spinach rots so quickly that within a few months no vestige of the tissue is left. It may be possible that the organism can live in the soil without the presence of the host. It perhaps can also live on dead stubble of other field crops, and thereby remain alive until the next spinach crop is available.

The fungus requires an excessive amount of moisture and apparently is favored by low temperatures. At least it is prevalent on plants after they have been stunted or otherwise injured by cold weather.

Control.

No definite control measures have been suggested. Thriftily growing plants on well drained soil are resistant to the leaf-spot. In addition to having such plants it may be profitable to spray with the copper-soap mixture recommended for the control of downy-mildew of spinach.

References

Clinton, G. P. Notes on plant diseases of Connecticut. Conn. Agr. Exp. Sta. Ann. Rept. **1909-10: 714.** 1910.

Cooke, M. C. On Heterosporium. Grevillea **5: 122-123. 1877.**

Jennison, H. M. A spinach disease new to Massachusetts. Mass. Agr. Exp. Sta. Ann. Rept. **23: 146-148.** 1911.

Reed, H. S. and J. S. Cooley. *Heterosporium variable,* its relation to *Spinacia oleracea* and environmental factors. Va. Agr. Exp. Sta. Ann. Rept. **1909-10: 78-99.** 1910; Centralbl. f. Bakt. u. Par. II **32: 40-58.** 1911.

Anthracnose of Spinach

Caused by *Colletotrichum spinaciæ* E. and H.

The anthracnose of spinach is evidently not of great importance, except in occasional fields. It was first observed in New Jersey in **1890** and since then been observed in Texas, Virginia, Italy, and probably in other places. The disease causes few to many small circular spots on the leaf, which at first are water-soaked in appearance but which later turn gray or brown. The lesions may become so numerous that the plant is made unfit for use. In five to seven days after the first indication of the spots, the affected parts on both sides of the leaf are bearing the acervuli of the parasite. Setæ are very evident among the conidiophores. The small oval unicellular spore sends out a branching hypha which grows into a host stoma, causing a new infection.

No control measures have been suggested. Perhaps the spray applications recommended for downy-mildew may be helpful, although this has not been proved.

References

Halsted, B. D. Some fungous diseases of the spinach. N. J. Agr. Exp. Sta. Bull. **70**: 1-15. 1890.

Reed, H. S. and J. S. Cooley. *Heterosporium variable,* its relation to *Spinacia oleracea* and environmental factors. Va. Agr. Sta. Ann. Rept. 1909-10: 78-99. 1911.

Voglino, P. Intorno ad un nuovo deperimento degli spinaci. Ann. R. Accad. Agr. Torino **56**: 377-379. 1913.

Rust of Spinach

Caused by *Puccinia subnitens* Dietel

In Oregon a rust with bright orange cluster-cups caused considerable damage to spinach in **1922**. As the disease had not been seen before, it was surmised that the fungus may have its telial stage on some local plant and spread to spinach. After a search it was found that the salt-grass or alkali-grass, which was very abundant about the spinach fields, was covered with the telial stage of *Puccinia subnitens* and that this rust was able to infect spinach and beet.

The only means of controlling the parasite is the destruction of the alkali-grass, or not growing spinach where such grass is present.

Reference

Barss, H. P. Destructive rust (*Puccinia subnitens* Dietel) on spinach in the Northwest. Phytopath. **12:** 446. 1922.

Cercospora Leaf-Spot of Spinach

Caused by *Cercospora flagelliformis* E. and H.

In New Jersey, Indiana, and Minnesota a Cercospora leaf-spot of spinach has been reported, but in most cases it is only a minor trouble. The disease begins as small dead areas on the leaf which increase in number so rapidly that the whole plant may turn yellow and be much stunted in growth.

The fungus is characterized by unusually long thin septate spores which are of the usual Cercospora shape and borne on short conidiophores. The fungus probably lives in old diseased plant refuse left in the field. It is not known whether it is ever disseminated with the spinach seed.

Spraying as suggested for downy-mildew of spinach may protect the leaves from infection, although this has not been proved experimentally.

References

Halsted, B. D. Spinach. N. J. Agr. Exp. Sta. Ann. Rept. **1890:** 354-355. 1891.

Plant Disease Survey Office. Plant Disease Survey Bull. Suppl. **26:** 161. 1923.

Scab or Black-Mold of Spinach

Caused by *Cladosporium macrocarpum* Preuss

Old leaves or those that have been injured by unfavorable environment may at times be attacked by Cladosporium, which causes a black-mold over the leaf surface. The fungus can hardly be classed as a parasite, since it does not attack healthy plants. After it once gains an entrance into the host it may cause considerable injury.

Reference

Halsted, B. D. Some fungous diseases of the spinach. N. J. Agr. Exp. Sta. Bull. 70: 1-15. 1890.

Smut of Spinach

Caused by *Entyloma ellisii* Hals

A number of years ago the smut of spinach was reported in New Jersey. It does not seem to have been found since then. It causes white spots on the leaf. The only interesting part of the disease is that the smut spores are borne in the tissue under the stomata and germinate while still grouped among the leaf-cells. The basidia are pushed out in clusters through the stomata and bear needle-shaped basidiospores, thus superficially appearing like fruit-bodies of some imperfect fungus.

Reference

Halsted, B. D. Some fungous diseases of the spinach. N. J. Agr. Exp. Sta. Bull. 70: 1-15. 1890.

Phyllosticta Leaf-Spot of Spinach

Caused by *Phyllosticta chenopodii* Sacc.

The Phyllosticta leaf-spot is an unimportant trouble that occurs occasionally on spinach. The lesions are circular brown spots with pink borders and marked by numerous black pycnidia. The one-celled hyaline spores are discharged in long tendrils.

No control measures are necessary.

References

Halsted, B. D. Some fungous diseases of the spinach. N. J. Agr. Exp. Sta. Bull. 70: 1-15. 1890.

Reed, H. S. and J. S. Cooley. *Heterosporium variable,* its relation to *Spinacia oleracea* and environmental factors. Va. Agr. Exp. Sta. Ann. Rept. 1909-10; 78-99. 1911.

Fusarium Wilt of Spinach

Caused by *Fusarium spinacæ* Hungerford

In Idaho and Texas a Fusarium wilt has recently caused injury in some spinach fields. The trouble is made evident by the yellow color and stunting of the plant and the rolling of the margins of the leaves. The plant finally dies. The root system is almost completely destroyed and the vascular tissue of the stem is darkened.

The causal organism is a Fusarium with mostly three-septate spores. It evidently lives in the soil and enters the host through the roots.

There is no practicable means of eradicating the fungus in the soil. In Texas where the fungus is supposed to be a different species, the New Zealand spinach has proved to be resistant and may be grown where the disease is destructive.

References

Hungerford, C. W. A Fusarium wilt of spinach. Phytopath. **12**: 447. 1922; 13: 205-209. 1923.

Taubenhaus, J. J. Studies on Fusarium wilt of spinach in Texas. Phytopath. 14: 29. 1924.

Crown-Rot of Spinach

Caused by *Phytophthora* sp.

A crown-rot of spinach has been discovered doing much damage in a few fields in New Jersey and Indiana. In the latter place, a Phytophthora was isolated from the diseased tissue. Excessive watering seems to favor the fungus.

Reference

Plant Disease Survey Office. The Plant Disease Bull. Suppl. **26**: 163. 1923; **22**: 414. 1922.

Damping-off of Spinach

Caused by various fungi

(See Damping-off of Tomatoes, page 546)

Root-Knot of Spinach

Caused by *Heterodera radicicola* (Greef) Müll.

(See Root-Knot of Tomatoes, page 550)

CHAPTER XVII

SWEET-POTATO DISEASES

THE sweet-potato crop, due to the encouragement the growers received from modern control methods, has increased enormously during the last ten years. According to the census report of the United States the 1909 acreage of sweet-potatoes and yams was 643,225, and in 1919, 803,727. The number of bushels increased even more rapidly, jumping from fifty-nine millions to seventy-eight millions during the same time. The interesting point, however, is the rapidity with which prices rose as the diseases were controlled and the quality of the sweet-potato improved. Instead of a 25 per cent increase, as was true of the acreage, the value of the crop increased 400 per cent. No doubt some of this enhancement was the result of war conditions, but certainly a large share of the increase was brought about by the confidence with which distributors could buy potatoes and hold them in a healthy condition until the market demanded the product.

All the states in which climatic conditions are favorable for the growth of sweet-potatoes devote a certain acreage to this crop. Following are the more important states in the order of their produciton as given by the census of 1920: Georgia, Alabama, North Carolina, Mississippi, Virginia, Texas, South Carolina, Louisiana, Tennessee, Arkansas, Florida, Oklahoma, New Jersey, Delaware, Maryland, Kentucky, Missouri, California, Illinois, Kansas, Ohio, West Virginia, Iowa, Indiana, and Pennsylvania.

Notwithstanding the efforts which have been made for the eradication of the sweet-potato pathogenes, there is still an excessively large loss from the many diseases. The Plant

Disease Survey Office estimates of the total losses are 32 per cent for 1917, 36 per cent for 1919, 26 per cent for 1920, and 28 per cent for 1921. These estimates are probably low, but show that more than one-fourth of the crop is sacrificed to the ravages of fungi.

The most noteworthy recent step in the progress of sweet-potato production is the organization of an inspection and certification service for stock which is to serve as seed the following year. In nearly all cases, this is wholly voluntary on the part of the grower, who makes a written application to the proper authorities, and signs an agreement to follow certain given directions. In return his crop is inspected once while in the seed-bed, once when in the field, and once after harvest. If his potatoes meet the pre-determined standards, he is given certification tags, which he is permitted to attach to his containers, and which testify to the good qualities of the crop. This makes it possible for the consumer to know what he is buying. As the standards become higher and the inspection service more general, there is sure to be an ever increasing acceleration of improvement, until every grower has caught the vision of what constitutes good cultural practices for the sweet-potato.

References

Plant Disease Bulletins Volumes 1-6 and Supplements 12, 18, 24, and 34. 1917 — 1923.

Fourteenth Census of the U. S.—1920. 6: part 1. 27, 73. 1922.

Black-Rot of Sweet-Potatoes

Caused by *Sphæronema fimbriatum* (E. and H.) Sacc.

The history of black-rot prior to 1890 is not well known, but at that time it was found by Halsted in New Jersey, Maryland, Delaware, and Virginia. It is now known to occur wherever sweet-potatoes are grown, and is one of the most important field and storage troubles with which growers have to contend. It not only attacks all varieties of cultivated sweet-potatoes

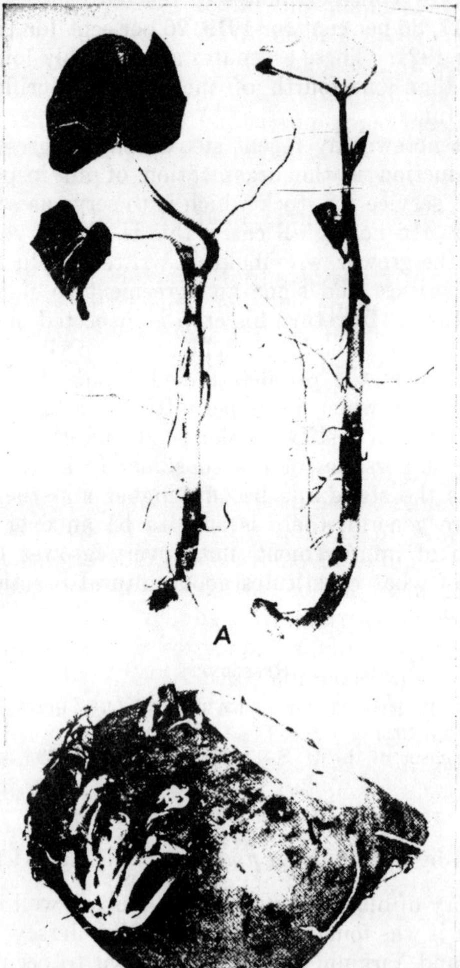

FIG. 132.—Black-rot of sweet potato. A, cankers on underground stem; B, rotting of young sprouts on a bedded sweet-potato.

but has also been found on the wild morning-glory (*Ipomœa purpurea*) and on the wild sweet-potato (*Ipomœa pandurata*).

Symptoms.

All underground parts of the plant are attacked (Figs. 132, 133), the most conspicuous lesions being on the roots where

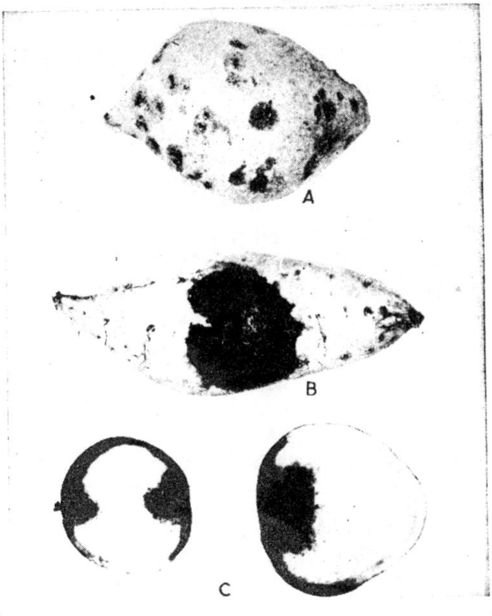

FIG. 133.—Black-rot of sweet-potato. A, a healthy root inoculated with spores of Sphæronema; B, a typical black-rot spot; C, cross-section of root showing depth to which the fungus penetrates.

dark circular areas may be observed varying in size from those which can just be seen to those which cover a large part of the potato. In the center or in the older portion of the lesion, small black bodies or pycnidia often develop. When the disease is not followed by saprophytic organisms the lesions usually extend inward only to the vascular ring, but Fusaria or other fungi often follow and rot the entire root. The diseased tissue is green in color and bitter in taste, the bitterness becoming more evident the longer the potatoes are kept in storage. The distasteful substance is soluble so that in cooking it is diffused throughout the healthy tissue.

In the hotbed, black spots occur on the young sprouts. These spots may eventually girdle the plant, causing dwarfing and yellowing of the foliage and finally death. The blackening involves the roots and may extend upward until an inch or more of the stem is included. This symptom is often referred to as "black-shank." The disease at times attacks the stems and petioles of the young sprouts, although this is not so common as is the infection on the small roots.

On the vines in the fields the symptoms are similar to those on the sprouts. The blackened part seldom includes the whole stem, being usually confined to the areas between two leaves. The part of the vine above the diseased tissue wilts and dies. In other cases, the vine may not be attacked above ground but have every root affected with black-rot. Ordinarily the color of the diseased roots is somewhat paler than that of the normal ones.

Cause.

The fungus causing the disease is *Sphærонema fimbriatum.* A perfect stage has been named but is now known not to be related to the Sphæronema.

Several spore stages have been described as occurring in the life history of the pathogene. The more common method of spore formation is in long-necked pycnidia found in the epidermis of the older parts of the lesions, from which globose spores are discharged in a gelatinous mass. Some of the long terminal cells of the hyphæ may each enclose a sheath in which are borne endogenously a chain of hyaline spores. Thick-walled olive-tinged chlamydospores may be borne singly or in chains on the hyphæ within the infected tissue. In addition the cells of the mycelium break apart and each cell functions as a spore.

The infected tissue of the roots is invaded by thick-walled olive-colored hyphæ. When the roots are planted in the hotbed either this mycelium grows up into the sprouts causing infection, or some of the various types of spores are disseminated in the seed-bed by insects, water, or mechanically and serve as the necessary inoculum. The sprouts which are not too seriously injured are often transplanted and thus convey the disease to the field. Undoubtedly also, the wild morning-glory and the wild sweet-potato may act as a harborers of the pathogene.

The black-rot organism may attack healthy roots but more often the disease follows injury. When the potatoes are put in storage where the conditions are favorable the parasite fruits abundantly, and the numerous mites which are generally present carry the inoculum to other potatoes. A few diseased potatoes, therefore, may cause infection throughout an entire bin.

Control of black-rot.

The first requisite is clean seed, which can be obtained only by making selections in the autumn and again in the spring from good lots of seed potatoes, or if this is not possible by buying certified seed. The second is seed treatment. One ounce of corrosive sublimate is dissolved in a little hot water, then enough cold water added to make eight gallons of the solution. After ten bushels have been treated, one-half ounce of the poison is added to each thirty-two gallons of the used solution. This is repeated until fifty bushels have been dipped, after which the old solution is discarded and a new one made. The solution should always be kept at its original volume. The seed potatoes are treated eight to ten minutes, after which they are bedded. The precautions suggested on page 622 about the use of corrosive sublimate should be kept in mind. The solution is weakened when placed in metal containers, or when gunny sacks are dipped into it.

The boards or cement used about the hotbeds should be scrubbed with blue vitriol solution (one pound in five gallons of water) or with formaldehyde (one part in fifty parts of water). If the soil is not renewed each year it is imperative that the bed be sterilized (see page 600 for methods). If manure is added to produce heat, it cannot be sterilized because the fermenting bacteria would then be killed. Only manure free from sweet-potato refuse can be used. If there is some other method available for heating, and manure is added merely as fertilizer, it should be applied before sterilization because manure is one of the chief causes of contamination, since much sweet-potato refuse is thrown on the manure heap.

In addition to seed selection, disinfection, and soil sterilization, as long rotations as possible should be practiced and the weed hosts eradicated.

The storage problem is another phase in the control of black-rot. It is necessary to take every precaution for the elimination of the disease in the bin, directions for which are given in the control of soft-rot.

References

Chester, F. D. The black-rot of the sweet potato. *Ceratocystis fimbriata* Ell. and Hals. Del. Agr. Exp. Sta. Ann. Rept. **3**: 90-91. 1890.

Elliott, John A. The ascigerous stage of the sweet potato black-rot fungus. Phytopath. **13**: 56. 1923.

Halsted, Byron D. Some fungous diseases of the sweet potato. The black-rot. N. J. Agr. Exp. Sta. Bull. 76: 7-14. 1890.

Halsted, Byron D. and D. G. Fairchild. Sweet potato black-rot. Jour. Mycology **7**: 1-11. 1891.

Harter, L. L. Control of the black-rot and stem-rot of the sweet potato. U. S. Dept. Agr. Bur. Plant Ind. Circ. 114: 15-18. 1913.

Harter, L. L. Notes on the distribution and prevalence of three important sweet potato diseases. Phytopath. **5**: 124-126. 1915.

Taubenhaus, J. J. The black-rot of the sweet potato. Phytopath. **3**: 159-165. 1913.

Taubenhaus, J. J. and Thomas Manus. The diseases of the sweet potato and their control. Del. Agr. Exp. Sta. Bull. 109: 1-55. 1915.

Weimer, J. L. Reduction in strength of the mercuric chlorid solution used for disinfecting sweet potatoes. Jour. Agr. Research **21**: 575-587. 1921.

Stem-Rot of Sweet-Potatoes

Caused by *Fusarium batatatis* Woll. and *Fusarium hyperoxysporum* Woll.

The first important description to be published of sweet-potato diseases was an article by Halsted in 1890. It is, therefore, impossible to obtain an accurate account of stem-rot before that time even though growers insist that it was present at least twenty-years prior to the above date. It is now known to be prevalent generally wherever sweet-potatoes are grown, being considered one of the most important problems of this crop. It is not a storage trouble, but affected roots decay more readily because of the ease with which storage-rot organisms can gain entrance. The biggest source of loss is the reduction in yield caused by weak and missing hills. Because of its various symptoms it is known locally under the names blue-stem, yellows, and wilt.

There are no immune strains of sweet-potatoes but certain varieties show much less susceptibility than others. Among the more susceptible ones are listed Early Carolina, Yellow Jersey, Red Jersey, and Nancy Hall while the more resistant are Dahoney, Red Brazilian, Pierson, Yellow Strassburg, and Miles Yam. The only host besides sweet-potato on which the disease has been found is *Ipomœa hederacea.*

Symptoms.

The first indication of the disease is a slight off-color or yellowish tinge of the vines, followed later by more pronounced discoloration, puckering of the leaves, and lastly by the wilting of the plant When the petioles of the older leaves are attacked they drop off leaving those near the tip of the vine which usually are the first to show the symptoms of yellowing. When young vines are affected many short stems may begin to grow at the center of the hill, giving a rosette appearance. Toward the last of the season the infested field can be recognized by the number of missing hills.

The term stem-rot is somewhat misleading since there is no actual rotting of the tissue unless the disease permits entrance for saprophytic organisms. If the cortex is peeled away from the diseased stem the exposed tissue will show a dark discoloration, which has led some growers to call the disease blue-stem. The discoloration of the vascular bundle may extend long distances up and down the stem and into the root where, by clipping the stem-end, it may be observed. In hot dry weather the base of the diseased stem may swell and the epidermis burst. If the vine dies and wet weather then occurs, the pathogene may produce a pinkish mass of spores on the dead stem.

Cause.

When the first work was done with stem-rot it was supposed that the fungus, *Nectria ipomoeæ* Hals. was the cause, but more recent investigations have shown that the two fungi, *Fusarium batatatis* and *Fusarium hyperoxysporum,* are responsible for this disease. They live in the soil as well as in the seed potatoes, and are disseminated with the soil by such methods as flooding, wind storms over the dry sand, on the feet of men and animals, with cultivating tools, and in any other way in which particles of soil from an infested field may be conveyed from one place to another. The transferring of the pathogene over long distances is most generally accomplished by infected seed potatoes. These when planted in the hotbeds may or may not produce diseased sprouts. In some cases the mycelium grows from the mother potato directly into the vascular bundle of the slips where the darkened streak may be seen through the hyaline tissue when held up to the light.

The life histories of the two fungi, either or both of which may be present, are apparently very simple. A large amount of mycelium is grown within or upon the host. During the rainy weather an immense number of long septate sickle-shaped spores are borne. It is not known in what form the pathogenes live over long periods of time when no susceptible crops are grown. It is conceivable that other hosts not yet discovered keep the fungi alive during the intervening periods, or it is possible that they live saprophytically in the soil almost indefinitely.

Control of stem-rot.

A soil-infesting parasite is very difficult to control when it attacks a field crop. About the only recourse is that of selecting or breeding resistant strains of the host. This is being done at present so that there are already certain varieties of sweet-potatoes which may be grown with a reasonable assurance of a healthy progeny. Until such strains become more common there are certain other practices well worth keeping in mind. All the points mentioned in regard to the control of black-rot are applicable to stem-rot. Long rotations are emphasized. In the fall before frost has injured the vines, seed potatoes should be selected from hills in which no blackening of the tissue shows when the stems are split. Too strong an endorsement of the use of certified seed for planting cannot be made. The inspections of certified seed may not in each case be perfect, but such seed certainly is much better than the average seed on the market.

In some localities an additional step has been taken in fighting the stem-rot. After the selection of the best potatoes, seed treatment, seed-bed sterilization, and long rotations, the young plants in the field and which have begun to grow well are inspected carefully. Cuttings are made from the stems of those plants which show no discoloration of the vascular bundles when the cortex is removed near the base of the plant. The vines are cut into lengths so that there are at least two buds on each piece, and are then transplanted in soil in an isolated seed-plot. The purpose of using the vine cuttings is defeated if they are placed in an infested field. If the proper care is taken the seed from such cuttings should be free from stem-rot and can be used for propagation the following year in a seed plot.

References

Cook, Mel T. and R. F. Poole. Stem-rot diseases of sweet potatoes in New Jersey. Phytopath. 12: 51. 1922.

Halsted, Byron D. Some fungous diseases of the sweet potato. The stem-rot. N. J. Agr. Exp. Sta. Bull. 76: 20-22. 1890.

Harter, L. L. Control of the black-rot and stem-rot of the sweet potato. U. S. Dept. Agr. Bur. Plant Ind. Circ. 114: 15-18. 1913.

Harter, L. L. Notes on the distribution and prevalence of three important sweet potato diseases. Phytopath. 5: 124-126. 1915.

Harter, L. L. and Ethel C. Field. *Fusarium batatatis* Woll. MS., not *Nectria ipomœcæ* Hals. the cause of the sweet potato stem-rot. Phytopath. 3: 68. 1913.

Harter, L. L. and Ethel C. Field. The stem-rot of the sweet potato (*Ipomœa batatas*). Phytopath. 4: 279-304. 1914.

Harter, L. L. and Ethel C. Field. Experiments on the susceptibility of sweet potato varieties to stem-rot. Phytopath. 5: 163-168. 1915.

Harter, L. L. and J. L. Weimer. Sweet potato stem-rot and tomato wilt. Phytopath. 10: 306, 307. 1920.

Poole, R. Frank. Recent investigations on the control of three important field diseases of sweet potato. N. J. Agr. Exp. Sta. Bull. 365: 1-39. 1922.

Taubenhaus, J. J. Some recent studies on new or little known diseases of the sweet potato. Phytopath. 4: 50, 51, 305-320. 1914.

Wollenweber, H. W. Identification of species of Fusarium occurring on the sweet potato, *Ipomœa batatas*. Jour. Agr. Research 2: 251-286. 1914.

Foot-Rot of Sweet-Potatoes

Caused by *Plenodomus destruens* Harter

Foot-rot or die-off was first recognized as a distinct disease in 1912 when Harter received specimens from Virginia. It probably was present many years before but had not been brought to the attention of the pathologists. It is now present in nearly all the states where sweet-potatoes can be grown, having been reported to the Plant Disease Survey office from Alabama, California, Florida, Iowa, Kansas, Maryland, Mississippi, Missouri, New Jersey, North Carolina, Ohio, Oklahoma, Tennessee, and Virginia. In the brief time in which the disease has been known it has progressed from relative unimportance to one of the most destructive field troubles of the sweet-potato. No variety of the host has been reported as being resistant. It has also been made to infect the closely related plant, *Ipomœa coccinea.*

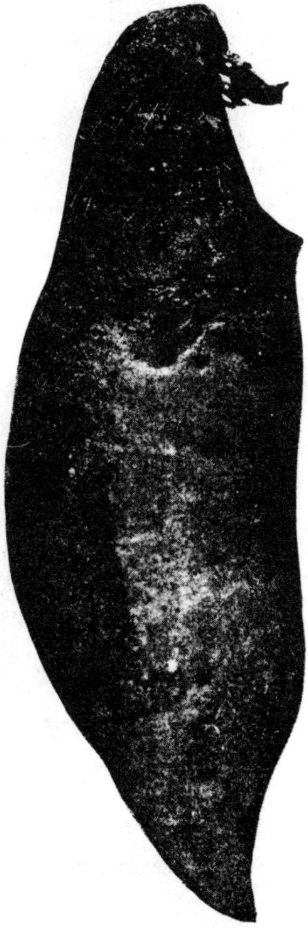

FIG. 134.—*Plenodomus destruens* causing rot of sweet-potato root.

Symptoms.

The base of the stem of a diseased plant turns brown from just below the surface of the soil to four or five inches above the soil line. The vines in low wet fields are sometimes affected several feet from the hill. Soon after infection takes place the lower leaves turn yellow and drop off. Unless adventitious roots are put out by the plant at the nodes of the stem to support itself, the vine wilts and dies. Even before the vine wilts fruit-bodies of the fungus begin to appear in the affected tissue, forming numerous dark papillæ-like bodies.

The disease is usually not conspicuous in the seed-bed and may even not be evident in the field until midseason. Consequently a number of affected vines may produce potatoes which in the course of time are attacked. The firm brown rot ordinarily does not include the whole root, but enough of it to make it worthless for food purposes or for sale (Figs. 134, 135). When these roots are kept in a moist place the conidia from the fruit-bodies embedded in the diseased tissue ooze out in viscous drops and collect on the surface of the root.

FIG. 135.—A, longitudinal section of sweet-potato showing rot caused by *Plenodomus destruens;* B, numerous small infections on the root caused by sweet-potato scurf; C, a root entirely covered with the scurf.

Cause.

The pathogene is *Plenodomus destruens,* a fungus with the pycnidial form of fruit-bodies. These pycnidia yield great numbers of spores, which not only help to disseminate the organism at once but are able to retain their viability through the winter and cause infection in the spring. In connection with the pycnidia chlamydospore-like bodies are borne within the host tissue but their exact function in the propagation of the fungus has not yet been determined. Apparently the fungus does not live in the soil unless parts of the host are present. When affected potatoes are bedded the sprouts growing from them are liable to have the disease. The symptoms are not always evident so that it is common to have the sprouts transplanted into the fields. Diseased potatoes are often thrown upon the manure pile and then spread out on the soil where they serve as a ready source of inoculum for the new crop. When the stem of the host becomes affected the parasite grows downward until it invades the roots, or the latter may become infected directly from the spores developed on the diseased stem.

The disease is more severe when warm weather is accompanied by high humidity. In the laboratory the spores germinate best at temperatures ranging from 75° to 95° F.

Control of foot-rot.

The recommendations suggested for the control of black-rot of sweet-potatoes will suffice for the foot-rot. Seed selection and treatment are of particular importance. It has been found, also, that more vigorous sprouts can be grown in beds under glass sashes than under cloth. The glass may not influence directly the presence or absence of the disease, but any method that gives greater vigor to the plant is to be commended. The faster plants develop the more chance there is of growing a crop before the disease, which is rather slow in developing, makes its appearance. The sprouts start about a week earlier under glass than under cloth.

It is maintained by some growers that common salt broadcast over the field at the rate of about 500 pounds an acre a month or more before planting will help in eradicating the organism. In Virginia it was proved that although this treatment may give a slightly greater yield in dry seasons, the results are not due to any deleterious effect which the salt has on the fungus.

References

Harter, L. L. Foot-rot, a new disease of the sweet potato. Phytopath. 3: 243-245. 1913.

Harter, L. L. Notes on the distribution and prevalence of three important sweet potato diseases. Phytopath. 5: 124-126. 1915.

Harter, L. L. The foot-rot of the sweet potato. Jour. Agr. Research 1: 251-274. 1913.

McClintock, J. A. Sweet potato diseases. Foot-rot. Va. Agr. Truck Exp. Sta. Bull. 22: 465-472. 1917.

Scurf of Sweet-Potatoes

Caused by *Monilochætes infuscans* Hals.

Scurf, also known as soil-stain, rust, and Jersey-mark has been present in the United States ever since the diseases of sweet-potatoes have been studied, and is now found wherever the host is grown. The loss incurred is not so great as is that caused by the more important maladies, but the shrinkage in storage and the reduction in price due to the unsightliness amount to a considerable sum each year. All varieties of the sweet-potato are more or less susceptible although the scurf is more pronounced on the white-skinned ones than on the yellow varieties, because the latter are normally more nearly the color of the stain caused by the fungus.

Symptoms.

In the field the lesion starts as a small brown speck on the root, and under very favorable conditions may continue to spread even after harvest. When there are many such small brown patches they may coalesce, forming a uniform rusting of a part or of the entire surface of the potato (Fig. 135). The disease is wholly superficial, seldom causing any direct damage to the potato, but the skin is injured enough so that the moisture can escape, permitting the potato to shrivel in storage

FIG. 136.—Sweet-potatoes affected with scurf-shrivel in storage.

(Fig. 136). When the disease is more virulent than usual, the potato may crack (Fig. 137) and in this way permit the entrance of decay organisms.

Cause.

The fungus is called *Monilochœtes infuscans,* which means the black bristly Monilia. It was so named because the black

Fig. 137.—Sweet-potatoes with scurf sometimes crack under conditions of excessive moisture.

conidiophores stick up from the surface of the lesion like minute bristles, and bear spores in chains like the common genus Monilia. The organism over-winters on the potatoes and in decaying vines in the field. When the diseased potatoes are placed in the hotbed the fungus grows on the sprouts and is carried to the field, where it passes down to the roots again. In this way the simple life cycle is completed. The fungus may survive for some time on humus, for it has been shown that the disease is more prevalent in heavy wet soil containing a large percentage of organic matter. An alkaline soil also favors the growth of the parasite.

Control of scurf.

The recommendations are the same as those given for the control of black-rot, which include seed selection, seed treatment, seed-bed sterilization, long rotation, and care in storage.

Recent work in New Jersey has shown that highly beneficial results are obtained by the application of 300 to 400 pounds an acre of inoculated sulfur, broadcast about one month before setting the crop. A grain drill will prove more satisfactory in making such applications than in trying to broadcast by hand, for it can be done more uniformly with the drill. After the application the sulfur is dragged into the ground with a harrow.

References

Halsted, Byron D. Some diseases of the sweet potato. The sweet potato scurf. N. J. Agr. Exp. Sta. Bull. 76: 25-27. 1890.

Harter, L. L. Sweet potato scurf. Jour. Agr. Research 5: 787-791. 1916.

Harter, L. L. Sweet potato diseases. Scurf (soil-stain, rust, Jersey mark). U. S. Dept. Agr. Farmers' Bull. 1059: 13-15. 1919.

McClintock, J. A. Sweet potato diseases. Scurf or soil-stain. Va. Truck Exp. Sta. Bull. 22: 461-463. 1917.

Poole, R. Frank. Recent investigations on the control of three important field diseases of sweet potatoes. N. J. Agr. Exp. Sta. Bull. 365: 1-39. 1922.

Taubenhaus, J. J. Soil-stain and pox, two little known diseases of the sweet potato. Phytopath. 4: 405. 1914.

Taubenhaus, J. J. Soil-stain, or scurf, of the sweet potato. Jour. Agr. Research 5: 995-1002. 1916.

Soil-Rot of Sweet-Potatoes

Caused by *Cystospora batata* Elliott

Like nearly all the other common diseases of the sweet-potato, soil-rot, or as it is more appropriately called pit or pox, was first described by Halsted in 1890. It is present in most of the sweet-potato-growing states, but evidently is most prevalent in New Jersey, Delaware, and Maryland. In certain fields where the soil is alkaline the loss is extremely heavy, although the average loss for the whole country is small. Soil-rot not only attacks the sweet-potato but can also infect the Irish potato and the turnip.

Symptoms.

The disease is present on the underground parts of the plants in the field, seldom having been observed in the seed-bed.

FIG. 138.—Sweet-potatoes affected with soil-rot compared with those that are healthy.

The lesions are at first small dark dry spots on the surface of the potato. After a time the affected tissue dries and falls out leaving a pit or pox mark, which resembles the scab-spot of the Irish potato (Fig. 138). While the root is still small the lesions begin to appear, especially when the moisture is conserved by a bend in the root or when two roots touch each other. On being affected such roots are checked in their growth under the pox spot, but continue to enlarge elsewhere, making a distorted unsightly root. Older roots when infected do not become misshapen, but are covered with the characteristic pits. Affected potatoes are often attacked through these pits by one or more of the numerous soil or storage-rot organisms.

Cause.

According to Elliott soil-rot is caused by a slime-mold for which he proposed the name *Cystospora batata.* This work has been confirmed by Taubenhaus, yet, because of their inability to duplicate the experiments, it has not been accepted by all the other pathologists.

Control.

Some interesting work has been done recently in New Jersey in checking up the control measures which Halsted suggested in 1898. It was found that by the addition of inoculated sulfur, applied by broadcasting 300 to 400 pounds an acre just previous to the making up of the ridges, gives very good results. It is also recommended that 4 or 5 per cent of nitrogen be applied to the infested soil instead of a lighter amount which is frequently used. In addition to the treatment of the soil, the same care should be given the potatoes as is advised for the control of black-rot.

References

Elliott, John A. The sweet potato "soil-rot" or "pox," a slime-mold disease. Del. Agr. Exp. Sta. Bull. 114: 1-25. 1916.

Halsted, Byron D. Some fungous diseases of the sweet potato. The soil-rot. N. J. Sta. Bull. 76: 14-20. 1890.

Poole, R. F. Recent investigations on the control of three important field diseases of sweet potatoes. N. J. Agr. Exp. Sta. Bull. 365: 5-39. 1922.

Taubenhaus, J. J. Pox, or pit (soil-rot), of the sweet potato. Jour. Agr. Research 13: 437-450. 1918.

Taubenhaus, J. J. Soil-stain and pox, two little known diseases of the sweet potatoes. Phytopath. 4: 405. 1914.
Taubenhaus, J. J. and Thos. F. Manns. The diseases of the sweet potatoes, and their control. Del. Agr. Exp. Sta. Bull. 109: 1-55. 1915.

Southern Sclerotium Rot of Sweet-Potatoes

Caused by *Sclerotium rolfsii* Sacc.

The Sclerotium rot was discovered on tomatoes in Florida in 1892. Even at that time it was destroying whole fields of the crop. Since then it has been found in nearly all the states south of Maryland and Illinois, and on many hosts. Some of the vegetables that are injured are tomatoes, sweet-potatoes, Irish potatoes, peppers, watermelons, muskmelons, radish, cabbage, young corn, bean, rhubarb, beet and eggplant, as well as many other cultivated and weed hosts. The fungus is most virulent on young succulent plants, although old stems are not immune. The parasite frequently is very destructive in storage on such vegetables as sweet-potatoes, pumpkins, squashes, cabbage, and Irish potatoes.

Symptoms.

Both in the seed-bed and in the field when plants are standing closely together, the disease has its origin at one to a few centers and from these spreads in all directions until a considerable area is invaded. A white mycelial weft spreads over the soil and the affected plant parts. Infection on the stem takes place just below the surface of the ground, then passes downward into the nearest roots and upward on the stem. The affected tissue exhibits a firm brown rot. On the seedlings the whole plant may be invaded, but on the older vines the base of the stem is girdled, while the parts above turn yellow, wilt and finally die. In the white mycelium covering the lesions are formed small brown sclerotia about the size, shape, and color of a mustard seed. When fruits of melons or squashes or the roots of sweet-potatoes are attacked, a soft watery rot is produced. Later the sweet-potatoes become hard and stringy.

Cause.

The fungus, *Sclerotium rolfsii,* bears much the same relation to vegetable diseases in the southern states as *Sclerotinia libertiana* does in the North. The mycelial growth of the two is similar in that both form large white masses of hyphæ. The sclerotia and method of reproduction of the two organisms are quite distinct as is their reaction to temperature. The Sclerotium has numerous small brown sclerotia which, so far as is known, never produce any fruit-bodies. After a period of inactivity, hyphæ again grow from the brown seed-like form. In fact, in no stage of its development does the parasite produce spores. The mycelial strands grow in profusion in the soil and after becoming well established are able to penetrate the host tissue and decompose them rapidly. Because of the clamp connection and binucleate condition of the mycelium, the fungus is supposed to belong to the Basidiomycetes, which includes the rusts, smuts, and mushrooms.

The parasite is most destructive on the light sandy soils when an excessive moisture supply is present. It is rarely found on the heavy waxy soils.

Control.

As the Sclerotium is a soil organism, no very satisfactory method of combating it has been demonstrated. Sanitation which includes the destruction of the diseased dead crop, eradication of possible weed hosts, the avoidance of contaminated manure, frequent cultivation to dry the top soil, and long rotations with the less susceptible crops, aids in reducing the amount of disease. In the seed-bed the soil may be sterilized (see page 600). In Florida good results were obtained on tomatoes when the base of the stem and the adjoining soil were wet with an ammoniacal copper solution (2 pounds of copper sulfate, 3 pints of ammonia, and 50 gallons of water). In

Georgia one to one and one-half tons of lime an acre harrowed into the plowed ground gave beneficial results and are recommended.

References

Gilbert, W. W. *Sclerotium rolfsii* on velvet beans. Phytopath. **11**: 378. 1921.

Harter, L. L. Rhizoctonia and *Sclerotium rolfsii* on sweet potatoes. Phytopath. **6**: 305, 306. 1916.

Higgins, B. B. Notes on the morphology and systematic relationship of *Sclerotium rolfsii* Sacc. Jour. Elisha Mitchell Sci. Soc. **37**: 167-172. 1922.

Taubenhaus, J. J. Recent studies on *Sclerotium rolfsii* Sacc. Jour. Agr. Research **18**: 127-138. 1920.

Texas Root-Rot of Sweet-Potatoes

Caused by *Ozonium omnivorum* Shear

Texas root-rot is restricted to Texas, southern Oklahoma, New Mexico, and Arizona, and well it is for no organism comes so nearly causing a total loss in the field where it occurs as does this virulent Texas root-rot fungus. It is found on many of the field crops as well as upon most of the common weeds. It, however, is not supposed to attack monocotyledons, such as cereals and grasses.

Symptoms.

The organism enters the plant below the surface of the soil, and not only grows downward throughout the root. but may invade the stem for a foot or more above the ground. Both on the stem and on the root a firm brown rot is produced. The root may be infected from the stem-end or through the side, in either case being completely destroyed. Not all the roots in the same hill may be decayed, nor are all the vines attacked necessarily. It is common to find part of a hill affected, and to have the disease appear only in spots in the field.

Cause.

The fungus has been known as *Ozonium omnivorum* and is characterized by dirty-yellow mycelium which permeates all parts of the host, as well as forms a yellowish weft of mycelial strands on the outside. The fungus is able to live over indefinitely in the soil or survive from fall to spring on the winter crops that are planted.

Duggar found a conidial stage which he believes is connected with this organism. His proof that the fruit-bodies are those of the parasite is based on the similarity of the mycelium, and on the fact that when the conidia are grown on media they produce a culture which apparently is identical with Ozonium. Inoculation experiments, however, were not successful. He changed the name to *Phymatotrichum omnivorum* (Shear) Dug. King later verified this conidial stage.

The Texas root-rot causes most injury in black poorly drained soil. It is very susceptible to low temperature, being killed by heavy freezing, and thus permanently barred from spreading northward.

As the fungus lives over in the soil it is difficult to control. Good cultural methods will eliminate much of the destruction. These consist in long rotations with corn and cereal crops, deep plowing or if possible plowing in the fall, frequent careful cultivations, destruction of weeds, proper drainage, and the selection of disease-free seed-tubers.

It has also been found that where the infection starts from a common center and progresses outward in an ever-widening circle, the growth of the fungus may be stopped by digging a trench around the area of infestation and soaking the soil to a width of three feet and to a depth of four feet with commercial formaldeyhde diluted one part in a hundred parts of water. This is not feasible under ordinary conditions.

References

Duggar, B. M. Root-rot of cotton and alfalfa. *In* Fungous diseases of plants, pp. 479-481. 1909.

Duggar, B. M. The Texas root-rot fungus and its conidial stage. Ann. Mo. Bot. Gard. 3: 11-23. 1916.
Harter, L. L. Sweet potato diseases. Root-rot. U. S. Dept. Agr. Farmers' Bull. 1059: 15-16. 1919.
King, C. J. Cotton root-rot in Arizona. Jour. Agr. Research 23: 525-527. 1923.
Shear, C. L. New species of fungi. Bull. Torrey Bot. Club. 34: 305-317. 1907.
Shear, C. L. and G. F. Miles. Texas root-rot of cotton. Field experiments in 1907. U. S. Dept. Agr. Bur. Plant Ind. Circ. 9: 3-7. 1908.

Sweet-Potato Mosaic

Cause undetermined

Mosaic on sweet-potato plants has been reported several times, and since nearly every cultivated crop is subject to a mosaic type of injury, it is easy to believe that such a disease exists. The point has hardly been proved, however, because the disease observed on this host, unlike a true mosaic, cannot be transmitted from one plant to another. The symptoms which have been described are mottling of the leaves, shortening of the internodes, and shortening and thickening of the petioles, which give a rosette appearance to the affected vine. No definite control measures are suggested.

References

Ensign, M. R. Sweet potato mosaic. Phytopath. 9: 180, 181. 1919.
Rosen, H. R. The mosaic disease of sweet potatoes. Ark. Agr. Exp. Sta. Bull. 167: 1-10. 1920.
Weber, Geo. F. Mosaic disease of sweet potato. Phytopath. 13: 42, 43. 1923.

White Rust of Sweet-Potatoes

Caused by *Albugo ipomϾ panduranæ* (Schw.) Swingle

As white rust of sweet-potatoes is so conspicuous on the leaves, it has been known for a long time and reported from all the sweet-potato-growing districts. It is usually of minor importance but occasionally may become severe enough to cause much defoliation and consequent weakening of the plant.

The white rust normally appears so late that the plants have made most of their vine growth. The older leaves near the center of the hill are affected first, and are also first to drop when defoliation occurs. The lesions on the foliage are irregular yellow areas, appearing on the lower surface slightly sooner than on the upper. As the affected tissue begins to turn brown, the white cheesy fruit-bodies, in the form of dots or slight patches, appear on the lower surface of the leaf. This stage is often referred to as leaf-mold.

The fungus is known as *Albugo ipomœœ panduranœ.* It is probably the same fungus which occurs also on wild morning-glory and other related plants, although the suggestion has been made that the strains on the two hosts are different physiologically. The mycelial threads penetrate all parts of the affected tissue, until finally the chains of conidia break through the lower epidermis and form the white pustules mentioned above. The conidia when mature are dust-like and are easily blown about by the wind. When the stem of the host is affected sexual spores or oospores are borne in the swollen distorted portion. These live over the winter in the host tissue, which on decaying in the spring liberates the parasite. Both the conidia and oospores germinate by means of swarm-spores.

High temperature is regarded as favorable for the growth of the organism, but pathologists differ in their opinions about the effect of moisture on the development of the fungus. Probably the weather requirements are much like those for Phytophthora and other Peronosporales.

No definite control measures are known.

References

Halsted, Byron D. Some fungous diseases of the sweet potato. The leaf mould. N. J. Agr. Exp. Sta. Bull. 76: 28-30. 1890.

McClintock, J. A. Sweet potato diseases. White rust leaf spots. Va. Truck Exp. Sta. Bull. 22: 458. 1917.

Stevens, F. L. Oogenesis and fertilization in *Albugo ipomœœ—panduranœ.* Bot. Gaz. 38: 300-302. 1904.

Taubenhaus, J. J. and Thos. J. Manns. The diseases of the sweet potato and their control. Del. Agr. Exp. Sta. Bull. 109: 1-55. 1915.
Taubenhaus, J. J. Some recent studies on new or little known diseases of the sweet potato. Phytopath. 4: 50, 51. 1914.

Septoria Leaf-Spot of Sweet-Potatoes

Caused by *Septoria bataticola* Taub.

The leaf-spot of sweet-potatoes is most common in the northern tier of sweet-potato-producing states, although not confined wholly to them. It is of no economic importance, even though it is widely distributed and plentiful in many fields.

The leaves are marked indiscriminately with varying numbers of minute white spots each bordered with a narrow reddish-brown zone. The older lesions contain one or more black pycnidia which are just visible to the unaided eye. These pycnidia belong to the fungus, *Septoria bataticola,* the cause of the disease. The spores, which are of the long septate Septoria-type, ooze out in tendrils when water is present in the leaf, then are splashed by the rain or carried by insects to neighboring leaves or plants. The parasite is able to live over winter in the old leaves on the ground, and then is ready in the spring to attack the new crop.

The organism is not of sufficient severity to demand control measures.

References

McClintock, J. A. Sweet potato diseases. Septoria leaf-spot. Va. Truck Exp. Sta. Bull. 22: 458. 1917.
Taubenhaus, J. J. Some recent studies on new or little known diseases of the sweet potato. Phytopath. 4: 50, 51. 305-320. 1914.

Phyllosticta Leaf-Blight of Sweet-Potatoes

Caused by *Phyllosticta batatas* (Thum.) Cke.

Phyllosticta leaf-blight is present in all the sweet-potato sections of the United States, but is more plentiful in the far

South than in the tier of states to which New Jersey and Delaware belong.

The lesions are similar to those caused by the leaf-spot fungus, except they are larger than those of the latter, and the pycnidia are more numerous and somewhat more conspicuous. The life history is similar to that of Septoria, the distinguishing difference being globose hyaline unicellular spores which ooze from the pycnidia instead of the long septate ones.

The loss is not sufficient to justify control measures.

References

Halsted, Byron D. Some fungous diseases of the sweet potato. N. J. Agr. Exp. Sta. Bull. 76: 27-28. 1890.

Harter, L. L. Sweet potato diseases. Leaf-blight. U. S. Dept. Agr. Farmers' Bull. 714: 18. 1916. (Same in Farmers' Bull. 1059: 17. 1919.)

McClintock, J. A. Sweet potato diseases. Phyllosticta leaf-spot. Va. Truck Exp. Sta. Bull. 22: 459. 1917.

Storage-Rots of Sweet-Potatoes

In the past from one-fourth to one-half of all the sweet-potatoes placed in storage decayed. This part of the industry was so discouraging that many growers turned to other, and possibly less profitable, crops in order to avoid the excessive loss. It is encouraging to know, however, that each year more modern storage-houses are being built and that the disease-tax is decreasing rapidly.

Many fungi are responsible for the loss. Harter and Weimer, who have done much work on this subject, list the following organisms as being able to produce rot in stored potatoes: Rhizopus species, *Sphæronema fimbriatum, Sclerotium bataticola, Diaporthe batatatis, Diplodia tubericola, Mucor racemosus* (Fig. 139), Alternaria species (Fig. 139), *Botrytis cinerea* (Fig. 139), Penicillium species (Fig. 139), Epicoccum species (Fig. 139), *Plenodomus destruens, Gibberella saubinetii, Fusarium culmorum,* and *F. acuminatum.* They were not able to cause a storage decay with *Fusarium*

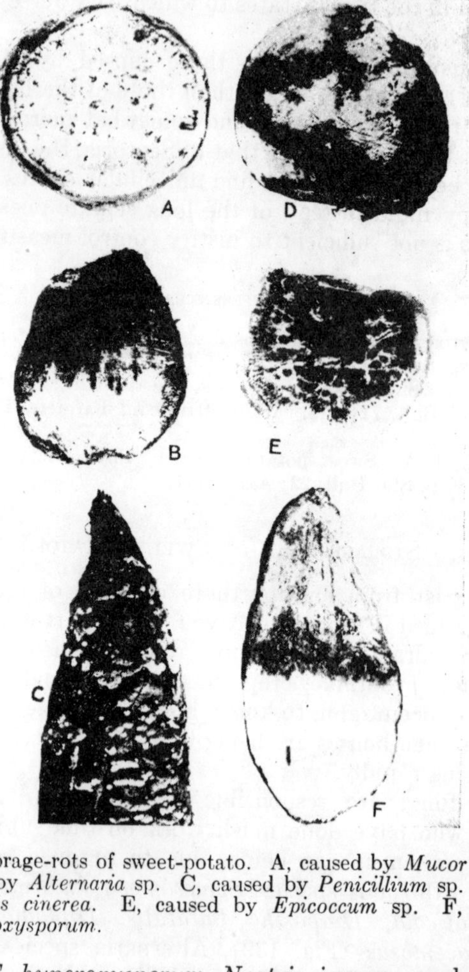

FIG. 139.—Storage-rots of sweet-potato. A, caused by *Mucor racemosus*. B, caused by *Alternaria* sp. C, caused by *Penicillium* sp. D, caused by *Botrytis cinerea*. E, caused by *Epicoccum* sp. F, caused by *Fusarium oxysporum*.

batatatis, F. hyperoxysporum, Nectria ipomœa, and *Fusarium solani.*

All of these organisms are nearly enough alike in their reaction toward environmental conditions so that they can be grouped together, when recommending control measures. Most of them will not thrive in a low temperature coupled with low humidity. Therefore, if the storage-house is handled correctly according to the directions given for the control of soft-rot, there should never be any great loss. It is also important that none of the field disease organisms is propagated and then carried with the potatoes into the bins. In order to avoid such a possibility, the measures suggested for the control of black-rot, foot-rot, stem-rot, and the other field diseases should be put into practice. The whole procedure of control then resolves itself into the formula of healthy potatoes for harvesting, extreme care in handling the crop, and relatively low temperature and dry atmosphere in the storage-house.

References

Elliott, John A. Storage rots of sweet potatoes. Ark. Agr. Exp. Sta. Bull. 144: 1-16. 1918.

Harter, L. L. and J. L. Weimer. Storage rots of sweet potatoes. Phytopath. 8: 73. 1918.

Harter, L. L. and J. L. Weimer and J. M. R. Adams. Sweet potato storage rots. Jour. Agr. Research 15: 337-368. 1918.

Manns, T. F. Sweet potato storage in Delaware. Del. Agr. Exp. Sta. Bull. 127: 1-64. 1920.

Taubenhaus, J. J. et al. Storage and disease of the sweet potato in Texas. Tex. Agr. Exp. Sta. Bull. 250: 1-41. 1919.

Soft-Rot of Sweet-Potatoes

Caused by *Rhizopus* spp.

Apparently it is easier to grow sweet-potatoes than it is to keep them afterwards. The most common form of decay is the soft-rot, which probably has been a limiting factor in storage ever since sweet-potatoes have become a commercial crop. Growers, feeling that they could not store their product successfully, were forced to market at harvest time when prices were extremely low. They were justified in this attitude for in the past it has not been uncommon to have 50 per cent of the crop lost by decay. Fortunately better storage conditions are

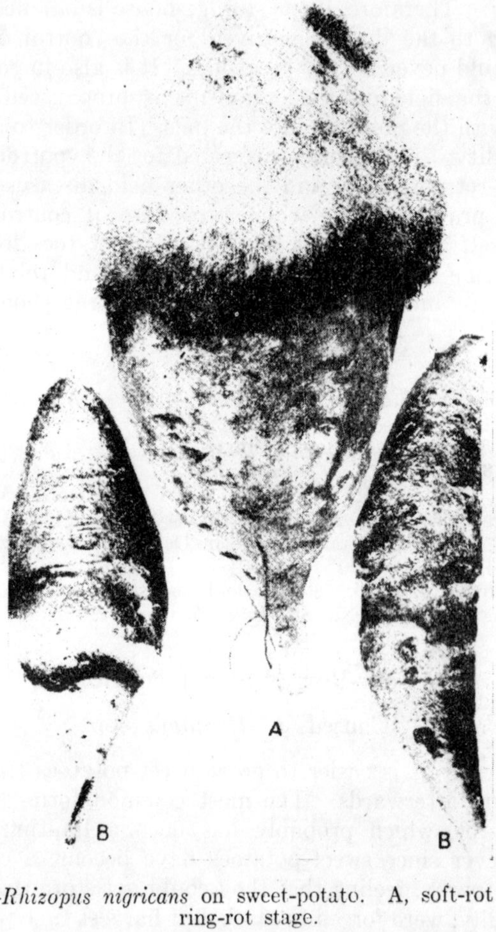

FIG. 140.—*Rhizopus nigricans* on sweet-potato. A, soft-rot stage. B, ring-rot stage.

decreasing this damage annually. Soft-rot also causes injury in the seed-bed and in the field before harvesting.

Symptoms.

Affected roots are so soft that they crush easily when handled. The decayed tissue is water-soaked and when the skin is broken a straw-colored liquid often drips out. If the root is left intact after the decay begins, it gradually loses moisture until it finally becomes a dry shriveled mummy. If, however, the skin ruptures while the disease is in progress, the pathogene at once begins to fruit abundantly covering the affected portion with a thick bushy growth of hyphæ and sporangiophores, on the tips of which are large spherical black sporangia common to the bread mold fungus (Fig. 140).

The entire root is not always affected; the disease may begin at any point and spread around the potato rather than lengthwise, thus giving rise to the "ring-rot" stage. If the storage conditions are unfavorable for the progress of the fungus, the girdle of tissue dries forming the dry ring-rot.

The rot itself has a mild pleasant odor, but fermentation organisms may follow and cause a soured smell, or other saprophytes may cause putrefaction.

Cause.

It has generally been considered that *Rhizopus nigricans* Ehrnb. was the cause of the soft-rot but later work has shown that other species of this genus may also be responsible. Among these are *R. reflexus* Bainier., *R. tritici* Saito, *R. artocarpi* Racib., *R. delemar* (Boid.) Wehmer and Hanzawa, *R. maydis* Bruderl., *R. nodosus* Namysl., *R. oryzæ* Went and Pr. Geerlings, and *R. arrhizus* Fischer.

The species are so nearly alike that for all practical purposes a description of *Rhizopus nigricans* will apply equally well to the others. The vegetative part is made up of root-like rhizoids, white wefts of mycelium, some threads of which act as stolons connecting older with newer growth centers, and the tall straight sporangiophores, on the tips of which are borne the well-known dark globose sporangia. The sporangia are filled with spores which when liberated form dark brown dust. An innumerable number are borne on a diseased root, and by handling are coated over the healthy tubers. They germinate readily but the germ-tube cannot enter the host unless the skin has been broken. The parasite secretes enzymes which bring about the dissolution of the starch grains and dissolve the middle lamellæ causing a softening of the tissue by the separation of one cell from the other.

Certain facts about the physiology of decaying sweet-potatoes have been worked out and are of much importance in connection with the proper storage conditions. First, the various species of Rhizopus have different optimum temperatures for growth. *Rhizopus nigricans*, which is the commonest species, together with *R. reflexus* and *R. artocarpi*, thrive best at temperatures from 60° to 70° F.; *R. tritici* and *R. delemar*, at 70° to 90° F.; and *R. arrhizus*, *R. maydis*, and *R. nodosus*, at 90° to 105° F. It has also been found that at a temperature of about 90° F. with a high humidity the injured surface of a sweet-potato protects itself from infection by means of a cork layer. Under ordinary storage conditions this cork layer is not developed but the exposed surface becomes dry and hard so that it retards materially the infection from Rhizopus. It has been considered that a high humidity was required by the fungus for entrance into the host, but it is now shown that when the temperature is 72° to 75° F. and the humidity 94 to 98 per cent it is very difficult to get infection with the pathogene. As the humidity is lowered the ease with which infection is obtained increases until at 72 to 84 per cent the optimum is reached. From this point the amount of infection decreases as the percentage of humidity is lowered.

Control of soft-rot.

The condition of the storage-house determines to a great extent the amount of loss due to any one of the numerous storage-rots. The house, after having been in use one season, should be cleaned by having the floors, walls, and partitions scrubbed with some disinfectant such as blue vitriol (1 pound in 5 gallons of water). The sweet-potatoes when harvested should be handled as carefully as possible in order to avoid bruising. Furthermore, it is wise to remove by sorting all diseased, injured, or otherwise objectionable potatoes. Since much of the rotting occurs following the harvesting in wet weather, it is advisable to do the digging when the sun is shining so that the potatoes can be dried quickly. It is not well, however, to let the roots lie too long in scorching sunlight. The potatoes should be placed carefully in open bushel crates or hampers and left in these when placed in the storage-house. This eliminates extra handling and permits the free circulation of air. The temperature of the house should be kept at about 80° F. for the first two weeks until the potatoes are well cured, then lowered gradually until it reaches 55° F. In regulating the humidity frequent inspections should be made and extra ventilation given when necessary. After the potatoes are placed in the bins it is well not to molest them by further sorting until time for their final removal, for the extra handling will only serve as a means for disseminating any spores of the fungus which may be present.

Chilling which can usually be avoided results in much soft-rot later. The chilling may take place either in storage or in transportation. Great care should be exercised in keeping the temperature fairly constant. The same measures recommended for the storage-house apply equally well to the car in which the crop is to be shipped to market.

References

Ames, Adeline. The temperature relations of some fungi causing storage rots. Phytopath. 5: 11-19. 1915.

Halsted, Byron D. Some fungous diseases of the sweet potato. The soft-rot. N. J. Agr. Exp. Sta. Bull. 76: 4-7. 1890.

Harter, L. L. Amylase of *Rhizopus tritici* with a consideration of its secretion and action. Jour. Agr. Research 20: 761-786. 1921.

Harter, L. L. and J. L. Weimer. A comparison of the pectinase produced by different species of Rhizopus. Jour. Agr. Research 22: 371-377. 1921.

Harter, L. L. and J. L. Weimer. Studies in the physiology of parasitism with special reference to the secretion of pectinase by *Rhizopus tritici*. Phytopath. 21: 609-625. 1921.

Harter, L. L., J. L. Weimer and J. I. Lauritzen. The decay of sweet potatoes (*Ipomœa batatas*) produced by different species of Rhizopus. Phytopath. 11: 279-284. 1921.

Lauritzen, J. I. and L. L. Harter. The relation of humidity to the infection of sweet potatoes by Rhizopus. Phytopath. 13: 56. 1923.

Taubenhaus, J. J. Some recent studies on new or little known diseases of the sweet potato. Phytopath. 4: 50-51, 305-320. 1914.

Weimer, J. L. and L. L. Harter. Wound-cork formation in the sweet potato. Jour. Agr. Research 21: 637-647. 1921.

Weimer, J. L. and L. L. Harter. Temperature relations of eleven species of Rhizopus. Jour. Agr. Research 24: 1-40. 1923.

Dry-Rot of Sweet-Potatoes

Caused by *Diaporthe batatatis* (E. and Hals.) Harter and Field

Although this disease was mentioned by Halsted in 1890, no further work was done with it for the next twenty years. It was then found to be quite prevalent, but so closely associated with other diseases such as black-rot and stem-rot that its true economic importance could not be determined. It has been demonstrated, however, that the disease causes much loss in storage even though the injury in the seed-bed and field is only slight.

Symptoms.

In the field the rot usually enters at the stem-end of the root, then later in storage involves the whole potato. The shrunken wrinkled roots, which may become mummified, are often covered with papillæ (Fig. 141). When the surface of the potato is removed it may be seen that under the papillæ are pycnidia massed in a coal-black stroma. The symptoms on the stem are more difficult to detect. The affected plant, weakened from other conditions, slowly succumbs to the disease, first becoming yellow then finally dying. It is not until after the death of the host that the fungous fruit-bodies appear on the stem, the upper surface of the leaf, and on the petioles.

Cause.

The perfect stage of the fungus, *Diaporthë bätatatis*, has been found only in laboratory cultures. The imperfect stage, which is the one on the host, was called *Phoma batatæ* E. and H. at first but later changed to *Phomopsis batatæ* (E. and H.) Harter and Field. When diseased potatoes are planted the

FIG. 141.—*Diaporthe batatatis* causing rot of sweet-potato.

sprouts become affected, and since the symptoms are so hard to detect on them, they are often transplanted. If the vines are healthy otherwise, the disease makes very little progress during the summer. There is just enough development to cause the infection of the potato, which when placed in the storage-house may rot if conditions are favorable for the fungus, or they may be only slightly affected and be used for seed in the spring. The fungus may live over winter in the fields on sweet-potato stems or roots.

The optimum temperature for the growth of the organism is 75° to 90° F., which is higher than in a good storage-house except during the curing period.

Control.

The instructions given for the control of black-rot and of soft-rot include the required precautions for this disease.

References

Halsted, Byron D. Some fungous diseases of the sweet potato. The dry-rot. N. J. Agr. Exp. Sta. Bull. 76: 23-25. 1890.
Harter, L. L. and Ethel C. Field. Diaporthe, the ascogenous form of sweet potato dry-rot. Phytopath. 2: 121-124. 1912.
Harter, L. L. and Ethel C. Field. A dry-rot of sweet potatoes caused by *Diaporthe batatatis*. U. S. Dept. Agr Bur. Plant Ind. Bull. 281: 1-38. 1913.

Surface-Rot of Sweet-Potatoes

Caused by *Fusarium oxysporum* Schlect.

The surface-rot of sweet-potato is found everywhere and in certain seasons does an appreciable amount of injury. In the past it was considered a stage of the stem-rot but this has been disproved. It also has symptoms which might be mistaken for black-rot. The spots on the potatoes never penetrate deeply and never attain a large size. In this respect as well as in being lighter in color, the lesions can be differentiated from those of the latter disease. The spotting causes loss, not only because of disfiguration but also because such spots are accompanied by shriveling and drying of the potato. The pathogene, *Fusarium oxysporum*, which also causes the wilt of the white or Irish potato, lives in the soil for an indefinite period apparently as a saprophyte. When the temperatures are low and the humidity high, the fungus is able to infect the sweet-potato. The amount of infection seems to be increased when the potatoes are dug in wet weather. In attempting to control the loss the crop should be dug when the sun is shining so that the potatoes will dry quickly. This practice together with the other measures suggested for the control of soft-rot will reduce the amount of surface-rot if it will not entirely eliminate the disease (Fig. 139).

Reference

Harter, L. L. and J. L. Weimer. The surface-rot of sweet potato. Phytopath. 9: 465-469. 1919.

Charcoal-Rot of Sweet-Potatoes

Caused by *Sclerotium bataticola* Taub

The charcoal-rot of sweet-potatoes, which was at first confused with black-rot, is of minor importance. Generally it is found only in storage-houses in which the temperature and humidity are regulated improperly. The roots, the only part of the plant which is injured, show no pronounced external symptoms, except a possible drying and shrinkage. If a potato which is lightly affected is cut open the tissue will show a darkened ring, the color varying from jet-black on the outside to an ashen-gray toward the center. The tissue at this time appears water-soaked and quite solid. The badly diseased root dries out until it becomes very light, and may easily be broken. The inner tissue then becomes a charcoal-black. Aside from sweet-potatoes, the parasite has been inoculated successfully into pepper, tomato, cucumber, apple, eggplant, and turnips.

The cause of the disease is a fungus known as *Sclerotium bataticola*. Infection takes place at a bruise on the surface of the root, and from there the pathogene slowly penetrates the host tissue, in which it produces a large number of sclerotia. Many of these sclerotia are large enough to be seen with the unaided eye when the potatoes are cut open. As the organism is disseminated by means of the discarded sweet-potatoes and other refuse, and since sanitation is a relatively rare condition on the average farm, there are many chances for the fungus to be returned to the fields where sweet-potatoes are grown or brought into the storage-house with the dirt and injured roots. If the control measures suggested for the soft-rot of sweet-potatoes are followed there will be no appreciable loss from charcoal-rot.

References

Taubenhaus, J. J. The black-rots of the sweet potato. Phytopath. 3: 70, 71, 159-166. 1913.

Taubenhaus, J. J. Some recent studies on new or little known diseases of the sweet potato. Phytopath. 4: 50, 51, 305-320. 1914.

Java Black-Rot of Sweet-Potatoes

Caused by *Diplodia tubericola* (E. and E.) Taub

In 1894 and again in 1895 sweet-potatoes were received at the Louisiana Agricultural Experiment Station from Java, which showed a peculiar type of decay. In 1910 the disease was discovered in Delaware, and since then has been reported from most of the states where sweet-potatoes are grown. Specimens also were received from Cuba, Isle of Pines, Philippine Islands, Japan, Porto Rico, and South America.

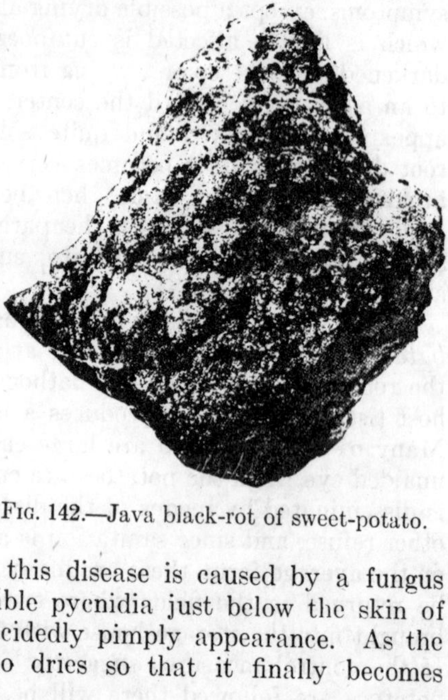

Fig. 142.—Java black-rot of sweet-potato.

This Java black-rot is strictly a storage trouble. The inner part of the tuber is black and brittle much like that described for charcoal-rot. When the latter is present, however, the tissue is filled with black sclerotia, which is not true of the Java black-rot. Furthermore, this disease is caused by a fungus which produces innumerable pycnidia just below the skin of the potato giving it a decidedly pimply appearance. As the decay spreads the potato dries so that it finally becomes mummified (Fig. 142).

The original specimens from Java were sent to Halsted who determined the pathogene as new to this country and called it *Lasiodiplodia tubericola*. When a more careful study was made later the genus was renamed, the fungus now being known as *Diplodia tubericola*. It was also discovered that the

Diplodiæ from cotton, dasheen, watermelon, and other hosts will produce rots similar to that caused by the fungus on the sweet-potato. Evidently the Diplodiæ attack a wide range of species of plants. This would seem to indicate that the type of black-rot under discussion was present in the United States long before the specimens were received from Java.

The rot caused by *Diplodia tubericola* progresses very slowly, normally beginning at one end of the potato and requiring four to eight weeks to destroy it entirely. After the decay begins almost a month elapses before there are any signs of pycnidial formation. As the pycnidia are borne not only beneath the skin of the tuber but also deeper in the tissue, they must await the rupturing of the skin or the partial disintegration of the host before the spores can be liberated. The spores are at first hyaline and one-celled, later becoming dark and one-septate. Pseudo-paraphyses may be present with the conidia.

No detailed control measures have been worked out experimentally but the same recommendations made for holding in check the soft-rot will prove beneficial in warding off Java black-rot. These include extreme care in handling the roots so that the skins are not bruised, proper curing after harvest, suitable temperature in a proper storage-house, and preferably storing in crates.

References

Harter, L. L., J. L. Weimer, and J. M. R. Adams. Sweet potato storage rots. Jour. Agr. Research 15: 337-368. 1918.

Reinking, Otto. Philippine plant diseases. Phytopath. 9: 114-140. 1919.

Taubenhaus, J. J. The black-rots of the sweet potato. Phytopath. 3: 70, 71. 159-166. 1913.

Taubenhaus, J. J. The nonvalidity of the genus Lasiodiplodia. Phytopath. 4: 47. 1914.

Taubenhaus, J. J. Some recent studies on new or little known diseases of the sweet potato. Phytopath. 4: 50, 51. 1914.

Taubenhaus, J. J. and Thos. F. Manns. The diseases of the sweet potato and their control. Del. Agr. Exp. Sta. Bull. 109: 1-55. 1915.

Trichoderma-Rot of Sweet-Potatoes

Caused by *Trichoderma koningi* Oudemans

The Trichoderma-rot of sweet-potatoes is of some importance in the Delaware-Chesapeake peninsula. It occurs as a dry rot of the tubers often completely girdling them, and forming a ring of discolored tissue. At first the spots are light brown with a tendency to wrinkle. The hard water-soaked affected part is bounded by a narrow black zone which marks the line between the healthy and diseased area. Under conditions favorable for the organism the entire root is finally included, the diseased potato often being covered with a luxuriant white mycelial growth.

The decay is caused by a soil-inhabiting fungus known as *Trichoderma koningi.* It is not able to infect a perfectly sound potato, so that it must depend on bruises or the different important root diseases to gain an entrance. If the control measures given for soft-rot are followed, there should be no trouble with Trichoderma.

Reference

Cook, Mel T. and J. J. Taubenhaus. *Trichoderma koningi* the cause of a disease of sweet potatoes. Phytopath. 1: 184-189. 1911.

CHAPTER XVIII

DISEASES OF TOMATO

NEXT to potatoes and sweet-potatoes the tomato is the most important vegetable grown for commercial purposes. More than three hundred thousand acres are planted to this crop each year. Although every state in the Union produces tomatoes, they are grown most intensively in the South Atlantic district. The important states in the order of their acreages of tomatoes are Maryland, New Jersey, California, Delaware, Virginia, Indiana, Florida, New York, Ohio, and Missouri.

The tomato has nearly as many diseases as the potato, and in many cases these are not successfully controlled as is shown by the estimated percentage of losses given in the Plant Disease Survey Bulletin. According to these compiled data, more than one-fourth of the total crop is sacrificed to disease.

It is difficult to place the many tomato diseases in any logical order, but six of the most important troubles are given first, followed by four diseases caused by bacteria, then the remaining stem and root troubles, the fruit-rots, and lastly the physiological injuries and rare degeneration diseases.

SEPTORIA LEAF-SPOT OF TOMATO

Caused by *Septoria lycopersici* Speg.

The leaf-spot or leaf-blight of tomato was first observed in Argentina in 1882. Although the disease was found a few years later in Italy, it was not discovered in the United States until 1896. Within the following ten years it was reported from a large number of states, suggesting that it was present in the United States for some time before being brought to the attention of plant pathologists It is now an important disease in the central eastern states, and probably is present in most parts of the world where tomatoes are grown commercially. In some localities it is considered of no economic importance, as on the Pacific Coast, the North Atlantic and the Gulf States, while in others the losses are very large. It has been estimated that during a season favorable for the fungus the losses in this country amount to five million dollars.

The leaf-spot organism attacks not only every variety of tomato, but has also caused infection when inoculated into other solanaceous plants such as the potato, eggplant, jimson-weed, and the horse-nettle. In some of the states where the fungus is most injurious, the horse-nettle is a common weed and apparently is a host on which the pathogene over-winters and multiplies. Attempts have been made to breed a variety of tomato immune to the disease, but so far they have not been encouragingly successful.

Symptoms.

The plant may be attacked at any stage of its life, from that of the seedling until the time when frost or disease kills the vine. The small circular lesions are first observed as water-soaked areas on the under surface of the lower leaves, although infection may take place on the upper surface as well, and quite often does on the leaves at the tip of the plant. The fungus quickly penetrates the leaf tissue causing spots with dark brown margins and sunken grayish centers to appear on both the upper and lower surfaces. The size of the lesion may vary from one which is just visible to the eye to spots more than a fourth of an inch in diameter. The average size is about an eighth of an inch. The number of spots on a single leaf varies with the virulence of the disease, although if the environment for the fungus is unusually favorable individual spots may increase in size so rapidly that the intervening leaf tissue is killed before many infections appear, and before fruiting bodies of the causal fungus are formed. Ordinarily, however, the foliage is disfigured with many lesions, each of which is marked by the presence of minute black pycnidia on the upper side of the leaf. These pycnidia and the small circular grayish areas comprise the distinctive symptoms by which this disease may be recognized and differentiated from the other tomato leaf-spots. No matter whether the spots are large or small the leaf usually dies, and is blown or knocked off. The disease spreads from the lower leaves to those higher on the vine, so that it is not uncommon to find a plant which is leafless except at the tip where the new foliage has not yet had time to become infected. Similar spots are sometimes borne on the calyx and on the stem, although the spots on these parts of the host are usually not so prominent. On the stems they are elongated rather than circular. Lesions have been reported on the fruit, but these are relatively rare.

Cause of leaf-spot.

The disease is caused by *Septoria lycopersici,* whose many-celled filiform spores are borne in great abundance in the comparatively sparse pycnidia of each lesion. During periods of high temperature and humidity the spores ooze from the pycnidium, and are carried to neighboring leaves or plants by insects, by splashing rain water, and on the hands and clothing of the pickers. As the spores exude only in the presence of water, the wind does not help in their dissemination. When the plant is once inoculated, the spore germinates quickly by means of a germ-tube, which penetrates directly through the cell-wall and in its growth ramifies all parts of the adjoining leaf tissue. Reproductive bodies are borne readily, so that within two weeks after spores are disseminated others are ready to repeat the cycle.

The diseased leaves and stems which drop to the ground, as well as affected portions of weed hosts, serve as a place where the mycelium and pycnidia hibernate. This refuse is left in the field or accidently even mixed with the seed-bed soil. If some of the seedlings are affected, the organism is scattered among the healthy plants at the time of transplanting. Such early infection does much damage if environmental conditions are favorable for the fungus. Infection late in the season is often looked upon as being beneficial, for by defoliating the plants, the fruit ripens more quickly and thus avoids injury from possible early frost.

Attempts have been made to inhibit the development of the pathogene by applying various combinations of fertilizers or fungicides to the soil on which the host is grown. None of these attempts has been successful. The only conditions which seem to have any effect on the virulence of the disease are the presence of medium temperatures and an abundance of rainfall. The optimum temperature is about 77° F., while the minimum and maximum for sporulation of the fungus are 59° and 80.5° F. respectively. These conditions are encountered in the middle Atlantic states where the disease does the most damage. On the other hand, the fungus is extremely sensitive to very high temperatures, and will not long survive heat at 98° F., which may be obtained in some of the southern states.

Control of leaf-spot.

As is true in the control of most other diseases, sanitation is important in dealing with Septoria leaf-spot. Inasmuch as infection may take place in the seed-bed, it is necessary to use every precaution in avoiding such early appearance of the disease. This may be done by scrubbing with a dilute solution of formaldehyde any boards or boxes surrounding the bed and by using new clean soil or sterilizing that which has been utilized before (see page 600). In addition to the care given the seed-bed, long rotations, together with the eradication of the horse-nettle, should be practiced in the field. The spores are spread in the presence of moisture, so that cultivating should be done only when the vines are dry. This is true particularly when the plants are small, for it is the early infections that reduce the yield. If only a small number of plants is handled, it is possible to remove at transplanting time the lower leaves, which are affected, and thus avoid carrying the fungus into the field.

Sanitation may be supplemented by spraying. Bordeaux mixture alone has usually not been satisfactory unless applied heavily and under high pressure. Success has been attained by adding resin fish-oil soap to the fungicide. Three pounds of the soap are dissolved in three gallons of hot water, and added to the forty-seven gallons of the spray mixture, after the copper and lime have been stirred together (see page 617). One or two applications of the soap bordeaux mixture are made to the plants in the seed-bed, and at least every ten days in the field. When the plants are large, three or four nozzles to the row are necessary. It is often convenient when intending to spray to place tomato plants close together in the row and in rows farther apart, so that the wheels of the spraying machine will fit between them. In this manner, an ordinary potato-sprayer with a boom of eight nozzles can be fitted to spray one row completely, and one-half of each adjacent row of tomatoes with four nozzles to the row. It is advisable to direct part of the spray from the ground upward, so that the lower sides of the leaves will also be coated with the poison.

In preliminary trials copper soap dust has been used instead of the liquid, and has given satisfactory results. This may become a popular form of fungicide for tomatoes.

In the northern states there is one objection to spraying tomatoes, because the application of a fungicide keeps the plants green longer, and consequently much of the fruit does not ripen before frost injures it. Occasionally the frost injury may be as severe as the disease would have been if left unchecked. Fortunately, in the cooler climate the disease is not often serious enough to call for control measures.

References

Fromme, F. D. Experiments in spraying and dusting tomatoes. Va. Agr. Exp. Sta. Bull. 230: 1-15. 1922.

Levin, Ezra. The leaf-spot disease of tomato. Mich. Agr. Exp. Sta. Tech. Bull. 25: 1-51. 1916.

Martin, W. H. Dissemination of *Septoria lycopersici* Speg. by insects and pickers. Phytopath. 8: 365-372. 1918.

Pritchard, Fred J. and W. Blair Clark. Effect of copper soap spray mixtures on control of tomato leaf-spot. Phytopath. 9: 554-564. 1919.

Pritchard, Fred J. and W. S. Porte. Use of copper soap dust as a fungicide. Phytopath. 11: 229-235. 1921.

Pritchard, Fred J. and W. S. Porte. Relation of horse-nettle (*Solanum carolinense*) to leaf-spot of tomato (*Septoria lycopersici*). Jour. Agr. Research 21: 501-506. 1921.

Pritchard, Fred J. and W. S. Porte. Effect of fertilizers and lime on control of tomato leaf-spot (*Septoria lycopersici*). Phytopath. 11: 433-445. 1921.

Pritchard, Fred J. and W. S. Porte. The relation of temperature and humidity to tomato leaf-spot (*Septoria lycopersici* Speg.). Phytopath. 14: 156-165. 1924.

Macrosporium Blight and Rot of Tomato

Caused by *Macrosporium* spp.

The organism causing leaf-spot of tomato is identical with the one which causes early-blight of potato; therefore, the history and distribution of the two diseases are similar (see page 360). The loss resulting from this leaf-spot of tomato and the nailhead-spot, as the injury on the fruit is known, reaches many thousand dollars each year. The injury is usually not severe to the crop in the northern tier of states, but may result in a complete loss in fields in the latitude of Indiana, New Jersey, and states farther south.

Symptoms.

The symptoms on the leaf are like those described for early-blight of potatoes. The fungi, in addition to their production of foliage injury, do much damage to the tomato fruit. The lesions which in some instances follow cracks or sunburn, and in others are found on the otherwise unblemished fruit are circular depressed areas which at first are brown, but which may become black in the center, due to the fruiting of the pathogene (Fig. 143). In typical nailhead lesions the diseased area never becomes much larger than is represented by the head of a nail, while in other types the spots may become relatively large, although they never attain the size of the Phoma-spot, nor do they ever show the presence of pycnidia which is one of the diagnostic signs of the latter disease. The lesions often crack open permitting entrance of soft-rot organisms. Even though the lesions may not break open, the flesh of the fruit is usually dark below the spot. This discoloration extends to the core and makes the tomato unfit for table consumption or for canning and thus necessitates its being discarded or sold at a discount to catsup manufacturers, who use the undamaged portions.

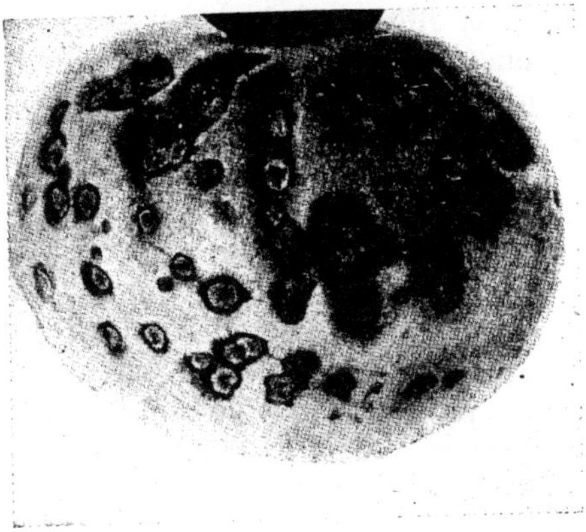

FIG. 143.—Tomato fruit spotted with Macrosporium blight.

It has recently been shown that the fungus may also cause a typical canker at the base of the stem. The affected tissue at first turns light brown, gradually taking on a darker hue, until the canker is almost black. The tissue shrivels, and the plant may fall over, or the leaves may wilt and become discolored. If attacked when young, new roots occasionally grow out above the canker, and permit the plant to survive (see collar-rot, page 576).

Both the leaf-spot and canker form of the disease may be found at any stage in the growth of the plant, from that of the seedling until the plant is mature. Late infections on the leaves would be of slight importance were it not for the fact that the fruit is susceptible to the pathogene.

Cause.

Until recently all the injury was attributed to the parasitism of *Macrosporium solani* E and M. The fruit-spot called nailhead and found mostly in the southern states is now known to be caused by a species with much smaller spores, and which evidently is identical with *Macrosporium tomato* Cooke. A species different from either of these has recently been reported from California. So far as the life histories are concerned, the three species may be treated as one. These have been included under the discussion on early-blight of potatoes (see page 360), so that only the matter pertaining to fruit infection need be added here.

It has often been observed that the southern nailhead-spot was very severe on many shipped tomatoes, and it was generally believed that most of the infection took place during transportation. Careful inoculation experiments, however, have proved that large tomatoes like those which are usually harvested are so well protected with a thick epidermis that the fungus cannot penetrate the tissue. Infection can take place only in the young stages of the fruit; consequently if sound large fruit is packed for market, the grower may feel assured that his product will not become disfigured with the nailhead spots.

When a tomato is badly infected, the mycelium enters the seed where it forms a mat between the seed covering and the endosperm. The hyphæ also mingle among the hairs on the surface of the seed. When such contaminated seeds are planted, the embryo is killed before it elongates, or the mycelium grows up with the young sprout.

Control.

The directions given for the control of Septoria leaf-spot of tomato apply also to the control of the Macrosporium blight and rot. In addition, the seed should be obtained from healthy fruit.

References

Douglas, Bruce. A new alternaria spot of tomatoes in California. Phytopath. **12**: 146-148. 1922.

Pritchard, Fred J. and W. B. Clark. Effect of spraying on early ripening of tomato fruit. Phytopath. **9**: 289-291. 1919.

Pritchard, F. J. and W. S. Porte. Collar-rot of tomato. Jour. Agr. Research **21**: 179-184. 1921.

Rosenbaum, Joseph. Studies with Macrosporium from tomatoes. Phytopath. **10**: 9-21. 1920.

Rosenbaum, Joseph. A Macrosporium foot-rot of tomato. Phytopath. **10**: 415-422. 1920.

Rosenbaum, J. and Sando C. E. Correlation between size of the fruit and the resistance of the tomato skin to puncture, and its relation to infection with *Macrosporium tomato* Cooke. Amer. Jour. Bot. **7**: 78-82. 1920.

Massee, I. On the presence of hybernating mycelium of *Macrosporium solani* in tomato seed. Roy. Bot. Gard. Kew. Misc. Inform. Bull. **4**: 145-146. 1914.

Leaf-Mold of Tomatoes

Caused by *Cladosporium fulvum* Cke.

The first specimens of leaf-mold of tomatoes were sent for identification from North Carolina to England, where the fungus was described in 1883. The distribution of the disease is now world wide, especially where the host is grown under glass. The pathogene, although not always appearing in epidemic form, may cause almost the complete loss of the crop in occasional greenhouses. All varieties of tomatoes which have been tested are about equally susceptible.

Symptoms.

The stems, leaves, blossoms, and occasionally the fruit are affected. The disease is present most often, however, on the leaves where the first sign of the trouble is a light green or yellowish area on the upper epidermis. Simultaneously with this discoloration, the organism fruits on the reverse side of the leaf, so that the corresponding part of the latter is covered with an olivaceous or purple mold. As the fungus becomes well established, the affected tissue is killed and takes on a brownish-yellow color until finally the whole leaf dies. The parasite, after attacking the older or lower foliage, rapidly advances to the leaves which are higher on the stem.

Lesions similar to those on the leaf are found on the young stems and fruit-pedicels, and the moldy growth of the fungus sometimes envelopes the blossoms, causing them to die before the fruit is set or the small fruit to drop. The fruit itself is rarely if ever attacked.

Cause.

The causal organism is *Cladosporium fulvum,* a fungus with dark colored mycelium and fruit-bodies. The conidiophores arise in clusters from the stomatal openings. The conidia are borne in large numbers on the tips of the irregularly shaped conidiophores. They are disseminated readily by air currents or by clinging to the tools, clothing, and hands of the gardeners. They may also be washed from one plant to another when the bed is watered. In a humid atmosphere, like that which the greenhouse often provides, the spores germinate readily, and by entering the stomata cause infection. As the skin of the fruit has no breathing pores, it usually is able to resist infection. When the old infected leaves drop to the ground, the mycelium in the dead tissues form sclerotial-like bodies that are able to live over winter and produce new conidiophores and conidia the following spring.

The variation in the morphology of the parasite has led to the recognition of a special variety known as *Cladosporium fulvum violaceum.*

The parasite is extremely susceptible to temperature and moisture. The spores germinate best when the thermometer ranges from 65° to 75° F. The spores do not germinate in a relatively dry atmosphere; the mycelium thrives in proportion to the amount of humidity in the air. Cloudy days are favorable for increasing the virulence of the organism, not only because cloudiness is usually accompanied by a higher humidity, but also because the germ-tubes grow more rapidly in the dark than they do in the light.

Control of leaf-mold.

Keeping the humidity in the greenhouse as low as is consistent with good growth of the tomato is the first requisite. This may be accomplished by furnishing slightly more heat if necessary and opening the ventilators so that the moisture given off by the plants may escape. If electric power is available, it may be economical at times to draw off the water-laden air by means of a fan. In addition to the decrease in humidity the plants should be provided with as much of the available sunshine as they will stand, and the temperature kept at the maximum for good growth of the host.

Supplementary to governing the environmental factors, sanitation and the use of a fungicide are also to be recommended. All of the diseased dead plants should be removed far enough from the greenhouse to eliminate the possibility of their serving as a source of inoculum for later crops; neither should they be placed on the compost heap unless the composting is complete enough to kill the conidia with the heat which is generated by the decomposing vegetable matter.

Bordeaux mixture has been suggested as the fungicide to use, possibly for the reason that it does not injure the host rather than for any fungicidal effect it may have on Cladosporium. Its use for the control of this fungus, however, has not been satisfactory unless applied very heavily and under high pressure. Applications are made at five-day intervals, beginning before there is any possibility of inoculation by the pathogene. Sulfur is much more toxic to the organism than is the copper in bordeaux mixture, but at the same time the former is injurious to tomato plants. Makemson recommends the fumigation with sulfur of each house just before the crop is planted. Under no condition should the fumigation be attempted after the plants are set into the beds. If it is necessary to make further applications of sulfur, the plants may be sprayed lightly with self-boiled lime-sulfur. This is made by slaking slowly eight pounds of stone lime, and when the bubbling begins sifting in the same amount of sulfur. As soon as the lime is slaked, enough water is added to make fifty gallons of the mixture. Applications are made sparingly at weekly intervals beginning when there is a possibility that the fungus may appear.

References

Anonymous. Brief notes on *Cladosporium fulvum.* Conn. Agr. Exp. Sta. Ann. Rept. **1889:** 171-174. 1890.

Makemson, Walter Kenneth. The leaf-mold of tomatoes, caused by *Cladosporium fulvum* Cke. Mich. Acad. Sci. Rept. **20:** 311-348. 1918.

Savelli, M. Sullo sviluppo del "*Cladosporium fulvum,* varietas *violaceum.*" Ann. R. Accad. Agr. Torino **56:** 63-66. 1913.

Mosaic of Tomatoes

Cause undetermined

In 1892 Bailey in New York described a disease of tomatoes which he named winter-blight. His description and illustrations are almost certain proof that he was dealing in part with the trouble now known as mosaic. Since then the disease has often been reported as present in greenhouses and more recently as the cause of great losses in field-grown tomatoes. Probably the disease is present wherever this crop is grown. Because of the various symptoms produced, the disease is known locally as mottle-top, grass-leaf, fern-leaf, and by other similar terms.

The tomato mosaic, aside from affecting all varieties of tomato, also occurs on tobacco, pepper, petunia and a number of related weed plants, especially ground-cherry (Physalis spp.) and horse-nettle (*Solanum carolinense*). Evidently the mosaic on potato is slightly different although successful inoculations can be made from tomato to potato. The symptoms, however, do not resemble the true potato mosaic.

Symptoms.

As the name indicates, one of the commonest symptoms is the mottling or the mosaic effect on the leaves (Fig. 144) resulting from irregular patches of light green or yellow commingling with similar patches of normal green. The lighter

FIG. 144.—Mosaic of tomato causing mottled leaves and dwarfed branches.

colored areas grow more slowly than does the remainder of the leaf, so that the normal green patches cup up, giving the affected foliage a warty appearance. The edges of the leaves turn downward and are less flexible than are those of the healthy leaf. The yellow areas may die and turn brown. Similar brown streaks may also occur on the stems and petioles. All stages of this distortion may be found in an infested field.

The new leaves which are formed after the plant contracts the disease often consist merely of the midrib with a narrow flange of adjoining tissue. These narrow terminal leaves are often very striking, and have earned for the malady such names as filiform-leaf and string-leaf.

The tomatoes when attacked late in the season may show mottling only on the upper leaves, while the remainder appear normal, thus the plant is not dwarfed nor its yield reduced. The earlier the infection the more profound will be the change, until, in severe cases, the host is much reduced in size, bears no salable fruit, and may even be killed outright. Occasionally the fruit is affected with a russeting, with streaking, or with raised greasy appearing blotches, which after drying out become sunken and cracked.

Cause of mosaic.

Like the mosaic of other hosts, the cause of this disease on tomatoes has not been determined, and until more is known of the causal factor very little of the probable life history can be sketched. Within the sap of the diseased plant there is a substance or potentiality known as the virus or contagium. Any means by which sap is transferred from a diseased to a healthy stalk will also function in the dissemination of the virus. The handling of a few diseased seedlings may be the means later of inoculating many healthy ones when the latter are conveyed by the same hands, which have not been sterilized. Similar contamination may take place when the plants are pruned or when the crop is harvested, although such late infection ordinarily does not result seriously. Like the virus of other mosaics the one in tomato is probably conveyed by aphids, white-flies, and other sucking insects. If an old diseased tomato plant or infected weeds are permitted to remain in or near the greenhouse in which seedlings are grown, inoculation may readily take place, presumably by means of aphids and other sucking insects which carry the sap from these old sources of inoculum to the young plants. Within ten days the new infection will begin to show. Although only a few leaves may become mottled, the contagium has penetrated every part of the plant including the roots. It, however, does not get into the seed and is thus not seed borne; neither does it persist in the soil.

It has recently been discovered that the virus lives over winter not only on old tomato plants in the greenhouse, but remains alive in the roots of perennial weed hosts such as the ground-cherry. When this weed begins to grow in the spring, the new leaves are infected and serve as a source of inoculum.

Control of mosaic.

From the above brief description it is evident that the first step in control is to eradicate all infected tomatoes or susceptible weeds in or near the greenhouse or hotbed where seedlings are to be grown. Furthermore, the greenhouse should be fumigated so that no aphids or other possible insect carriers remain alive. As an extra precaution, the hands should be washed repeatedly with soap if there is any possibility that diseased plants are being handled when pulling the seedlings and transplanting them. The field should be isolated from others in which are planted potatoes, petunias, eggplants, peppers and other susceptible crops. The most important step in the control of tomato mosaic, however, is the destruction of all ground-cherries, horse-nettles, jimson-weed, nightshade, bittersweet, matrimony vine, and other weeds which might act as sources of the virus. All such plants growing within the field or within a zone of several hundred feet about the field should be pulled up by the roots and carried away or burned. Spraying the vines in the field with bordeaux mixture combined with an insecticide is also a wise precaution.

References

Allard, H. A. The mosaic disease of tomatoes and petunias. Phytopath. **6**: 328-335. 1916.

Chapman, G. H. "Mosaic" and allied diseases, with special reference

to tobacco and tomatoes. Mass. Agr. Exp. Sta. Ann. Rept. **25:** Part 2. 41-51. 1913.
Gardner, Max W. and James B. Kendrick. Tomato mosaic. Ind. Agr. Exp. Sta. Bull. 261: 1-24. 1922.
Westerdijk, Johanna. Die mosaikkrankheit der tomaten. Meded. Phytopath. Lab. "Willie Commelin Scholten" **1910:** 1-19. 1910.

Damping-off of Tomatoes

Caused by various fungi

There is scarcely a cultivated or weed plant which, in its most delicate seedling stage, is not susceptible to damping-off. Even certain tree seedlings are affected, although the most severely injured group of plants is the garden vegetables when grown intensively or under forcing conditions. As soon as they are large enough to contain woody supporting tissue in the base of the stem, the plants no longer succumb to damping-off. The further attack of the same fungus, however, may produce a foot-rot, wilt, or stem-canker.

Each year enormous losses are sustained by growers on account of the damping-off organisms. The entire crop of seedlings may be destroyed and, when the disease comes late in the spring, it is often impossible to replace the dead plants by another seeding. Some of the vegetables that show most injury are tomatoes, cucurbits, eggplant, beet, garden-cress, peppers, lettuce, and celery.

Symptoms.

The disease may appear as soon as the seedlings are above ground and continue its devastation until there is sufficient growth of the woody tissue to hold the stems erect. The susceptible period usually lasts one to two weeks; for certain plants and under conditions favorable for the fungus, this time may be extended considerably. Seedlings that appear healthy one day may have collapsed by the next morning. The tissue of the stem near the surface of the soil becomes water-soaked or discolored, and exhibits wilting and constriction or a soft-rot. The invaded portion quickly becomes too weak to support the plant, which falls prostrate, often while the leaves are still green and turgid. In this stage the base of the stem frequently is disintegrated to such a degree that it will fall apart when handled. The root system also decays and the whole plant dies.

Cause.

The damping-off is usually attributed to *Pythium debaryanum* Hesse, although there are other fungi that probably are more responsible for the disease. The Pythium is similar in its life history to the downy-mildew fungi. It has coarse much-branched usually non-septate mycelium. The hyphæ grow through the soil in profusion, and when in contact with a host enter either the stomata or directly through the epidermis. Three types of spores are produced. The tip of a hyphæ or an intercalary cell may enlarge, and after being cut off from the remainder of the thread by a wall, it proceeds to function as a reproductive body. If it germinates with a germ-tube, as it frequently does, the body is known as a conidium. Often, however, it functions as a zoosporangium by forming a vesicle suspended on a beak and in which are developed five to eighteen kidney-shaped biflagellate zoospores. These motile spores swim about for a time in the soil-water, then lose their cilia and germinate by means of a germ-tube. The third type of spore is the sexual oospore resulting from the nuclear fusion of an antheridium and oogonium. The thick-walled oospore undergoes a period of rest before it germinates. These numerous spore stages, together with the rapid development of the mycelium and the dissemination of the fungus with loose soil, adapt the fungus to invasions of epidemic proportions.

Another fungus whose destructiveness and general presence make it a dreaded enemy of vegetable seedlings is Rhizoctonia. Still other damping-off organisms are various species of Phytophthora, Sclerotinia, Botrytis, Fusarium, Sclerotium, Phoma, and many soil-inhabiting fungi of other genera. The life history of each genus may be obtained by referring to the preceding discussions of the different diseases caused by the respective fungi.

Having so many possible parasites to consider, it is impossible to predict the most favorable temperature for the production of damping-off in any particular seed-bed. Rhizoctonia has been found to cause the disease more readily in the lower temperatures ranging from 60° to 75° F., while Pythium is most destructive at higher temperatures represented by 68° to 86°F. Therefore, the temperatures that may cause the avoidance of one organism will prove ideal for the development of another. The presence of water is a different matter. As a rule, damping-off is found when the plants are crowded closely together, heavily watered, in soil that dries slowly, or in greenhouses and coldframes in which poor ventilation permits a high humidity of the air. All these conditions are conducive to the preservation of moisture.

Control of damping-off.

Since some of the parasites may be carried on the seed and others are in the soil and all are favored by the continuous presence of water, the three control measures for damping-off are seed treatment, soil sterilization, and keeping the top soil and plants reasonably dry. Tomato seeds may be treated in the same manner as outlined for the control of bacterial spot (see page 557). Most other vegetable seeds may be given the same treatment, yet proper precautions must be taken to avoid injury, especially if the room temperature is above 70° F. when the treating is being done.

The soil may be sterilized either with steam or formaldehyde (see page 600). Changing the soil has also been suggested, but at best it is a rather doubtful remedy, since few soils are free from all the organisms that cause injury to seedlings. It is better, however, than using the same unsterilized soil continuously for growing young plants.

If, before the plants are up, there seems to be a possibility of damping-off of the new crop, treating the soil once with a copper sulfate solution (1 pound in 25 gallons of water) has proved beneficial. The solution is applied at the rate of one pint to the square foot, and in order to rinse the copper farther into the soil a similar amount of water is applied immediately. The copper cannot take the place of soil sterilization but its use is commendable when steam or formaldehyde are not easily available.

The most important practice in preventing the trouble is in eliminating surface moisture of the soil. This is relatively easy if the greenhouse and hotbeds are so constructed that the plants will get the full benefit of the sun at all hours of the day, and comparatively dry air can be supplied by a proper system of ventilation. In addition, the plants should stand far enough apart in the bed so that they will dry quickly. If grown in coldframes they should be protected from fog or excessive rain. The top soil in the seedling-bed should not be heavy or slow-drying. If the available soil is not suitable its quality may be improved by mixing it with a small amount of sand, or at least sprinkling sand over the surface. It may also be advisable to practice shallow planting in the heavier soil. The watering should be done in the morning and preferably on bright days. If the subsoil can be wet down with a heavy application of water at long intervals, such watering is more desirable than frequent sprinklings in which the plants and surface soil are kept moist while the subsoil dries out. Stirring the surface soil after an application of water will help it to dry, and thereby reduce the chances for injury to the seedlings.

References

Atkinson, G. F. Damping-off. N. Y. (Cornell) Agr. Exp. Sta. Bull. 94: 233-272. 1895.

Butler, E. J. *Pythium debaryanum*. Mem. Dept. Agr. India Bot. ser. 5: 262-267. 1913.

Clinton, G. P. Damping-off, *Pythium debaryanum* Hess. Conn. Agr. Exp. Sta. Bull. 222: 475-476. 1920.

Hemmi, Takewo. On the relation of temperature to the damping-off of garden-cress seedlings by *Pythium debaryanum* and *Corticium vagum*. Phytopath. 13: 273-282. 1923.

Johnson, James. The control of damping-off disease in plant beds. Wis. Agr. Exp. Sta. Res. Bull. 31: 29-61. 1914.
Peters, L. Über de erreger des wurzelbrandes. Arb. aus d. Biol. Anst. f. Land. u. Forstw. 8: 211-259. 1911.
Sherbakoff, C. D. Report of the associate plant pathologist. Fla. Agr. Exp. Sta. Ann. Rept. 1915-16: 80R-98R. 1917.
Ward, H. M. Observations on the genus Pythium. Quart. Jour. Micros. Sci. 23: 487-497. 1883.

Root-Knot of Tomato

Caused by *Heterodera radicicola* (Greef) Müller

Root-knot has been reported on every vegetable grown, as well as on many other cultivated plants and weeds. Among the most seriously affected garden crops are cucumbers, muskmelons, squash, watermelons, pumpkins, tomatoes, eggplant, lettuce, spinach, parsley, beets, carrots, parsnips, celeriac, and celery. Asparagus, beans, cabbage, kale, onions, peas, potatoes, and radishes are less susceptible, while only sweetpotatoes show any marked resistance. In one respect the slightly injured crop is as much of a menace as is the one that is seriously damaged, for the former is grown year after year and furnishes an ever-present host for the nematode. Cereals are the most resistant to root-knot of any of the cultivated plants.

The disease is present in all parts of the world where the temperature does not drop too low during the winter months. It is not present under field conditions in the northern part of the United States; the trouble is common, however, in the South, especially in the sandy coastal regions from Maryland to Texas and in some of the southwestern irrigated sections. In the North it is the most serious malady in greenhouses. The average loss for all crops has never been estimated accurately, but it probably would not be an exaggeration to say that 5 per cent of all susceptible crops are taken by the nematode. Ten to 50 per cent reductions in yield from this trouble are not rare, and complete crop failures are often reported when control methods are not followed.

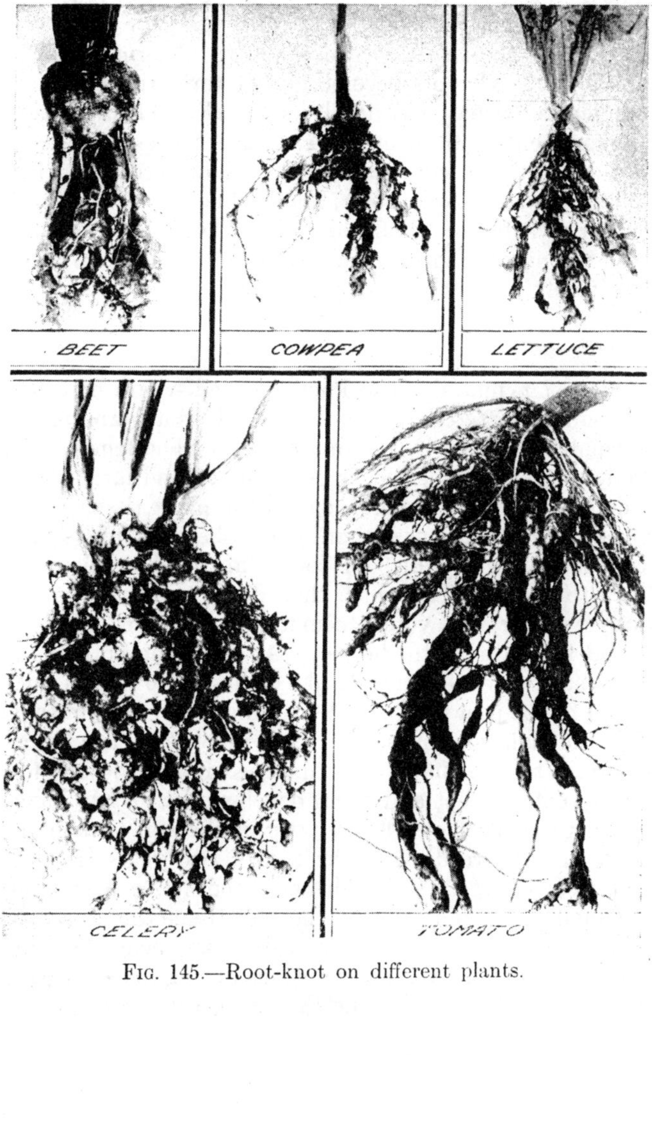

FIG. 145.—Root-knot on different plants.

Symptoms.

The disease has been confused with club-root of crucifers and legume nodules, even though it does not resemble either closely. In root-knot the tissue is stimulated to abnormal growth, so that the invaded roots become two or three times as large in diameter as normal, and finally decay (Fig. 145). The top of the plant being robbed of the food which goes to the swollen roots, and at the same time deprived of feeding surface, is stunted in size, pale in color, and inclined to wilt. The fruit is either dwarfed or wholly lacking. The affected plants may linger through the growing season, although most of them die prematurely.

Cause.

The root nematode, also known as gall-worm or eel-worm, is classified as *Heterodera radicicola*. It is a worm, ordinarily too small to be seen by the unaided eye. The female, just before her egg-laying period, is pear-shaped and large enough to be detected without a hand-lens in the plant tissue where she appears as a minute pearly granule. She may lay as many as five hundred eggs, which of course are microscopical in size. When the eggs hatch, which they do in two or three days in warm weather, the young nematodes may feed in the roots or find their way into the soil. At this time both the males and females have thin thread-like bodies, 12 to 15 μ in thickness by 375 to 500 μ in length. The female then grows in thickness while the male continues to increase in length. The larvæ may remain in the soil for a long time, seemingly without food, or quickly enter a root and begin development. They have been known to survive at least a year in the soil without a host plant, but all are dead at the end of two years if the soil is kept free of vegetation. The nematodes live over winter in perennial roots, and when in the root tissue die soon after the host. In the South where the seasons are long, there may be as many as ten or twelve generations during the growing period of the host, but farther north the shorter seasons and the colder climate retard the development of the parasite. The Heterodera is disseminated by means of infested soil, flowing water, and with diseased root-crops.

Most of the root-knot is found in light sandy soil. It is almost impossible to obtain infection in heavy clay. The organism thrives best where there is a medium amount of moisture, and is killed when the soil is dust dry for more than several months, or when land is flooded part of the season. As was stated before, they survive only the southern winter, therefore are susceptible to low temperatures. The larvæ are inactive at 50° F. and many of them, if not all, are killed by the alternate freezing and thawing of the average winter in the northern half of the country. Their thermal death point is between 120° and 140° F.

Control of root-knot.

Much of the injury to vegetables takes place in the greenhouse where soil sterilization may be practiced. Notwithstanding the rather general application of steam and formaldehyde, the nematodes have not been satisfactorily controlled in many instances, due to the fact that the sterilization was conducted carelessly or with ineffective apparatus. It is possible to kill the parasite by means of steam, applied preferably by the drain-tile method (page 602). Growers who do not have steam available under pressure, may use boiling water for flooding the soil (page 612). Changing the soil is also effective, although it usually is expensive.

When the trouble occurs in the field, the problem of control is much more complicated. The organism is disseminated so readily that when it once enters a community every field is soon infested. Consequently it frequently is impossible to find clean soil in which to grow susceptible crops. Where irrigation is practiced, or in some of the river bottom land, flooding for a month or more may be practicable. In the much larger proportion of fields in which flooding is not possible, the grower must rely on proper rotations of crops and the growing of those plants which are least susceptible. In crop rotations it has been found that the following hosts are very resistant and should be grown as often as possible when attempting to starve the Heterodera: barley, broom-corn, millet, pearl millet, corn, cowpea (certain varieties), nearly all grasses, red-top, timothy, peanuts, sorghum, rye, velvet beans, wheat, and winter oats. It is not enough merely to change crops, but it is also necessary to keep down the susceptible weeds. It is for this reason that plants which may be cultivated, as corn, are often preferable to cereals sown broadcast.

A rotation has been suggested for the southern gardener who has only a small space and who is bothered with root-knot. The lot is divided into three strips, each inclosed with wire netting. One strip is used for a chicken pasture, the second for the garden, and the third for growing sweet-corn, soybeans, or other crops which are immune. The second year the chickens are pastured in the lot where the garden was, while the vegetables follow the sweet-corn or other immune crops and the latter are planted in the old chicken run. A similar shifting is practiced each year with the result that a comparatively healthy crop may be grown.

In Florida commercial calcium cyanamide applied dry at the rate of one to three tons an acre has been suggested. Such a treatment has a limited application since the large quantity necessary may injure the plants, and the material is too expensive for use on extended areas.

References

Atkinson, G. F. A preliminary report upon the life history and metamorphoses of a root-gall nematode, *Heterodera radicicola* (Greef) Müller, and injuries caused by it upon the roots of various plants. Ala. Agr. Exp. Sta. Bull. 9: 1-54. 1889.

Bessey, E. A. Root-knot and its control. U. S. Dept. Agr. Bur. Plant Ind. Bull. 217: 1-89. 1911.

Byars, L. P. Experiments on the control of the root-knot nematode, *Heterodera radicicola* (Greef) Mueller. Phytopath. 9: 93-104. 1919.

Byars, L. P. and W. W. Gilbert. Soil disinfection with hot water to

control the root-knot nematode and parasitic soil fungi. U. S. Dept. Agr. Bull. 818: 1-14. 1920.
Cobb, N. A. Nematodes and their relationships. U. S. Dept. Agr. Year book 1914: 457-490. 1915.
Godfrey, G. H. Root-knot: its cause and control. U. S. Dept. Agr. Farmers' Bull. 1345: 1-26. 1923.
Stone, G. E. and R. E. Smith. Nematode worms. Mass. (Hatch) Agr. Exp. Sta. Bull. 55: 1-67. 1898.
Watson, J. R. The control of root-knot by calcium cyanamide and other means. Fla. Agr. Exp. Sta. Bull. 136: 147-160. 1917.
Zimmerley, H. H. and H. Spencer. Hot water treatment for nematode control. Va. Truck Exp. Sta. Bull. 43: 267-278. 1923.

Bacterial-Spot of Tomato

Caused by *Bacterium vesicatorium* Doidge

The bacterial-spot of tomatoes and peppers, although recognized only ten years ago, is now known to be widespread, reports of its presence having been received from South Africa, Florida, Georgia, Tennessee, Indiana, Illinois, Iowa, Kansas, Michigan, and probably other states. It is not a serious leaf disease except in wet seasons, but does much damage on tomato and pepper fruits, not only in disfiguring them but in giving access to rot-producing saprophytes. In some counts made at a canning factory in Indiana, the amount of injured fruit ran as high as 5 per cent, and in a few shipments from Florida the losses equalled those of the nailhead disease in severity.

Symptoms.

Small water-soaked areas, the centers of which later turn black and parchment-like, appear on the leaf. The spots are usually slightly angular and may be greasy on the upper surface. The lesions never become large, but several spots on a single leaflet may cause it to turn yellow and drop. The lesions resemble the Septoria leaf-spot very closely, although the latter is characterized by the presence of pycnidia and never has the yellowish halo about the lesion and the irregular outline as is typical of the bacterial-spot. On young plants the organism may attack stems and petioles with enough virulence to cause considerable injury, and may even form cankers on older injured stems.

The most important lesions are on the fruit. They first appear as minute black raised spots surrounded by a narrow water-soaked zone. The protrusion of the darkened area has earned the name, scab, for this stage of the disease. The spots enlarge until they reach six to eight millimeters in diameter, and extend some distance into the flesh of the fruit. As the affected inner tissue disintegrates, the raised portion sinks until it becomes a depressed cavity bordered by the ruptured epidermis. When several adjacent spots coalesce, a large blotch-like lesion is formed.

Cause.

The parasite, *Bacterium vesicatorium,* is a rod-shaped yellow uniflagellate organism that gains entrance into the leaves through the stomata, and into the fruit through insect punctures. It is not known in how many ways the bacterium is able to survive the winter but it has been shown that seeds may act as conveyors of the organism and undoubtedly furnish one means by which the new crop may be inoculated. As in other seed-borne diseases, the seedlings become infected and being in close proximity to each other, the few invaded plants serve as centers of infection for their neighbors. The result is an increased number of unhealthy plants set into the field.

The prevalence of the disease is usually in direct proportion to the amount of rainfall; the most favorable temperature is between 75° and 85° F. Thus a warm wet season may inaugurate an epidemic of bacterial-spot.

Control.

If the clean seed is planted in soil that has not borne tomatoes for at least three years, and the plot is isolated from infested fields, it is fairly certain that the crop will be free from the bacterial-spot. It is usually easy enough to rotate and to isolate the crop, but the difficulty lies in procuring clean seed. The latter may be accomplished if the grower selects his own tomato seed by marking the most prolific and otherwise desirable vines, and picking the unblemished fruit from them. Before the seed is extracted the tomatoes are placed in a solution of corrosive sublimate, 1-1000 (1 tablet in each pint of water) for ten minutes, then rinsed in running water. As little of the pulp as possible is taken with the seed which may be washed free from the flesh at once or permitted to stand for a day or two in the watery solution before the seeds are finally separated, cleaned with water and dried. Such seed does not need to be treated.

If seed, bought on the market, is not known to be free from the bacterial-spot, it should be treated with corrosive sublimate, 1-3000 (1 tablet in 3 pints of water) for five minutes, rinsed in clean water and dried, or better still planted before it is thoroughly dry. The disinfecting solution should be used only once before being discarded and should always be placed in wooden, glass, or glazed crockery containers, for metal causes it to weaken. The seed is tied very loosely in cheesecloth bags, and submerged in the solution. It is well to stir the bags with a stick, making sure that the air within the cloth is replaced by liquid, so that the seeds will be in direct contact with the fungicide.

References

Doidge, E. M. A tomato canker. Ann. Appl. Biol. 7: 407-430. 1921.

Gardner, Max W. and J. B. Kendrick. Bacterial-spot of tomato. Jour. Agr. Research 21: 123-156. 1921.

Gardner, Max W. and J. B. Kendrick. Tomato bacterial-spot and seed disinfection. Ind. Agr. Exp. Sta. Bull. 251: 1-15. 1921.

Garden, Max W. and J. B. Kendrick. Bacterial-spot of tomato and pepper. Phytopath. 13: 307-315. 1923.

Grand Rapids Disease of Tomatoes

Caused by *Aplanobacter michiganense* EFS.

The Grand Rapids disease of tomatoes was first recognized in 1909 by Smith, who received affected plants from Michigan.

He later obtained specimens from western New York, from Massachusetts, and doubtfully from Texas. He is of the opinion that the organism described by Spiekerman as *Bacterium sepedonicum* is probably the same or a closely allied organism, and that the disease, therefore, may be found in Europe. He was able to induce infection not only on the tomato, but also on a West Indian plant, *Solanum mammosum.* The trouble is found both in the field and in the greenhouse, and may cause considerable injury as was evidenced by the fact that at Grand Rapids the loss was eight thousand dollars in one season.

Symptoms.

According to Smith, the malady differs from brown-rot in being highly infectious through the parts of the tomato above ground, producing less brown stain in the vascular bundles, and not developing in the host such a strong tendency to force incipient aerial roots. The leaflets shrivel slowly one by one rather than exhibiting a sudden general wilting. The parasite disorganizes the phloem more than any other tissue in the stem, and oozes to the surface of affected leaves, fruits, and shoots through fissures caused by the disease where it causes a slight slimy coat.

Cause.

The rod-shaped non-motile yellow bacterium, bearing the number 211.2222522, is named *Aplanobacter michiganense.* It is carried by the splashing of water or by insects to tomato leaves and stems where it enters through the stomata. It may also gain admission through broken roots. In either case it causes a very destructive disease. The over-wintering of the bacterium has not yet been determined in detail, but it is probable that it is carried on or in the seed.

Control.

Although no control methods have been evolved, it is well to take the precaution of burning all diseased vines, rotating crops if the trouble is in the field, or sterilizing the soil if in the greenhouse, and obtaining seed from healthy plants or treating the seed as suggested for control of bacterial-spot (see page 557).

REFERENCES

Smith, E. F. Bacteria in relation to plant diseases 3: 161-165. 1914.
Smith, E. F. Bacterial canker of tomato. *In* An introduction to bacterial diseases of plants, pp. 202-222. 1920.

THE STREAK DISEASE OF TOMATOES

Cause undetermined (possibly *Bacillus lathyri* M. and T.)

For thirty years or more a rather serious disease of greenhouse tomatoes, known as streak or winter-blight, has been present in this country. More recently a disease with the same symptoms has been named stripe or black-stripe in England, and spotted-wilt in Australia. It is generally considered that these three differently designated troubles are identical, but up to the present time the American workers have not been able to duplicate successful inoculations with the causal organism suggested by the investigators in England. This failure may be due to the difference in environmental conditions in the two countries, or it may be ascribed to the fact that, although the symptoms are alike, the causal factors are not the same in the two cases. Still other reasons for the discrepancy, which are not yet evident, may be involved.

Symptoms.

The streak disease usually is found in greenhouses, although it may occasionally be present on tomatoes grown in the field. Often it is associated with mosaic and its symptoms on the leaves may be confused with this disease. It seldom is noticeable until the plants are fairly well grown. The symptoms may appear on leaves, stems, and fruits, all of which may be affected at once or any one part be injured while the remaining parts still seem healthy. The diseased plant is dwarfed and

FIG. 146.—Streak showing on the leaves of tomatoes.

given a wizened appearance because of the sparseness and the peculiar rolling and withering of the foliage (Figs. 146, 147). In advanced stages the plant may die.

The disease usually begins on the young leaflets at the tip of the vine and advances to those which are lower and older. The first visible effects on the leaf are irregular light yellow areas, which generally change to a chrome-yellow, and finally turn brown and die. The dead areas not only include the mesophyll cells but also part of the smaller veins, and react on the remaining healthy portion in such a manner that the edges of the leaf are curled downward and inward in a very characteristic manner. The green areas may be cupped upward giving the foliage a roughened appearance.

Brown lesions in the form of longitudinal streaks are very common on the petioles and large veins. The discoloration of the tissue ordinarily is superficial, but in extreme cases it may extend inward beyond the vascular bundles. Such tissue is very brittle so that the stem can be snapped off readily. As the diseased stem does not develop as fast as that of the normal plant, it is weak and spindling.

Light brown sunken spots are also found on the fruit. They are irregular in outline, variable in size, and may appear on any part of the surface. Frequently the spots coalesce, producing large blotches. Sometimes the affected epidermis is cracked, giving the lesion a scabby appearance. If the immature tomatoes are affected they fail to color normally, and remain hard and green. Frequently the diseased fruit falls before it begins to ripen. There is no rot unless secondary organisms enter through the injured tissue, but the discoloration may extend to the center of the fruit and make it worthless for the market.

Cause.

The investigational work conducted in Canada led to the conclusion that no organism was connected with the injury, and that it probably was the result of an unbalanced fertilizer.

FIG. 117.—Streak injuring tip of tomato plant, and breaking the leaf-petioles.

In England it is considered a settled fact that *Bacillus lathyri*, the organism that causes streak of sweet-peas, is also the parasite associated with the streak of tomatoes. The result of the work in the United States has been so conflicting that the pathologists are unwilling to deny or affirm either of the above statements.

The bacillus is a yellow rod-shaped actively motile organism, which is supposed to live over winter both in the soil and to a slight extent on the seed. It gains entrance into the host through wounds made by insects, or in cuts by pruning. It readily passes to all parts of the plant, causing the types of lesions described under symptoms. The diseased tissue left on the ground serves to inoculate the soil for another year according to the English view.

All investigators agree that an over supply of nitrogenous fertilizer causes the disease to appear in epidemic form. When the nitrogen is balanced properly with acid phosphate and potash, the disease is eliminated or much reduced. Poor ventilation, irregular watering, and fluctuating temperatures are favorable for the production of streak.

Control of streak.

Evidently the first requirement is normally growing robust plants. These are procured by good ventilation, uniform heating and watering, and feeding with a balanced fertilizer. Plant-food with an excessive amount of nitrogen is to be avoided. In Canada three ounces of acid phosphate and one-fourth ounce of potassium sulfate added to each plant when the disease began to show prevented any further development of the trouble. The proper environment is usually sufficient to insure a crop free from streak, but in order to be safe from the ravages of a possible bacillus, it is well to take the added precautions of sterilizing the seed (page 557), selecting a new site for the seed-bed, sterilizing the greenhouse soil (page 600), transplanting only healthy plants, and avoiding inoculation with contaminated clothes or tools.

References

Bailey, L. H. Some troubles of winter tomatoes. N. Y. (Cornell) Agr. Exp. Sta. Bull. 43: 149-158. 1892.
Brittlebank, C. C. Tomato diseases. Jour. Dept. Agr., Victoria 17: 231-235. 1919.
Howitt, J. E. and R. E. Stone. A troublesome disease of winter tomatoes. Phytopath. 6: 162-166. 1916.
Orton, C. R. and W. H. McKinney. Winter blight of the tomato. Penn. Agr. Exp. Sta. Ann. Rept. 1914-15: 235-246. 1916.
Paine, S. G. and W. F. Bewley. Studies in bacteriosis IV. "Stripe" disease of tomato. Ann. Appl. Biol. 6: 183-202. 1920.
Paine, S. G. and W. F. Bewley. Studies in bacteriosis VIII. Further investigations of the "stripe" disease of tomato. Ann. Appl. Biol. 10: 89-95. 1923.
Stone, R. E. and J. E. Howitt. Experiments with winter-blight or streak of tomatoes. Phytopath. 12: 41. 1922.

Bacterial Brown-Rot of Tomatoes

Caused by *Bacterium solanacearum* EFS.

(See page 401 for a more detailed discussion)

As stated under the discussion regarding brown-rot of potatoes, the causal bacterium attacks many kinds of plants, and among these is the tomato. In some of the southern states it is quite destructive on the latter host, but is of little importance farther north.

Symptoms.

In nearly all respects the symptoms on tomatoes are the same as those on the potato. It may at times be difficult to distinguish between this trouble and the Grand Rapids disease. The fruit seldom, if ever, is affected. The large leaves on injured vines may curl downward, giving a characteristic appearance to the plant. Great numbers of adventitious roots are forced into growth by the disease, but finally remain as small nodules on the side of the stem. The cut surface of the stem will frequently show oozing of the dirty-colored bacteria.

Cause.

The pathogene is *Bacterium solanacearum,* having the descriptive number $21\frac{1}{2}$.3333823.

Control.

It is advisable to sterilize the seed-beds with steam or formaldehyde (page 600) or if this is impossible it is recommended that soil used in the beds be taken from an uncontaminated source. In transplanting the seedlings, care should be exercised in keeping the roots uninjured or free from nematode infection, for it is through these injuries that the bacterium enters the plant.

References

Johnson, James. The relation of air temperature to certain plant diseases. Phytopath. **11:** 446-458. 1921.

Orton, C. R. and W. H. McKinney, Jr. Notes on some tomato diseases. Penn. Agr. Sta. Ann. Rept. **1915-16:** 285-291. 1917.

Sherbakoff, C. D. Solanaceous blight (Brown-rot). Fla. Agr. Exp. Sta. Ann. Rept. **1917-18:** 71R-74R. 1919.

Fusarium Wilt of Tomatoes

Caused by *Fusarium lycopersici* Sacc.

Fusarium wilt of tomatoes, also known as sleepy-disease, summer-blight, or merely as wilt, is present wherever the host is grown, although the fungus is not an active parasite in the northern tier of states and in other equally cool areas. The pathogene was first described in Italy in 1882, but the wilt disease was not mentioned until 1896 when it was discovered in England. It is probable that the malady was present in Europe and North America for a number of years before the latter date. The greatest loss in the United States from this disease occurs in the states bordering both sides of the lower Mississippi and Ohio rivers, where a reduction of 10 to 20 per cent of the crop is not unusual. The annual loss is estimated at more than one hundred thousand tons of tomatoes.

Symptoms.

The symptoms of the disease are those of a typical wilt. Even before any outward indications of the trouble are apparent, a cross-section of the stem will show the usual darkening of the vascular bundles. When the environment is very favorable for the fungus, the seedling plants may become infected, wilt or turn yellow, and soon die. Under normal conditions, however, infection takes place later or the early inoculated plant continues to live until, a least, the first fruit begins to form. In the meantime the injured stalks do not grow so rapidly as do those that are healthy. As the disease becomes more acute, the leaves, beginning with the lower ones, turn yellow and wither until finally the entire plant wilts and succumbs. It is possible for some of the more resistant vines to remain alive throughout the season even though they appear to be generally infected.

The roots as well as the stems are invaded, the symptoms showing as a black-rot particularly on the smaller laterals. This severing of the water supply hastens the withering of the parts above ground. After the vines are dead they also turn black and the fungus fruits on their surface. Under moist conditions the dead stem may be covered with the pale pink mycelium of the parasite.

Cause.

In America the causal fungus is probably in all cases *Fusarium lycopersici,* while in England Bewley has found that wilt caused by *Verticillium alboatrum* is much more common. Both organisms live over in the soil and in diseased plant refuse; and except in size, shape, and manner of bearing spores, their life histories are similar.

The plant may become infected at any stage of its growth. The fungous hyphæ grow from the soil into the small roots, and after penetrating to the vascular bundles extend upward through the xylem ducts until the whole plant is invaded.

When the host dies the mycelium pushes outward and fruits on the surface of the stem. In case of the Fusarium, the sickle-shaped septate spores are borne in great quantities, which reinfest the soil to the detriment of the succeeding crop of tomatoes. Being a soil organism, it can readily be carried from one plant to the next, or from one field to an adjoining one, by washing of water or in any other way in which particles of infested soil may be conveyed from row to row or from field to field. It has been proved quite definitely that seed bearing spores or possibly seed containing bits of mycelium furnish the means for distant dissemination. The soil in the northern states has become inoculated by the introduction of early seedlings from the South. When the parasite once becomes established, it will live for more than three years without the presence of the tomato crop. It is not known just how much longer it can survive away from the host, but experiments have shown that even one year's rotation will reduce materially the amount of inoculum in the soil, even though a ten-year lapse may not destroy all of it.

There is a difference in the ecological requirements of *Verticillium* and *Fusarium lycopersici*. The latter does not develop well in heavy soils, reaching its greatest growth in well aerated light soils that are neither extremely wet nor unusually dry. The maximum amount of infection is obtained when common garden soil is mixed with leaf-mold. In addition to the soil requirement, a high temperature is essential. The most favorable amount of heat is 85° F., which is reached as early as the middle of May in the extreme South, but which is attained only in an abnormally warm season in the farthest northern states.

The Verticillium parasite, on the other hand, thrives best in clay soils, which because of their greater water-holding capacity are cooler than those containing much sand. Its optimum temperature is approximately 70° F. as compared with 85° F. of the Fusarium. The Verticillium when once present could very readily do much damage to tomatoes in the cooler sections of the country.

Control of wilt.

In nearly all field diseases caused by soil organisms, the grower does not place enough emphasis on the care of the seedlings. If a tomato seed-bed is placed on soil which is known to be contaminated or which has an opportunity to become infested, there is no reason why the seedlings will not be diseased. Isolated seed-beds on clean soil are the first requirement for the control of wilt; long crop rotations in the field are the second. The seed-bed, if necessary, can be sterilized either with steam or formaldehyde (see page 600). In order to make sure that the bed will not be reinfested, the seed should be procured from healthy ripe tomatoes, or if the origin of the seed is unknown, it should be sterilized with 1-3000 corrosive sublimate for five minutes (1 tablet in 3 pints of water). The seed is tied loosely in cheese-cloth bags, dipped in the disinfectant for the required length of time, then rinsed in running water. The seed should then be partly or wholly dried and planted. The northern grower should not buy southern seedlings unless a guarantee of health can be obtained for them.

The most satisfactory method of control is the use of resistant varieties. In nearly every state where wilt is a menace some progress has been made in selecting strains that will withstand the effects of the parasite. Some of the most successful work has been conducted in Louisiana, Maryland, and at Washington, where such varieties as the Louisiana Red, Louisiana Pink, Marvel, Norton, Columbia, and Arlington have been produced.

References

Bewley, W. F. "Sleepy disease" of the tomato. Ann. Appl. Biol. 9: 116-134. 1922.

Clayton, E. E. The relation of soil moisture to the Fusarium wilt of the tomato. Amer. Jour. Bot. 10: 133-147. 1923.

Edgerton, C. W. and C. C. Moreland. Tomato wilt. La. Agr. Exp. Sta. Bull. 174: 1-54. 1920.

Edgerton C. W. and C. C. Moreland. Tests of the wilt-resistance of different tomato varieties. La. Agr. Exp. Sta. Bull. 184: 1-24. 1921.

Elliott, J. A. and R. F. Crawford. The spread of tomato wilt by infected seed. Phytopath. **12**: 428-434. 1922.

Pritchard, F. J. Development of wilt-resistant tomatoes. U. S. Dept. Agr. Bull. 1015: 1-18. 1922.

Western Yellow-Blight of Tomatoes

Caused by *Fusarium* et al.

As early as 1896 the western yellow-blight of tomatoes became so serious in Oregon that an investigation of its cause was begun at that date. It is now present in Washington, Oregon, Idaho, Utah, just across the boundary in Canada, and possibly in several other western states. No doubt the malady described as summer-blight or wilt in California is identical with yellow-blight. The disease is of considerable importance, a loss of 50 per cent of the crop not being uncommon, and losses ranging from 1 to 5 per cent for entire states have been reported. Apparently it affects all varieties of tomatoes.

Symptoms.

The first symptoms are commonly not apparent until the host is in blosson , and in many instances not until the fruit has attained half its growth. During the remainder of the season the disease increases in prevalence and in severity. There is never a typical stage of wilt in the progressive series of symptoms, but the foliage becomes more brittle and droops. The affected plants are usually stunted in growth, are spindling, and do not bear heavily unless inoculation takes place late in the season. The symptoms begin as a slight upward rolling of the leaves, and a characteristic purpling of the leaf-veins, followed by a peculiar sheen of the foliage that gives a grayish cast to the whole plant. Soon thereafter the host turns yellow, the fruit even though only half mature ripens, and finally the plant dies.

The roots also are invaded, especially the tips of the small feeding roots, which are often discolored and decayed. The cortical tissue becomes loose so that it can be slipped off leaving the central woody cylinder. This brown soft tissue is in marked contrast to the white firm roots of the healthy plant.

Cause.

There are so many conflicting statements regarding the cause of yellow-blight that it is impossible to weigh all the evidence accurately. Humphrey believes that *Fusarium orthoceras* and *Fusarium oxysporum* are the two causal organisms, while Heald is of the opinion that Rhizoctonia may be responsible for the trouble, although he also was able to isolate a Fusarium. It may be that there is first a physiological disturbance of the roots, followed by various soil organisms, thus accounting for the conflicting results so far obtained. No life history can be suggested until the parasite is determined definitely.

The disease is most common in the districts of the Northwest where the summer temperature is high during a long period. The optimum temperature seems to be approximately 85° F., so that light quickly heated soils are favorable for increasing the trouble. In some manner the velocity of the wind and the intensity of the sunlight also accelerate the progress of the disease. Where plants have a windbreak and are slightly shaded, the number of diseased plants is lessened

Control.

No sure methods have yet been devised for the control of yellow-blight. The fungus, whatever it may be, is evidently a soil inhabitant and as such must be warded from the seedling-bed. If clean soil cannot be obtained, sterilizing with steam or formaldehyde is necessary (see page 600). It has also been shown that when the young plants are grown under glass or other wind protection until they have attained maximum size for transplanting, the amount of infection will be reduced. In a like manner care in keeping the roots uninjured during transplanting into the field will result in increased healthy plants. Such a precaution may be exercised most easily if the dirt in the pot is transferred with the plant, leaving the roots in it undisturbed. Windbreaks, also, are important aids in fighting the ravages of the parasite. In places where natural windbreaks in the form of hedges are not available, use may be made of brush fences. In extreme cases young plants can be protected by furnishing them with individual windbreaks made by setting on edge two short boards eight inches wide and nailing them together in a V-shape, with a thin brace over the open part of the V. These boards are left about the plant during the entire season and serve later in keeping the fruit from touching the ground when the vine becomes over-laden.

Rotation with non-susceptible crops, late planting when possible, and the avoidance of infested soil, are also recommended although these practices alone cannot be relied on when other control measures are neglected.

References

Heald, F. D. Tomato blight. Wash. Sta. Hort. Assoc. Rept. 12: 35-42. 1916.

Humphrey, H. B. Studies on the relations of certain species of Fusarium to the tomato blight of the Pacific Northwest. Wash. Agr. Exp. Sta. Bull. 115: 1-22. 1914.

McKay, M. B. Western yellow tomato blight. Ore. Crop. Pest and Hort. Rept. 3: 174-178. 1921.

Rogers, Stanley S. The culture of tomatoes in California with special reference to their diseases. Calif. Agr. Exp. Sta. Bull. 239: 591-617. 1913.

Shapovalov, M. Effect of environmental conditions on western yellow blight of tomatoes. Phytopath. 14: 120, 121. 1924.

Phytophthora Foot-Rot of Tomato

Caused by *Phytophthora cryptogea* Peth. and Laff.

Phytophthora foot-rot of tomato was discovered recently in Ireland and England. So far as is known this fungus has not yet been introduced into the United States, although a closely related organism has been isolated from greenhouse tomatoes in New York. By artificial inoculation the foot-rot

FIG. 148.—Reddick's Phytophthora causing wilt of tomato plant in the greenhouse.

organism has been shown to infect asters, petunias, wall-flowers, *Gilia tricolor*, and seedlings of beech.

Symptoms.

When the plants are attacked in their seedling stage, typical symptoms of damping-off result. The stem is infected at the point where it emerges from the soil, and quickly becomes so flabby that the plant falls over and withers away (Fig. 148). When the host is more mature, the thick epidermis and woody structure of the stem aid in warding off the infections of the parasite, but the latter gains an entrance through the roots and works upward until the stem is decayed up to the surface of the soil. As the rot advances the tissue turns brown and wastes away until it can no longer hold the plant upright. Before the plant finally falls over, the foliage may show the rolled and yellow appearance that ordinarily is manifested in wilt diseases; but quite often the foliage remains green and turgid even after the plant is prostrate.

Cause.

The causal organism, *Phytophthora cryptogea,* is closely related to that large group of parasites which produce downy-mildews on nearly every cultivated crop. The genus is known for the rapid growth of its non-septate mycelium through the host tissue, and the production of lemon-shaped conidia, the contents of which break up into many swarm-spores. The mycelium may also bear thick-walled sexual spores which are able to live through the winter. These spores together with mycelium in old diseased tissue insure an abundance of inoculum for the new crop. The rains often wash the conidia into nearby ponds or open wells, the water of which is used to irrigate the tomato beds. No surer method of inoculation could be provided. With a sufficient amount of contaminated water and the temperature permitted to attain 77° F., the ideal conditions are available for the germination of the conidia and the penetration of the mycelium into the host.

Control.

When a greenhouse becomes infested it is not enough that the old soil be exchanged for that from cultivated fields or even from the woods, for it is possible that such soil may also be contaminated. The whole greenhouse should be sterilized thoroughly with steam or formaldehyde (see page 600). In addition, it is a wise precaution to have the sprinkling water analyzed for the presence of any Phytophthora spores, particularly if there is any possibility of the source of the water having become contaminated from flooded fields or from seepage. Water from deep wells is desirable.

As a further protection for the seedlings, it is better not to water the plants during the middle of the day. Such watering does not give the plant an opportunity of drying in the sunlight and, therefore, encourages the attack of the parasite.

Subirrigation has proved more satisfactory than overhead sprinkling when damping-off is to be combated. The use of potash fertilizer has also proved beneficial. When damping-off is imminent its attack may be delayed by sprinkling a thin layer of pure sand over the bed to prevent the collection of moisture at the surface of the soil. According to Bewley one part of finely ground copper sulfate is mixed with ten parts of hydrated lime and sprinkled over the bed between the seedling rows. This will check damping-off even after the disease has made its appearance. In a similar manner the bed may be wet down with a copper sulfate solution just after the seed is sown. One pound of the fungicide is dissolved in twenty-five gallons of water and a pint of this solution is applied to each square foot. In order to distribute the poison in the soil, the bed is immediately wet down with an equal amount of water. This will not injure the germination of the seed, but is sufficient to check the growth of Phytophthora which is susceptible in the highest degree to copper poisoning.

English workers have suggested the "Cheshunt Compound," made of copper sulfate and ammonium carbonate for the control of damping-off and similar diseases. It is not supposed to injure young plants even when poured over them. This solution, when made according to directions and applied to seed-beds in America, did not give favorable results.

References

Bewley, W. F. "Damping-off" and "foot-rot" of tomato seedlings. Ann. Appl. Biol. **7**: 156-172. 1920.

Bewley, W. F. and W. Buddin. On the fungus flora of glasshouse water supplies in relation to plant disease. Ann. Appl. Biol. **8**: 10-19. 1921.

Bewley, W. F. "Damping-off" and "foot-rot" of the tomato. Exp. and Res. Sta. Nurs. and Mark. Gard. Indus. Cheshunt, Herts. Ann. Rept. **7**: 38-39. 1922.

Pethybridge, G. H. and H. A. Lafferty. A disease of tomatoes and other plants caused by a new species of Phytophthora. Sci. Proc. Roy. Dublin Soc. n. s. **15**: 487-505. 1919.

Spinks, G. T. Damping-off and collar-rot of tomatoes. Agr. and Hort. Res. Sta. Univ. Bristol Ann. Rept. **1917**: 25-27. 1918.

Ascochyta Blight of Tomatoes

Caused by *Didymella lycopersici* (Brun.) Klebahn

The chief interest in Ascochyta blight of tomatoes lies in the difference of opinion regarding the nomenclature of the fungus, which was first described in France in 1887. It has since been found in a number of places in America and Europe, mainly on tomatoes, although it has been recorded on eggplant.

Symptoms.

The parasite occurs on the stems as well as on the leaves. On the latter the spots are brown with concentric rings and resemble very closely the lesions produced by Macrosporium. By close examination, however, the dead tissue is found to be covered with minute black pustules, the embedded pycnidia. A brown canker is formed on the stems of young plants, at the surface of the ground causing them to wilt or fall over.

Cause.

The perfect stage of the fungus has recently been discovered and named *Didymella lycopersici*. The more common stage is still known as *Ascochyta lycopersici* Brun., although several late writers have suggested that it is preferable to place this parasite in the genus Diplodina. Some of the earlier writers thought the organism identical with the Mycosphærella on cucurbits but this has now been disproved.

Under moist conditions, the spores are discharged in long whitish tendrils from the pycnidia. They are disseminated by splashing water and on tools and in the presence of high humidity, as is generally found in greenhouses, again cause infection. The fungus lives over winter either in the perithecial or pycnidial stage in the diseased tissue on the ground. So far as is known, it is not seed-borne. It is such a weak parasite that it can gain entrance into the host only when the latter is weakened by abnormal conditions, or when the fungus is encouraged by the most favorable environment for its development.

Control measures are not necessary.

References

(See also Mycosphærella wilt and black-rot of cucurbits.)

Brooks, F. T. and S. R. Price. A disease of tomatoes. New Phytol. **12**: 13-21. 1913.

Brooks, F. T. and O. Searle. An investigation of some tomato diseases. Trans. Brit. Myc. Soc. **7**: 173-197. 1922.

Cook, Mel T. and C. A. Schwarze. Two interesting diseases on greenhouse tomatoes. Phytopath. **6**: 104, 364-366. 1916.

Klebahn, N. Der pilz der tomatenstengel-krankheit und seine schlauchfruchtsform. Zeitschr. f. Pflanzenkr. **31**: 1-16. 1921.

Schoevers, T. A. C. De tomatenkanker, een voor nederland nieuwe, ernstige tomatenziekte. Tijdschrift over Plantenziekten **25**: 174-192. 1919.

Collar-Rot of the Tomato

Caused by *Verticillium lycopersici* P. and P., et al.

The collar-rot of tomatoes is most prevalent in New Jersey, Delaware, and Maryland, but probably is widely distributed in other parts of the country. A similar trouble has been described in Europe and Australia. It may become of great importance as was shown in Delaware in 1919 when almost a third of the seedlings were lost because of this disease.

Symptoms.

The trouble is confined almost wholly to the seedling stage of the host, which seems to become immune after it reaches a height of six inches or more. The trouble is not a damping-off even though the plant is attacked at a point near the surface of the ground and usually by the same organisms that cause the latter disease. The lesion is a brown sunken area that girdles the stem, and of varying width depending on the severity of the attack. The tissue becomes brittle so that the plant may fall over, but even if it does not the girdling kills it. Before death occurs, the leaves may roll upward, and their edges turn purple, or the whole plant take on an unhealthy yellow color. There is little or no root infection.

Cause.

The disease is caused by at least four fungi, namely, *Verticillium lycopersici, Macrosporium solani, Rhizoctonia solani,* and *Phytophthora cryptogea* (see Phytophthora foot-rot, page 571). Apparently the environmental factors determine which parasite will be the most destructive in any given area. In the three states mentioned above where the trouble causes marked losses, the Verticillium and Macrosporium are the chief offenders, while Rhizoctonia proves to be only a wound parasite. This last named fungus may be the same as the one causing a disease of potatoes (see page 367), but possibly may be a different race as has been proved for the one on cabbage. The life histories of Verticillium and Macrosporium are similar to those on pages 235 and 360 respectively.

The Verticillium grows at slightly lower temperatures than do the other two, although this is not much of an advantage to it, as its minimum temperature is lower than that at which tomatoes thrive. However, it reproduces so rapidly that it is a greater menace under conditions favorable for its growth than are the two other parasites.

Control.

As collar-rot is a seed-bed trouble and the four causal fungi hibernate either in the soil or on diseased tissue in the soil, the most effective means of control is sterilization of the seed-bed with steam or formaldehyde (see page 600). It is not always safe merely to change the soil or the location of the bed, for in localities where the disease is prevalent the parasites are widely disseminated.

References

Brittlebank, C. C. Tomato diseases in Victoria, Australia. Jour. Dept. Agr. Victoria, Aust. **18:** 413-416. 1920.

Cook, Mel T. The Alternaria fruit-rot and Rhizoctonia stem-rot of tomatoes. Phytopath. 10: 59. 1923.

Pritchard, F. J. and W. S. Porte. Collar-rot of tomato. Jour. Agr. Research 21: 179-184. 1921.

Timber-Rot of Tomatoes

Caused by *Sclerotinia libertiana* Fckl.

The timber-rot of tomatoes is one of the minor troubles affecting this host. The tomato apparently is more resistant to the Sclerotinia than many other vegetables, so that usually it is not attacked unless grown in soil that has become heavily infested from some previous crop, such as lettuce. Most of the injury occurs when tomatoes are grown in the greenhouse, although the disease has been observed in the fields.

The trouble is made manifest by the decay of the outer stem tissue at or just above the surface of the soil. The injured part is soon covered with a white weft of mycelium, in which later are embedded black sclerotia. If the stem is broken open, the pith is found to have vanished and the cavity lined with hyphæ and sclerotia. When the rot is severe, the entire plant wilts and dies.

Since the fungus can attack the tomato only under the most favorable conditions, which include heavy soil infestation and the presence of much water in the soil and air, obviously the method of control consists in altering the environment. If the disease has appeared in the greenhouse on any host, the soil should be sterilized before being cropped again (see page 600). In the field tomatoes should never follow a badly diseased crop of lettuce, cabbage, or other vegetable. If these rather elementary precautions are practiced, the grower may be fairly certain that the disease will not appear. (See lettuce-drop for a more detailed discussion of Sclerotinia.)

References

Dickson, B. T. Sclerotinia-wilt of green-house tomatoes. Phytopath 10: 500, 501. 1920.

Humbert, J. G. Tomato diseases in Ohio. Ohio Agr. Exp. Sta. Bull. 321: 159-196. 1918.
Pape, H. *Sclerotinia libertiana* Fckl. als schädling der tomatenpflanze. Gartenw. **26**: 309-312. 1922.

Botrytis Stem-Rot of Tomatoes

Caused by *Botrytis* sp.

The tomato is only one of the many hosts on which Botrytis will cause infection if the environment is ideal for the growth of the fungus. It is never important where conditions suitable for healthy plants are maintained.

The stem-rot ordinarily has its origin at some old leaf-scar, from which it spreads slowly until a large lesion, extending up and down the stem and girdling it, may be formed. The spot is slightly sunken, grayish in color, and the inner tissue is soft. When there is abundant moisture, the organism fruits on the lesion, and the ashen-gray mold thus formed is a valuable diagnostic sign. The same type of lesions appear also on the leaf-petioles, fruit-pedicels, calyces, and fruit. In extreme cases the plant wilts and dies.

The fruit with a roughened surface or with the old blossom still clinging to it is more liable to infection than are the smooth tomatoes. When the fungus has caused a decay of the calyx, the fruit on the same pedicel is almost sure to become diseased. In fact, the organism does not seem to be able to enter the host unless given the advantage of starting on dead or injured parts, then transferring to the adjacent healthy tissue. The numerous smoke-colored spores of the fungus are wafted about in the greenhouse by air currents, splashed by streams of water, or carried by insects until every vine is inoculated.

Fortunately for the grower, spore germination and penetration of the germ-tube do not take place when the air is moderately dry. The Botrytis thrives best under a cloudy sky and in a humid atmosphere at a temperature of 68° to 75° F. In greenhouses the temperature for tomatoes cannot be varied enough to control the disease by regulating the heat, but it is possible to give more ventilation so that the air will remain reasonably dry, and to see that each plant receives as much sunshine as the weather will permit.

References

Cook, Mel T. and C. A. Schwarze. Two interesting diseases on green-house tomatoes. Phytopath. 6: 104, 364-366. 1916.

Dickson, B. T. Stem end-rot of green-house tomatoes. Phytopath. 10: 398-500. 1920.

Stone, R. E. A new stem-rot and wilt of tomato. Phytopath. 9: 296-298. 1919.

Thomas, Roy C. Botrytis-rot and wilt of tomatoes. Ohio Agr. Exp. Sta. Mo. Bull. 6: 59-62. 1921.

Southern Sclerotium Rot of Tomato

Caused by *Sclerotium rolfsii* Sacc.

(See Southern Sclerotium Rot of Sweet-Potato, page 510)

References

Edgerton, C. W. and C. C. Moreland. Diseases of the tomato in Louisiana. La. Agr. Exp. Sta. Bull. 142: 1-23. 1913.

Rolfs, P. H. The tomato and some of its diseases. Fla. Agr. Exp. Sta. Bull. 21: 25-36. 1893.

Rolfs, P. H. Tomato diseases. Fla. Agr. Exp. Sta. Bull. 117: 33-48. 1913.

Sherbakoff, C. D. Tomato diseases. Fla. Agr. Exp. Sta. Bull. 146: 119-132. 1918.

Blossom-End-Rot of Tomatoes

Cause physiological

Blossom-end-rot or point-rot has been present no doubt ever since the modern varieties of tomatoes first came into existence, but the disease does not seem to have been described in literature until 1888. It is now known to be of common occurrence wherever the host is grown, and under certain combinations of environmental conditions may cause the loss of more than half the crop. Aside from the environment, the susceptibility of the variety has an influence on the amount of the disease present. Unfortunately the more commonly grown strains are rather susceptible, and only those which are not of economic importance, such as the red cherry and red currant, are immune.

Symptoms.

As the name of the disease implies, the injury is always at the blossom-end of the fruit, and in this manner differs from

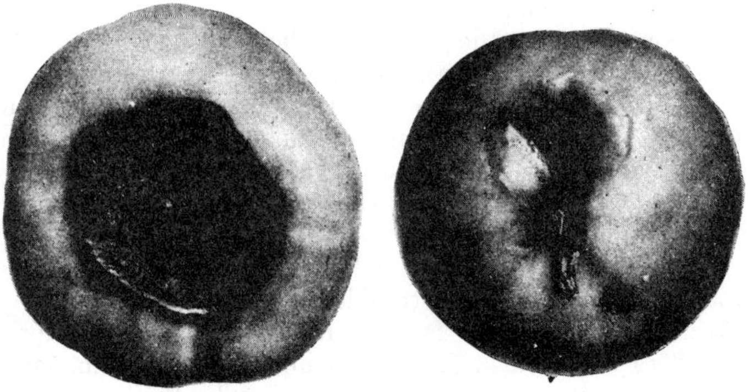

FIG. 149.—Blossom-end-rot of tomato.

Phoma-rot, which most often originates at the growth cracks radiating from the stem. The first indication of any abnormality is a slight water-soaked area about the blossom-end. The lesion soon darkens and enlarges rapidly in a constantly widening circle until the fruit begins to ripen (Fig. 149). The spot may be merely a speck or it may involve half or even more of the tomato depending on the time the fruit first begins to show symptoms of decay. As the lesion increases in size the flesh shrinks so that the affected blossom-end is flattened. The skin over the affected portion becomes black and leathery, unaccompanied by any soft-rot of the tomato unless secondary organisms gain an entrance.

It has been observed that the outward symptoms of the disease may at times be suppressed almost entirely, while the inner tissues near the blossom-end are wholly discolored and collapsed.

Cause.

There has been much controversy in regard to the cause of blossom-end-rot. Many investigators have felt sure that an organism was the causal agent, but no one has been able to isolate a pathogene that would consistently produce typical lesions when inoculated into the fruit. It is now rather generally conceded that the trouble is wholly physiological. The leaves have been shown to have a greater osmotic pressure than does the fruit. This indicates that the foliage is able to take up more water than does the fruit, and under conditions of emergency the water that rightfully should go to the tomato is transferred to the leaf, and thus the fruit-cells which are farthest from the roots are the first to suffer from the effect of drought. The condition is analogous to that in tip-burn of potatoes, and other similar troubles.

Although lack of water is the principal contributing factor, it alone does not account for all the injury. If the amount of moisture is reduced gradually, and uniformly, the rot may not appear at all. On the other hand, a period of over-watering may result in much blossom-end-rot, if later the amount of water is lessened suddenly. Manure and other ammonium-carrying fertilizers increase the trouble. Some investigators insist that nitrate of soda is detrimental, but Brooks, who has conducted extensive experiments with the disease contradicts this assertion. He concludes that the regularity of the water supply, lime added to the soil, and well aerated and cool soil retard the trouble, while the opposite factors aid in increasing the rot. Forcing the plants with large amounts of fertilizers is undesirable. Increased amounts of phosphate have little or no effect, while additional potash increases the amount of blossom-end-rot materially.

Control of blossom-end-rot.

The enumeration of the causes contributing to the development of point-rot suggest certain measures that may be applied for its prevention. The young plants should be given enough room and grown slowly enough so that they become well hardened and stocky when ready to be transplanted. It is difficult to govern conditions in the field, but it should be remembered that fresh barnyard manure, acid-producing chemicals and abnormal quantities of fertilizers are to be avoided. On the other hand, humus, such as green cover-crops or straw added to the soil, gives greater aeration, helps to retain the moisture, and keeps the temperature more uniform, and for these reasons should be used in growing tomatoes. In the greenhouse a straw mulch over the soil, subirrigation of the tomato bed and uniform sparing overhead watering will decrease the amount of injury. The disease is usually more severe near a steam pipe, or when the sunlight is too bright, thus showing that lowering the temperature of the soil and reducing the intensity of sunlight tend to lessen the amount of rot.

Spraying with bordeaux mixture and tying the vines on stakes may sometimes give beneficial results, but experiments have shown that these practices cannot be relied on on all occasions for the control of blossom-end-rot.

References

Brooks, Charles. Blossom end-rot of tomato. Phytopath. **4**: 345-374. 1914.

Galloway, B. T. Notes on the black-rot of the tomato. U. S. Dept. Rept. **1888**: 339-346. 1888.

Smith, E. H. Blossom end-rot of tomatoes. Mass. Agr. Exp. Sta. Tech. Bull. 3: 1-19. 1907.

Stuckey, H. P. Tomatoes. Ga. Agr. Exp. Sta. Bull. 112: 211-248. 1915.

Stuckey, H. P. Transmission of resistance and susceptibility to blossom end-rot in tomatoes. Ga. Agr. Exp. Sta. Bull. 121: 83-91. 1916.

Phoma Rot of Tomatoes

Caused by *Phoma destructiva* Plowr.

The Phoma rot of tomatoes is co-extensive with the host, and for a long time has been a serious disease of ripe tomatoes, having first been described as early as 1881. It is most destructive on the winter crop in the southern states. On several occasions it has also been found on fruits of pepper.

Symptoms.

The lesions are most common on ripe fruit although it is not unusual to find green tomatoes affected, and in severe cases

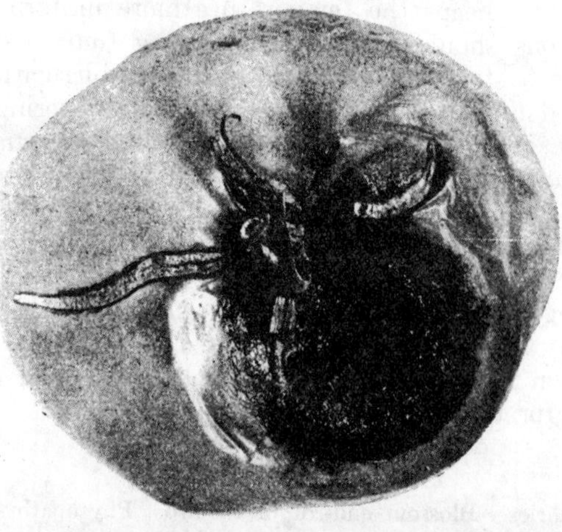

Fig. 150.—Phoma fruit-rot of tomato.

the leaves and stems are also invaded. On the foliage and stems, the lesions are irregular in outline, rather large, and brown or black in color. They are often located along the margin of the leaf. A few small black pycnidia may form in the dead tissue. If the spot is large or if several spots are on the same leaf, the latter turns yellow and drops. Ordinarily the infection first appears on the lower part of the tomato plant.

The lesions on the fruit begin most often near the stem-end along the growth-cracks (Fig. 150), but by no means are limited to this area. They may arise wherever the skin has been punctured either mechanically or by insects. The first indication of the disease is a small water-soaked area, which quickly enlarges and turns black on the surface. The diseased tissue may form a depression by shrinking, but frequently remains comparatively firm. The completely rotted areas vary in size from one-fourth inch in diameter to one-third or more of the tomato surface, and extend a considerable depth into the flesh. The darkened portion is thickly studded with small black pycnidia. When the epidermis included in the darkened portion is cracked, the parasite grows in white wefts along the margins of the opening.

Cause.

The pathogene is one of the imperfect fungi, named *Phoma destructiva.* In some of the older literature it is known as *Phyllosticta lycopersici* Pk. Recently a supposedly new species with a secondary spore stage like that of Alternaria has been listed as *Phoma alternariaceum,* but needs further investigation before becoming wholly acceptable. The Phoma probably lives over winter with diseased material in the soil, or possibly may even cling to the surface of the seed. The small hyaline unicellular spores are washed and splashed to the various parts of the host, where if enough moisture is present the germ-tubes gain an entrance into the tissue through wounds or stomata. The mycelium grows very rapidly, forming a visible lesion in four days, and a large decayed area in one to two weeks. Soon after the lesions turn black, numerous pycnidia begin to form, from which during periods of high humidity great masses of spores ooze in long tendrils. Such spores may infect the tomatoes on the vines, or cling to the surface of the picked fruit until the epidermis is injured by handling, and then cause infection. This often happens when tomatoes are shipped long distances. Much of the diseased material is thrown away or drops to the ground where it serves as the source of inoculum the following year.

The minimum temperature at which the fungus grows is about 43° F.; the optimum, 82° F.; and the maximum, 92° F. A high humidity is favorable for its development on the fruit.

Control.

No satisfactory methods of control have been suggested. The rotation of crops and the procuring of clean seed as recommended for wilt (see page 568) are worthwhile precautions. Some varieties of tomatoes have more growth-cracks than others, therefore if the market will accept the smoother type it is an advantage to grow the latter. The farther the growing fruit can be held from the ground, the less chance there will be for contamination. Thus supports for the vines will be of some aid in reducing infection. Careful handling while the fruit is being picked and crated will reduce the number of bruises through which the fungus enters. After the fruit is picked it should be marketed as soon as possible, and during transportation the temperature in the car should be kept as low as possible with iced bunkers.

References

Brick, C. Die Schwarzfleckenkrankheit der tomatenfrūchte durch *Phoma destructiva* Plowr. Zeitschr. f. Pflanzenkr. 29: 20-29. 1919.

Brooks, F. T. and O. Searle. An investigation of some tomato diseases. Trans. Brit Mycol. Soc. 7: 173-197. 1922.

Higgins, B. B. The diseases of pepper. Ga. Agr. Exp. Sta. Bull. 141: 48-75. 1923.

Jamieson, Clara O. *Phoma destructiva*, the cause of a fruit-rot of the tomato. Jour. Agr. Research 4: 1-20. 1915.

Link, Geo. K. K. and F. C. Meier. Phoma-rot of tomatoes. U. S. Dept. Agr. Circ. 219: 1-5. 1922.

Buckeye-Rot of Tomatoes

Caused by *Phytophthora terrestria* Sherb.

Although the first written description of buckeye-rot of tomatoes did not appear until 1917, specimens of diseased fruit were collected in Florida in 1911. The trouble has since been reported from Arkansas, Texas, Louisiana, Porto Rico, Cuba, and Virginia, and possibly is present in other southern states. The disease is of special importance to the tomato crop grown for northern markets. Not only is much of the fruit destroyed in the field, but there is also a large loss during transportation.

Symptoms.

The fungus affects only the fruit of the mature tomato plant. The invaded tissue remains firm while its color changes from the normal green to grayish or greenish-brown. The lesions often are characterized by distinct zonate markings, which resemble the outline of an eye and suggest the name applied to the disease. The tomato may become infected at any stage of its growth. Ordinarily the rot begins at the blossom-end because the inoculum in the soil reaches the tomatoes that touch the ground. For this reason the decayed area might be mistaken for blossom-end-rot were it not for the presence of the zonation, and under wet conditions the superficial growth of the causal fungus. The parasite may also cause a damping-off of seedlings.

Cause.

Buckeye-rot is caused by *Phytophthora terrestria,* a soil fungus which thrives best in the presence of abundant moisture. The mycelium in the soil or in the discarded diseased tissue produces conidia, which in turn give rise to swarm-spores. These swim or are splashed to the tomato on or near the ground, and by means of a germ-tube enter directly through the epidermis of the host. The parasite reenters the soil when the decaying fruit falls from the vine. Some tomatoes are picked after inoculation by the swarm-spores has taken place but before any visible signs of infection are present. In addition, the fungus is able to grow through the wrappers during shipment, and to infect tomatoes which were not inoculated at picking time. When such fruit is shipped to northern markets, the decay during transportation reduces the value of the shipment. The English workers think that the parasite is identical with *Phytophthora parasitica* Das.

Control.

Tying the vines to stakes will keep most of the tomatoes from touching the soil and consequently will decrease the amount of rot. This one measure, however, cannot always be depended on to eliminate the entire loss. It has been demonstrated that the fruit may be dipped for one and one-half minutes in water heated to 140° F. without shriveling the skin or injuring the tomato in any way. This high temperature is sufficient to kill not only the mycelium and spores that may cling to the surface of the fruit, but will also destroy the parts of the fungus which have penetrated the tissue, provided infection is still in the incipient stage. The treatment is successful even up to the time when the lesions become visible to the unaided eye. Therefore, if staking the vines does not provide adequate protection, the grower may resort to dipping in hot water. The treatment requires extreme care in regulating the temperature unless special apparatus is available to govern automatically the supply of heat.

References

Rosenbaum, Joseph. Infection experiments on tomatoes with *Phytophthora terrestria* Sherb. and a hot water treatment of the fruit. Phytopath. 10: 101-105. 1920.

Rosenbaum, J. The origin and spread of tomato fruit-rot in transit. Phytopath. 8: 572-581. 1918.

Sherbakoff, C. D. Buckeye-rot of tomato fruit. Phytopath. 7: 119-129. 1917.

Anthracnose of Tomato

Caused by *Colletotrichum phomoides* (Sacc.) Chester

Anthracnose or ripe-rot of tomatoes has been known since 1878 when the fungus was first observed in Italy. At present it is as wide-spread as is the culture of tomatoes, but is of minor economic importance. Occasional reports have been made in which the loss from the disease in individual fields is given as 25 per cent. This condition, however, is in marked contrast to the average case in which only a trace of the rot is found on late picked fruit.

Symptoms.

The disease is confined to the ripening tomato. The first symptom is a small translucent spot which while enlarging turns black and shrinks. Ordinarily the lesions are circular in outline, but by uniting with adjoining lesions may become irregular in shape. The larger spots may show very evident zonations, and when this is true the fruit-bodies of the fungus are borne in concentric circles. The soft decayed portion extends deeply into the flesh of the tomato and makes it valueless. The rot often follows sun-scald or other injury to the fruit and may be accompanied by Oospora or a yeast, which ferments the sap of the tomato.

Cause.

The parasite, *Colletotrichum phomoides,* can enter the host only through wounds. The mycelium quickly invades the fruit tissue, and later forms numerous yellow-lined acervuli, from which the spores ooze in pink tendrils. They are splashed by the rain or otherwise conveyed to neighboring fruit when the latter is wet. The fungus apparently lives from season to season on diseased tissue in the soil, its life being prolonged by the definite stromæ the parasite forms within the host.

Control.

In many localities no control measures are necessary, but where the ripe-rot is troublesome during successive years, it may be held in check by spraying with bordeaux mixture as suggested for the control of Septoria blight (see page 535).

References

Arthur, J. C. Rot in ripe tomatoes. N. Y. (Geneva) Agr. Exp. Sta. Ann. Rept. 3: 380-382. 1885.

Brooks, F. T. and G. O. Searle. An investigation of some tomato diseases. Trans. Brit. Mycol. Soc. 7: 173-197. 1922.

Chester, F. D. The ripe-rot or anthracnose of tomato, *Colletotrichum phomoides* Sacc. Del. Agr. Exp. Sta. Ann. Rept. 6: 111-115. 1894.

Cook, Mel T. and W. H. Martin. Diseases of tomatoes. N. J. Agr. Exp. Sta. Circ. 104: 1-15. 1918.

Pool, Venus W. Some tomato fruit-rots during 1907. Neb. Agr. Exp. Sta. Ann. Rept. 21: 1-30. 1908.

Oospora Rot of Tomato

Caused by *Oospora lactis parasitica* Pritchard and Porte

In 1885 Arthur observed the Oospora rot of tomato in New York, while Stoneman recorded it in 1898. Since then it has been found a number of times in this country and at least mentioned once in Italy. It probably is distributed widely, especially in the southern states.

Symptoms.

The organism attacks only the fruit, forming a watery soft-rot which involves the entire inner tissue but which usually does not decompose the epidermis. The ripened fruit that hangs near the soil or touches it is most often infected; the green fruit, however, is not wholly immune. Under humid conditions, the fungus produces a velvety or granular covering over the fruit surface, or makes a fluffy growth along the margins of the cracks in the epidermis. Fermentation of the sap often is produced either by the parasite or by a concomitant yeast that causes a juice to drip from the injured tomato, and results in a disagreeable odor.

Cause.

The organism has been known as *Oidium lactis* or *Oospora lactis,* but Pritchard and Porte, having found that there are physiological strains which differ in their ability to cause tomato-rot, have designated this particular strain as *Oospora lactis parasitica.* It is a weak parasite entering through wounds or injuries. It grows rapidly between and through the cells, decomposing the middle lamellæ, and soon produces chains of hyaline oval spores on the ends of the aerial hyphæ. Evidently it is saprophytic enough in its nature to live indefinitely in the soil, from which it is splashed to adjoining tomatoes.

Control.

Poole has shown that spraying or dusting five to seven times, as suggested for the control of Septoria blight (see page 535) reduces the number of rotted fruit. Keeping the vines from lying on the ground, and care in handling the picked fruit should also prove benficial.

References

Arthur, J. C. Rot in ripe tomatoes. N. Y. (Geneva) Agr. Exp. Sta. Ann. Rept. **3:** 380-382. 1885.

Poole, R. F. A new fruit-spot of tomatoes. Bot. Gaz. **74:** 210-214. 1922.

Pritchard, F. J. and W. S. Porte. Watery-rot of tomato fruits. Jour. Agr. Research **24:** 895-906. 1923.

Stoneman, Bertha. A comparative study of the development of some anthracnoses. Bot. Gaz. **26:** 69-120. 1898.

Isaria Rot of Tomato

Caused by *Isaria clonostachoides* Prit. and Porte

The Isaria rot of the tomato fruit is so rare that it has been reported only from the district about Washington, D. C., where it was first found in 1919. It differs from the many other fruit-rots in that the chief sign of the disease is the peculiar appearance of the fungus growing over the surface of the host. The tomato is covered with a white downy mycelium, which later assumes a pink or orange tinge or even becomes greenish-yellow. The amount of fungal growth on the tomato depends on the degree of humidity present; in a dry atmosphere there is very little development of the aerial hyphæ.

The causal organism, *Isaria clonostachoides,* bears its oval pinkish spores on branched conidiophores. The spores are so numerous that they are often grouped in globular masses or in dense angular spikes. They are not able to infect the stem or roots of the tomato, but apparently gain entrance through wounds into the fruit, which serves as an excellent substratum for nearly every organism that can gain lodging there. Presumably the fungus may live in the soil as a saprophyte until time for infection of the new crop.

The parasite thrives best in a high temperature, causing infection most easily at 84° F. and becoming harmless below 57° or above 90°.

No control measures have been necessary.

Reference

Pritchard, F. J. and W. S. Porte. Isaria-rot of tomato fruits. Phytopath. 12: 167-172. 1922.

Melanconium Fruit-Rot of Tomatoes

Caused by *Melanconium* sp.

This type of fruit-rot is of very slight importance. It attacks both ripe and green fruit, causing numerous minute black spots each surrounded by a slightly sunken zone. On ripe fruit the lesion may be bordered by a yellowish epidermis, or the natural color of the fruit may still prevail. The affected tissue is tough and corky. Acervuli are sometimes formed in the old spots and appear as a mass of white mycelium discharging dark spores.

The spores of the fungus are not only borne on the surface of the fruit but also in the seed cavities. Presumably the spores are scattered when the plants are watered; besides it has been shown that white flies may act as carriers. The germ-tube enters through insect punctures or other wounds, and if moisture conditions are favorable may penetrate even the uninjured epidermis.

No control methods are necessary.

Reference

Tisdale, W. H. A Melanconium parasitic on the tomato. Phytopath. 6: 390-394. 1916.

A Yeast Rot of Tomatoes

Caused by *Nematospora lycopersici* Schneider

For many years investigators have found a yeast associated with rots of ripe tomatoes, but for the most part they have considered that the organism followed other fungi. It has recently been shown that a certain yeast, when permitted by wounds to enter the ripening fruit, is able to cause decay even when other pathogenes are not present. This is true particularly of the yeast found on the host in southern California, Cuba, and Mexico. The lesions consist of sunken reddish-brown spots a few centimeters in diameter and not penetrating deeply into the flesh. The causal organism is made up of unicellular bodies that increase in number by having a new cell grow from the end of each old one. This process is called budding. Under certain conditions the ends of two cells grow together and their contents fuse, forming an ascus with eight spores. Rain, insects, and tools are means by which the yeast is disseminated to the recipient host.

No control measures are suggested.

References

Schneider, Albert. A parasitic Saccharomycete of the tomato. Phytopath. 6: 395-399. 1916.

Schneider, Albert. Further note on a parasitic Saccharomycete of the tomato. Phytopath. 7: 52-53. 1917.

Frost-Injury to Tomatoes

It is well known that the tomato is one of the first vegetables to be injured by frost. Its susceptibility to cold is of importance during transportation since many of the southern-grown tomatoes are shipped northward while winter weather still prevails. It is, therefore, of interest to know the lowest temperatures the fruit will bear before it is injured. Harvey made a number of tests on both mature green and ripe fruit of the most common varieties. He found that the green fruit is slightly more susceptible to cold injury than are those which have ripened. If a fruit has reddened at the blossom-end but is still green about the stem, the latter part is chilled more quickly. When tomatoes are lying on the ground they are usually not injured so soon as are those hanging on the vine unless the latter are well screened by foliage. The fruit covered with drops of water, as dew, will freeze more quickly than will one which is dry. It was found that tomatoes with uninjured skins are more resistant to cold than those with growth cracks or abrasions.

As was shown in the discussion on the freezing of potatoes (see page 457), plant tissues will bear temperatures lower than the freezing point of water. As the temperature falls below 32° F., there is a constantly increasing probability that ice crystals will be formed, but if the tissues are held perfectly still, crystallization will be delayed considerably. A slight jar during the under-cooling is sufficient to start the formation of ice, and the freezing of the tissue follows. The average temperature at which ripe tomatoes are injured is given as 30.46° F. with only slight variations for the different varieties.

Reference

Harvey, R. B. Frost injury to tomatoes. U. S. Dept. Agr. Bull. 1099: 1-10. 1922.

Sun-Scald of Tomatoes

Sun-scald may affect leaf, stem and fruit of tomatoes when certain combinations of environmental influences are present. For example, in the greenhouse the tomato leaves have been observed to have large dead areas following bright sunshine which was preceded by a prolonged period of cloudiness. When very tender large plants are transplanted into the field, a brown longitudinal streak of dead tissue may be produced on the sun-exposed side. Similar streaks are formed if the stems are chilled in a frosty night, and then subjected to an unusually bright and warm morning sun.

The spots on the fruit, which are caused by the hot rays of the sun and aggravated by the lack of soil-moisture, are not of much consequence in themselves, but are detrimental in that they serve as an open doorway for the entrance of Alternaria, Colletotrichum, Phoma, and other rot-producing fungi. Usually the injured surface is shrunken, the skin is dry and papery, and does not redden normally when the tomato ripens, but retains a yellowish color. If the injury is followed by an invasion of Alternaria, small brown specks, resembling freckles, form over the affected epidermis, which later becomes black and ruptures.

References

Britton, W. E. Blight, burn, or scald of tomato plants. Conn. Agr. Exp. Sta. Ann. Rept. **1896**: 232-234. 1897.

Cook, Mel T. Sunburn and tomato fruit rots. Phytopath. **11**: 379-380. 1921.

McCubbin, W. A. The diseases of tomatoes. Canada Dept. Agr. Exp. Farms Bull. 35: 1-16. 1918.

Œdema of Tomatoes

Cause non-parasitic

(See Œdema of Cabbage and Cauliflower, page 168, for additional discussion)

Tomatoes grown under glass during the winter may sometimes be affected by œdema. Numerous small intumescences appear on the veins, midribs, petioles, and stems. The swellings are elongated and have a frosty-white appearance at first, later they may turn yellowish or brown. The whole leaf may die.

The cause of the trouble is over-heating and over-watering of the soil, excess humidity in the air, and poor lighting of the greenhouse. Proper ventilation, heating and watering will overcome the difficulties sufficiently to avoid any loss that might otherwise occur.

References

Atkinson, G. F. Œdema of the tomato. N. Y. (Cornell) Agr. Exp. Sta. Bull. 53: 77-108. 1893.

Orton, C. R. and W. H. McKinney, Jr. Notes on some tomato diseases. Penn. Agr. Sta. Ann. Rept. 1915-16: 285-291. 1917.

Leaf Chlorosis of the Tomato

Cause undetermined

Leaf chlorosis of tomatoes was observed in a greenhouse in North Dakota, and was pronounced of no importance. It is of interest only in that occasionally it may occur in other localities, and also that it may prove to be stage of some other disease such as mosaic.

Not all the leaves of a given plant are affected. Beginning near the midrib or near one of the larger veins, the tissue of the leaf-blade dies and dries until it is white in color and parchment-like. These chlorotic areas may be preceded by a yellowing of the lesion or the dying may begin within a spot that has not yet lost its green color. The parchment-like area may shade gradually into the healthy tissue or show a sharp line of demarcation between the two. It is common for the affected leaf to roll upward or curl downward or become sickle-shaped by growing faster on one edge than on the other.

All attempts at determining the cause of the disease resulted negatively.

Reference

Reynolds, Ernest Shaw. Two tomato diseases. Phytopath. **8:** 535-542. 1918.

Curly-Top or Rosette of Tomato

Cause undetermined

The rosette of tomato has been mentioned several times in literature, but it is not definitely established that each investigator had the same disease in mind when he used this term. It is not a serious trouble, having been reported only from Pennsylvania, Ohio, and New Jersey.

The name curly-top describes the symptoms accurately. The plant does not show any trace of the disease until it is a foot or more in height, after which it grows more slowly and becomes dwarfed. The stalk is thicker at the tip and most of the foliage grows at the upper half of the plant. The leaves, although no more numerous than on the normal plant, have a massed appearance because of their shorter petioles. The leaves are smaller and exhibit a decided downward curling at their tips. The affected plants rarely bear blossoms and never set fruit.

In Ohio the malformation was attributed to the action of Rhizoctonia on the plants, but investigational work in Pennsylvania failed to connect any organism with the trouble. It does not seem to spread from one vine to another. It is possible that the abnormal condition is brought about by injury to the roots when the seedlings are transplanted in a careless manner.

Until more is known concerning the cause, no control methods can be suggested.

Reference

Orton, C. R. and W. H. McKinney. Notes on some tomato diseases. Penn. Agr. Exp. Sta. Ann. Rept. **1915-16:** 285-291. 1917.

White Leaf-Curl of Tomato

Cause undetermined

The white leaf-curl, which was first observed in Pennsylvania in 1914 on the Lorillard variety of tomatoes, is evidently closely related to mosaic, yet the symptoms are not the same. Instead of the leaves becoming filiform as is true of mosaic, in white leaf-curl they taper at each end and remain fairly broad in the center. The plants take on some symptoms of leaf-roll, except that the rolling begins at one leaflet and advances progressively to the other leaflets on the same petiole. Frequently the rolling of the edges is accompanied by a similar rolling of the tips of the leaves. In addition to the rolling and narrowing of the foliage, there are certain green spots among the faded areas on the leaf which bear a resemblance to mosaic. The most striking symptom, however, is the grayish cast of the affected plant. This probably is due to the loss of chlorophyll, the green giving way to a grayish color. The disease appears after the plants have begun to bear. As soon as the plant is affected, the blossoms drop and the calyx-lobes curl.

Evidently the malady is caused by a virus which may be transmitted to healthy plants. Until more is known regarding the trouble, it is well to apply the same control measures as are suggested for the mosaic of tomato (see page 545).

Reference

Orton, C. R. and W. H. McKinney. Notes on some tomato diseases. Penn. Agr. Exp. Sta. Ann. Rept. 1915-16: 285-291. 1917.

Leaf-Roll of Tomato

Cause physiological

When the excitement of finding a leaf-roll disease of potatoes was at its height, it was natural for similar symptoms on the tomato to receive the same nomenclature, and be attributed to a similar cause. More recent writers have shown that the rolling of the tomato leaves is not the result of a transmissible disease, nor is it related in any way to the degeneration troubles of potatoes, but is brought about by an over-supply of soil-moisture, or by pruning the plants too closely.

References

Gussow, H. T. Leaf-curling in tomatoes. Phytopath. **11**: 380-383. 1921.

Maupas, Albert. Sur la maladie de l'enroulement des feuilles de tomates. Rev. Hort. **94**: 52-54. 1922.

Schoevers, F. A. C. Het krullen van tomaten bladeren. Tijdschr. Plantenz. **25**: 11-12. 1919.

"Bull-Plants" Among Tomatoes

Cause undetermined

In Pennsylvania certain tomato plants in greenhouses have been observed as much larger than are the neighboring healthy ones, and were therefore named "bull-plants." Not only do the affected stems grow ranker, but the leaves are larger in every dimension and have a darker green color. Such vines do not blossom nor bear fruit. Although the cause of this gigantism has not been determined, it is known that the disease does not spread and is not of importance. In permitting the growth of such abnormal plants much of the soil food is used wastefully; thus it is better to remove them as soon as they are evident.

Reference

Orton, C. R. and W. H. McKinney. Notes on some tomato diseases. Penn. Agr. Exp. Sta. Ann. Rept. **1915-16**: 285-291. 1917.

Crown-gall of Tomatoes (Fig. 151)

Caused by *Bacterium tumefaciens* EFS. and Townsend

(See Crown-gall of Beet, page 71)

CHAPTER XIX

SOIL STERILIZATION

Among all the different control measures which the vegetable-grower must put into practice, one of the most common and effective is soil sterilization. No other method needs as much care in its application, and probably no other is conducted in such a half-hearted or slip-shod manner. Consequently much effort that is put into soil sterilization is spent in vain, with the result that the grower gets the impression that it is worthless, and either gives it up entirely or carries it on merely because others are doing it. Notwithstanding these difficulties, it is an extremely important means of disease control and when used correctly repays the labor and time many fold. An added advantage of sterilizing the soil is that weed seeds are killed. This item alone is often important enough to pay for the soil treatment.

Certain precautions should be taken into consideration when sterilizing soil, the most important of which relates to recontamination. If all the organisms are killed by a disinfectant and then the pathogene is introduced into this sterilized substratum, it usually develops faster than it would have done if it had to compete with the innumerable soil inhabitants which were there before the application of the disinfectant. In other words, it is worse than useless to sterilize the plant bed, then walk over it with contaminated boots, handle the dirt with infested tools, or water the plants with infested water from a nearby pond or open well. If the soil is sterilized thoroughly, the boards or cement about the edge of the bed should be equally well cared for, the footwear should be used only for the particular beds or disinfected before beginning work, and the tools treated in the same manner.

Some soils will be improved physically by steam sterilization while others are injured by the process. Heat will often make available certain salts that were in the ground before but had been insoluble or at least not available to the plant. Thus nitrogen, phosphorus, and probably potash are liberated, and when in the correct amounts will stimulate plants on sterilized soil. This has been observed commonly. In fact, gardeners have felt that some plants were made to grow too luxuriantly on sterilized soil. On the other hand, there are soils in which toxic substances are liberated and the succeeding crop is poorer than on non-treated soil. Fortunately, this latter case is rare.

FIG. 151.—Crown-gall at the base of a tomato plant. (See page 599.)

At times it is necessary to disinfect hotbeds in which the

source of heat is a base of fermenting manure. In order to have the fermenting process carried on, it is necessary to retain the organisms found naturally in the manure. In such cases it is well to sterilize all parts of the hotbed before the manure is applied, then procure the latter from some locality known to be free from the undesirable organisms.

Steam Sterilization

Steam is the most effective sterilizing agent in soil. There are various ways in which it may be applied. Some of the more common methods include the use of drain tile, covered steam pipes, inverted pan, and steam rake.

Drain tile.

There are certain advantages in using drain tile to conduct the steam through the soil. They need not be removed during the season, and can be retained a number of years without additional attention. Since their removal is unnecessary, the possibility of recontaminating the soil is reduced greatly. Other steam apparatus requires the presence of men in the greenhouse while steam is being applied, and the heat at times is almost unbearable; the drain tile eliminate this close indoor supervision. Aside from being used for sterilizing, they may be utilized for subirrigation, and for washing out an oversupply of harmful soluble salts from the soil. Furthermore, the tile are readily obtainable, are easy to put into place, and are not expensive.

Any ordinary three-inch tile may be used. It does not matter particularly whether they are made of cement or of clay, or whether they are glazed or not. The three-inch tile are probably as convenient in size as are the larger or smaller ones, and under average conditions are recommended. Tile with perforations for the escape of steam are on the market and may be used satisfactorily, but under ordinary conditions the spaces between the tile when laid end to end are sufficiently large to permit passage of steam into the soil. There are other tile with close jointed ends, so that particles of dirt cannot enter into the drain. Such a precaution may be desirable in loose sandy soil, but is not needed in the average greenhouse.

The tile are laid at various depths and distances apart, depending on the nature of the soil, depth of cultivation, and other local factors. Some experimentation on this subject may

FIG. 152.—Drain tile connected with the steam pipes for soil sterilization.

be necessary in individual houses. The tops of the tile usually are laid twelve to fifteen inches below the surface of the soil. It is a common practice to dig a trench eighteen inches deep, although this may be too deep in certain cases when only three-inch tile are used. The centers of the trenches may be spaced from fifteen to twenty inches apart, although the latter distance is too great to secure the best results under all conditions. The trenches are dug the entire length of the house if the distance is not too great. In a long house it may be desirable to lay the tile parallel with the width rather than the longer distance. In either case the ends of the drains are brought close to the inner walls of the house, and the first trench is dug close to the side wall so that every inch of ground may be sterilized. The tile are covered with the spaded-up dirt as is done in outdoor drainage. The end tile is connected with the surface of the soil by a closed rubber hose.

The heating pipes along the ends or sides of the house may be used for carrying the steam from the boiler to the drains (Fig. 152). A short branch pipe is extended downward from a joint directly over the end of each string of tile into the soil to the level of the drain, then turning at right angles is inserted several feet into the tile. Valves should be placed at intervals in the conducting pipe, so that steam can be directed simultaneously into just as many strings of tile as the capacity of the heating plant will bear. Some boilers may permit the application of steam to ten or twelve lines at once, while others can generate enough pressure for only two or three lines. The number of conducting drains and the length of the house will determine the pressure at the boiler, as well as the size of the pipe that leads from the heating plant to the head of the bed. For example, in one house seventy-five feet long, a two and one-half-inch pipe and a pressure of one hundred pounds was required to supply steam for ten lines simultaneously. There will also be some extra strain on the boiler itself, when such a large amount of water is converted into steam. If rain or other soft water is available, the clogging of the boiler and pipes with lime is lessened and considerable trouble avoided. The only other requisite is that the water gauge be watched closely. In all other respects the heating and firing is conducted as usual.

The disinfected soil retains the heat somewhat longer when covered with the ordinary roofing paper, boards, or straw mats. Such a covering, however, is not absolutely essential. Steam will permeate the soil better, also, if the latter has been plowed or spaded previously. It is suggested that the doors and ventilators be closed during the steaming; the temperature of tne whole greenhouse will then rise high enough to kill insects, and even fungi that float in the air or cling to the walls.

No definite length of time can be recommended for applying the steam. A few good soil thermometers are valuable in determining the thoroughness with which the sterilizing is done. The opposite end of each drain tile should be left open for a brief period after the valves are opened, so that the steam will start circulating freely without being hindered by a pocket of air under compression, and also to make sure that the drain tile are not clogged at any place. When these two points have been ascertained, the openings at the ends farthest from where the steam is introduced are closed securely. The steaming should be continued until the soil temperature reaches above 140° F. and then held at that heat for at least two hours.

Steam pipes.

One of the more popular methods of soil sterilization has been the use of steam pipes (Fig. 153). These are of varying lengths but usually are approximately twelve feet long and about three inches in diameter. On one side are one-fourth-inch holes placed at six-inch intervals along the entire length of each pipe. Specially prepared pipes may be bought for this purpose, or holes can be punched into the ordinary galvanized fluted pipes that are employed to conduct the water from eaves-troughs on buildings. The pipes are buried about a foot below the surface of the soil in parallel trenches, and with the small holes downward. They are placed from twelve to eighteen inches apart, and connected to a buried three-inch iron head-pipe long enough to reach across the width of the bed. At regular intervals in the iron pipes are openings supplied with flanges projecting three or four inches, over which the ends of the galvanized steaming pipes are fitted. At the middle of the head-piece is a place where the conducting pipe from the boiler may be attached. When the apparatus is put into place, the soil is smoothed down and boards or some other covering are fitted over it. To the end that the oppressive heat may be avoided, three or four groups of pipes with their head-pieces are buried one after the other in the greenhouse in the morning, then the steam is turned on these in succession during the re-

FIG. 153.—Galvanized pipes and iron head-pipes used for soil sterilization.

mainder of the day. The next day the apparatus is taken up and moved farther along, and the process is repeated. The steam pressure and length of application is the same as when the drain tile are used.

The steam pipes are an inexpensive arrangement for applying steam, and are much more effective than the steam rake. They have the same objection as does the steam pan, that of affording an opportunity for reinoculation when moved. They are less desirable than drain tile.

Steam pan.

The position of many seed-beds is changed often enough to make it impracticable to use the tile-drain method of sterilizing.

FIG. 154.—Steam pan in place for soil sterilization in the greenhouse.

In such beds and in older houses in which tile have never been installed, a different method for applying steam has been in use for many years. A shallow pan of varying dimensions is inverted over the soil (Fig. 154), and steam under pressure is forced into the pan.

If the bed to be sterilized is ten or twelve feet wide, the pan can be made just to fit. With this length, the pan should not be over six or seven feet wide, for it is difficult to sterilize much more than seventy or seventy-five square feet of surface at once. When one space has been sterilized the pan is moved forward, care being taken that the pan overlaps slightly the area that previously was sterilized. After the pan is pressed down firmly in its position, soil is banked around its edges. As soon as the pan is removed, the heated soil is covered with mats or boards to retain the high temperature longer. The usual time for sterilizing under the pan is from one-half to one and one-half hours when the steam gauge records from eighty to one hundred pounds pressure. It is a fairly common practice to bury a potato four inches in the soil under the pan, and then steam until the tuber is well cooked.

FIG. 155.—Steam pan attached to movable boiler for sterilizing outdoor seed-beds.

The steam pan may be constructed of sixteen or eighteen gauge galvanized iron or of boards. The iron pan is lighter and may be moved more easily, but it does not retain the heat as well as does the wood. It is made by pressing into shape a pan of the required length and width, and five and one-half inches in depth. The bottom inch and one-half of iron are then bent upward to form a flange along the lower edge, leaving a final depth of four inches. The pan may be used in this form or preferably reinforced by thin strips of iron riveted against the upturned flange on the edges, as well as having similar ribs of iron across the top of the pan, two in the middle and one on each end. Since the pan is hot when it must be moved from one position to another, it is well to fasten a large ring to each of the four corners on the top side. Pieces of wood may then be held in readiness for inserting into each ring and serving as handles for lifting the apparatus.

The wooden pan can be constructed cheaply with materials that are usually at hand. The same dimensions are adhered to as when using the galvanized iron. The frame is made of two-by-four-inch studding covered by thin narrow matched boards. When laying the boards it is well to place white lead in the grooves so that the box will be as nearly steam-proof as possible. Inasmuch as the boards swell when steam is applied, it is necessary to nail them securely in the groove to prevent drawing. In addition, pieces of strap iron long enough to reach from edge to edge over the top of the box are bolted or nailed into place at right angles to the boards. Four such strips are desirable, two in the middle and one at each end. A similar strip is nailed to the bottom edges on the inner side of the box in such a manner that the iron will protrude an inch or more and will serve for cutting into the soil when the box is dropped on the ground. This arrangement aids in sealing the inclosed chamber so that steam cannot escape without first passing through a layer of soil. Iron rings are bolted into each of the four top corners, and are used for lifting the box as was suggested in the case of the galvanized steam pan.

The pan is connected to the boiler (Fig. 155) with a three-fourths-inch or one-inch heavy rubber hose, long enough to permit considerable pan movement while the boiler remains stationary. The attachment is made at the end of the pan, where a short piece of gas-pipe has been fitted into an opening by a tightly adjusted gasket. The lighter pans must be weighted down, so that when the steam is turned on they will not be lifted off the ground. Bags of sand or stones may be piled on top for this purpose.

Before placing the pan into position, the soil should be

spaded up or loosened in some other way so that the steam can penetrate it readily. The soil, also, must be comparatively dry, for the heat passes slowly through that which is saturated with water. For the same reason, no attempt should be made to sterilize soil that is frozen. In order to overcome the handicap of wetness and freezing, some growers are tempted to sterilize late in the autumn. Such a procedure has not proved satisfactory, for the washing rains of spring, as well as many other causes, may afford ample opportunity for the seed-bed to become recontaminated before planting time the following season.

Steam rake.

The rake is shaped somewhat like the single section of a steel harrow. The cross pieces are made of hollow iron pipes to which are attached hollow teeth eight or ten inches in length with small holes at the bottom. The rake is forced into place until the teeth are completely buried. The steam which is injected through an attached pipe is forced into the soil through the lower openings. Most of the steam, however, escapes up the side of the teeth rather than permeating the soil. For this reason, the steam rake is a very poor makeshift in soil sterilization. A steam pan is sometimes inverted over the rake to make the latter more efficient. However, there is no point in combining the two methods when the steam pan will answer the purpose just as well when used alone.

Formaldehyde

An application of formaldehyde may be employed when there are no means of generating steam, and has proved effective against most, if not all, of the pathogenic soil bacteria and fungi. It, however, does not kill all the nematodes and usually is more expensive than steaming.

The quantity of formaldehyde and the method of application differs with nearly every worker. Nearly all the later experiments have shown that 1-100 or 1-200 dilutions are too weak to have a sufficient fungicidal value. The more recent recommendations are that one part of the commercial formaldehyde be diluted in fifty parts of water and that one-half to one and one-half gallons of this solution be applied to each square foot of soil. Another method is that of testing a square foot of soil to see how much water is necessary to saturate it, then adding a pint of formaldehyde to enough water to saturate from six to ten square feet. The solution is applied with a sprinkling can, the soil having first been loosened so that the solution can penetrate easily. After the bed has been wet down with the fungicide, it is covered with boards or canvass for twelve hours or longer. Following the removal of the cover, the soil is stirred several times to permit the fumes of the formaldehyde to escape. Work done lately in Massachusetts indicates that the covering is not absolutely essential, yet past experience leads to the belief that retaining the fumes makes results more certain. It has also been shown that when large areas are involved the amount of water can be reduced by more than one-half, and fairly good results be obtained. This, however, is not recommended when the full amount of water is available, for the deeper the saturation the more will the disinfectant be diffused. Seeds or plants must not be placed in the treated soil until ten days or two weeks after the application.

Steam formaldehyde.

When the application of steam is made hurriedly, it may at times be desirable to combine it with a slight dosage of formaldehyde. The latter is applied in the manner described above. The steam pan or drain tile is then put in operation. The steam is turned on for about half the length of time ordinarily recommended. Since the expense for steaming in this manner is almost as great as if no formaldehyde were applied, the addition of the liquid fungicide under these conditions is warranted only in rare cases.

Other Methods of Heating the Soil

Hot water.

Small flower-pots filled with soil may be freed from pathogenic organisms by dipping in boiling water for five minutes. Larger pots will take a relatively longer time for the heat to penetrate all parts of their contents. The hot water dipping may be used when there are only a few pots to sterilize, and may be employed even on a larger scale if a suitable tank is available. A false wooden bottom is placed in the tank so that the tops of the pots will just be submerged but will not be low enough for the soil to wash out into the water. The tank is kept over an open fire so that the water will remain at the boiling point. Pots can be set in as fast as high temperature will permit.

Large pots, flats, and even shallow benches may be partly sterilized, at least, when sufficient boiling water is poured over them. An eight-inch pot will require a gallon poured uniformly over all parts of the surface of its soil. Four and one-half gallons are needed for a flat 14 x 30 x 3 inches, or seven gallons of the boiling water for each square foot of soil in shallow benches.

The hot water is more effective in deep soil if the space is first covered with an inverted wooden pan similar to that used for steam sterilization. The water is applied through a hole in the center of the pan until at least seven gallons of the boiling water are added for each square foot. Before the water is applied, the soil should be warm, fairly dry, and well spaded. Good drainage is necessary to avoid puddling of the soil. After the water has been applied, the pan is moved to an adjoining section and the heated soil covered with burlap or other material to retain the high temperature as long as possible.

The hot water will kill not only bacteria and fungi, but is also effective against the nematode. For all that, it is not a popular form of sterilization because the cost of keeping the temperature up on such a large amount of water is excessive.

Surface burning.

When the timber was plentiful, it was not unusual to pile brush and logs on a given area and then by burning the wood sterilize the ground below. It was a cheap method at one time, but at present would be very expensive. Another very apparent disadvantage is that the humus in the soil is destroyed.

Soil baking.

A slightly more desirable process of sterilization than surface burning is that of baking the soil. A shallow iron or heavy tin pan is placed over a fire and a layer of soil shoveled into it. When this has been heated above 140° F. for a half hour, it is dumped and another layer placed in the pan. The method necessitate handling the soil twice, and is slow and cumbersome.

Miscellaneous Disinfectants

A large number of fungicides have been added to the soil, but only a few have shown any marked ability to hold in check the attack of the different parasites. Among the more successful ones are sulfuric acid, copper sulfate, and corrosive sublimate. Bleaching powder and a copper mixture named Cheshunt Compound have been recommended highly in England, but in America these two materials did not prove to be disinfectants of any particular value even when applied in large enough dosages to injure the host. Other substances that have been tried by the various investigators are lime-sulfur, copper acetate, potassium permanganate, toluol, carbon bisulfide, tobacco stems, moth-balls, ammonia, kerosene, gasoline, ammoniacal copper carbonate, zinc chloride, nitric acid, ether, chloroform, benzene, thymol, sodium cyanide, picric acid, and sundry other chemicals. None has proved effective enough to be admitted to the vegetable-grower's pharmacopœia.

Sulfuric acid.

This chemical is not of value to the vegetable-grower but is mentioned because it is used successfully in combating damping-off of forest nursery seedlings. From three-sixteenths to five-eighths of a fluid ounce of the commercial sulfuric acid, depending on the previous alkalinity of the soil, are added to each square foot of seed-bed at the time of planting. The beds are then watered twice a day with enough water to equal a three-tenths-inch rainfall, the special watering being continued until the seeds have germinated. The acid is helpful in killing weeds, and persists in the soil long enough to protect even such slowly germinated seedlings as the pine.

Copper sulfate.

When steam or formaldehyde have not been used to protect the seed-bed, a less complicated control measure as well as a less sure one may be found in the application of one pound of copper sulfate dissolved in twenty or twenty-five gallons of water. It has no value in controlling nematodes but does check the growth of damping-off fungi. The solution may be put on the seed-bed after the seed has been planted. One pint is applied to each square foot of space, and then to make sure that the disinfectant is well distributed, an equal amount of water is sprinkled over the beds. The copper sulfate, although harmless at this dilution to seeds, is not supposed to take the place either of steam or formaldehyde. It is recommended as of considerable protection against loss only when the more efficacious control measures have not been practiced.

Corrosive sublimate.

While using corrosive sublimate (mercuric chloride) at the Geneva Experiment Station for the control of cabbage maggot, it was discovered that the poison inhibited club-root infection and wire-stem caused by Rhizoctonia. It does not sterilize the soil, but disinfects merely a small area directly about the roots of the plant. The concentration of the solution is one ounce of the poison dissolved in ten gallons of water. One gallon of the solution is sufficient to treat from twenty to forty feet in a row of seedlings, depending on the size of the plants. The first application is made a week after the first plants appear above the ground, and three or four additional applications are given at weekly intervals. The poison can be applied with a sprinkling can from which the rose at the end of the spout has been removed. The chemical corrodes metal and should not be left longer than necessary in contact with this receptable.

Freezing.

Many growers still believe that freezing soil during the winter sterilizes it. The truth is that nearly all fungi and bacteria can remain alive in extremely low temperatures, although some of them may be killed, particularly when there is alternate freezing and thawing. The one great benefit derived from freezing is the killing of nematodes. The winter is severe enough in any of the northern tier of states, so that if the unused greenhouse is left open during the coldest weather, the soil will freeze deep enough to eliminate most if not all of these destructive ell-worms.

Changing soil.

In greenhouses or hotbeds where neither heat nor chemicals are applied to the soil, it is well to change the soil even though the practice is an expensive one. The used dirt is carted away, the empty beds are thoroughly cleaned, after which soil is procured from an uninfested field. Woodland soil may contain damping-off organisms, and nematodes and other organisms may be so generally distributed that no benefit is derived from the renewal of the substratum. If care is taken, however, the renewal may be an important means of avoiding plant diseases.

References

Barker, B. T. P. and C. T. Gimingham. Experiments on the treatment of Rhizoctonia disease of asparagus. Agr. and Hort. Res. Sta. Univ. Bristol, Ann. Rept. **1916**: 39-40. 1917.

Beinhart, E. G. Steam sterilization of seed-beds for tobacco and other crops. U. S. Dept. Agr. Farmers' Bull. 996: 1-15. 1918.

Bewley, W. D. Report of the work of the W. B. Randall assistant. Exp. and Res. Sta. Nursery and Mark. Gard. Indus. Cheshunt, Herts, Ann. Rept. **6**: 45-51. 1921.

Brown, H. D., I. L. Baldwin, and S. D. Conner. Greenhouse soil sterilization. Ind. Agr. Exp. Sta. Bull. 266: 1-27. 1922.

Byers, L. P. and W. W. Gilbert. Soil disinfection with hot water to control the root-knot nematode and parasitic soil fungi. U. S. Dept. Agr. Bull. 818: 1-14. 1920.

Darbishire, F. V. and E. J. Russell. Oxidation in soils and its relation to productiveness. Part II. The influence of partial sterilization. Jour. Agr. Sci. **2**: 305-326. 1907.

Gifford, C. M. The damping-off of coniferous seedlings. Vt. Agr. Exp. Sta. Bull. 157: 143-171. 1911.

Gilbert, W. W. The root-rot of tobacco caused by *Thielavia basicola*. U. S. Dept. Agr. Bur. Plant Ind. Bull. 158: 34-35. 1909.

Gloyer, W. O. and H. Glasgow. Effect of the mercuric chloride treatment for maggot on Rhizoctonia and club-root of cabbage. Phytopath. **14**: 1924. N. Y. (Geneva) Agr. Exp. Sta. Bull. 513: 1924.

Hartley, Carl. Injury by disinfectants to seeds and roots in sandy soils. U. S. Dept. Agr. Bull. 169: 1-35. 1915.

Hartley, Carl and T. C. Merrill. Preliminary tests of disinfectants in controlling damping-off in various nursery soils. Phytopath. **4**: 89-92. 1914.

Johnson, James. The control of damping-off in plant beds Wis. Agr. Exp. Sta. Res. Bull. 31: 29-61. 1914.

Lyons, T. L. and J. A. Bizzell. Effect of steam sterilization on the water-soluble matter in soils. N. Y. (Cornell) Agr. Exp. Sta. Bull. 275: 129-155. 1910.

Melchers, L. E. A method of steam sterilization of soil for controlling nematodes. Phytopath. **9**: 294-296. 1919.

Russell, E. J. and H. B. Hutchinson. The effect of partial sterilization of soil on the production of plant food. Jour. Agr. Sci. **3**: 111-144. 1909.

Winston, J. R. Effect of the steam-formalin treatment on certain soil organisms. Phytopath. **3**: 74. 1913.

CHAPTER XX

FUNGICIDES

THE discovery of bordeaux mixture in 1882 was the beginning of practical plant pathology. This mixture and lime-sulfur, which was first used as a summer spray in 1906, are still the only two fungicides that are applied generally as sprays. Lime-sulfur is injurious to most vegetables or does not control the parasites of the garden crops, therefore needs no further mention in this discussion. The copper-lime mixture with its many variations is the generally accepted fungicide for vegetables. The copper is the fungicidal agent in ordinary bordeaux mixture, in copper-lime dust, in copper sulfate solution, in ammoniacal copper sulfate, in ammoniacal copper carbonate, in copper-soap spray, in copper-soap dust, and in many of the proprietary compounds on the market.

Bordeaux mixture.

This spray is a precipitate obtained by mixing a solution of copper sulfate (blue vitriol) with a suspension of lime. The copper sulfate ordinarily used for the purpose is in the form of blue crystals but it may be obtained in a granulated form. The latter dissolves slightly more readily than do the crystals. When copper sulfate stands for some time in a dry place it becomes white. The change in color, however, does not indicate any loss in strength as is sometimes erroneously believed.

The best lime to use in making bordeaux mixture is lump lime or burnt lime. Air-slaked lime cannot be used, and for this reason it is desirable that the lump lime be as fresh as possible. Hydrated lime, if fresh may be employed if lump lime cannot be obtained, although bordeaux mixture made of it does not stay in suspension quite as long as does that made from the burnt lime.

When much spraying is to be done, it is well to make stock solutions of the blue vitriol and the lime. Fifty pounds of the copper crystals are placed in a sack and suspended in the top of a barrel holding fifty gallons of water. The warmer the water the more rapidly the crystals will dissolve, but they will dissolve in moderately cold water in five or six hours. As the lump lime needs slaking, a little warm water is poured on the lumps until they begin to generate heat. Care must be taken in not adding more water to the lime than is required for a smooth paste by the time it is through slaking. As many gallons of water are then added as there are pounds of lime, and the suspension of lime stored until needed for spraying. If hydrated lime is used, fifty pounds of the powdered form are worked into a paste and then stirred into as many gallons of water in a barrel.

After the stock solutions are made, they may be kept as long as desired if the water lost by evaporation is replaced. The bordeaux mixture should not be prepared until needed for it begins to deteriorate as soon as the ingredients are mixed together. In preparing the mixture, as much of the stock solution of blue vitriol as is required is placed into the sprayer. The tank is then nearly filled with water, the liquid being well stirred in the meantime so that the copper sulfate will become well diffused. If a 5-5-50 bordeaux mixture is desired, five gallons of the stock solution are dipped out for each fifty gallons of liquid that the sprayer holds. As soon as the tank is nearly filled with water and the copper sulfate well diluted, the same amount of lime-water as was used of the stock blue vitriol is added, the mixture being stirred violently during the process. As there are always some coarse particles in the lime-water, it is necessary to run this through a strainer as it is poured into the solution in the tank. A copper or brass wire strainer with eighteen meshes to the inch is best for this purpose, although several thicknesses of cheese-cloth or other similar fabric may be used in an emergency. The strainer may be fastened to a frame which fits into the opening of the sprayer.

As the lime in the stock solution settles to the bottom of the barrel and is difficult to get into suspension again, it has been found convenient to test the lime requirement of the bordeaux mixture rather than to attempt to get in the exact amount of lime by dipping up the solution. The material used for making the test is a solution of potassium ferrocyanide (yellow prussiate of potash). This material is a yellow waxy crystalline substance, which may be purchased at a drug store. A few crystals are placed in a half pint bottle with distilled or rain water. After the bordeaux mixture is made with what is considered sufficient lime, a small amount of the mixture is placed in a shallow receptable, as a saucer, and a few drops of the potassium ferrocyanide solution is added. If there is no discoloration, there is enough lime, but if there is a brown discoloration, as soon as the solution reacts with the bordeaux mixture more lime is required to neutralize the copper sulfate solution. Lime-water is added to the sprayer tank from the stock barrel until the test shows no discoloration of the mixture when potassium ferrocyanide is added.

There are nine different ways of mixing the solution of copper sulfate and milk of lime to make bordeaux mixture. The most practical and satisfactory method is to dilute the required amount of stock blue vitriol in the spray tank and add the necessary milk of lime from the stock barrel as already described; that is, adding strong lime suspension to a dilute copper sulfate solution. Both solutions might be diluted and poured together, or the lime-water could be diluted after which the concentrated copper solution is added. Thus by varying the methods, nine combinations are possible. It does not matter which combination is employed as long as both substances are not concentrated when they are poured together. If the two stock solutions are mixed together before either are diluted, the resulting spray consists of a coarse heavy precipitate that settles rapidly to the bottom and is worthless as a fungicide.

It is sometimes desirable to add a soap sticker to the bordeaux mixture. Three pounds of resin-fish-oil-soap are dissolved in warm water, and the solution poured into fifty gallons of the freshly made bordeaux mixture while the latter is being stirred. The formula may then be represented by the figures 4-4-3-50.

Even with the best made bordeaux mixture there is sure to be some sediment. When the spraying is finished each day, the nozzles should be removed and cleaned, the sprayer washed out, and water pumped through the pipe and hose in order to flush them.

Burgundy mixture.

Occasionally when it is difficult to procure good lime, the latter may be replaced in the spray mixture by sal soda (sodium carbonate). The spray is made in the same manner as is bordeaux mixture, except that the formula in the case of burgundy mixture is one pound of copper sulfate, one and one-half pound of sal soda and fifty gallons of water, or 1-1.5-50. It is more caustic in its action on foliage, and somewhat less effective in its fungicidal properties than is bordeaux mixture.

Copper-lime dust.

Sometimes it is more convenient to dust plants than to spray them, especially since the time and labor are much reduced when dusting is practiced. In making the dust, four-fifths of the water of crystallization of ordinary copper sulfate is removed by heating the crystals. The chemical is then known as mono-hydrous copper sulfate. This is ground as finely as it can be made by machinery, after which it is mixed with a good grade of hydrated lime. The ratio of copper and lime varies in different dusts, ordinary mixtures having either fifteen or twenty pounds of copper dust and eighty-five or eighty pounds of lime. These mixtures are designated by the formulæ 15-85 and 20-80, and are sometimes known as Sander's dust. Five to ten pounds of the lime may be replaced by an insecticide. The copper-lime dust may be bought already prepared, or the different ingredients may be purchased and mixed at home. In the latter case, it will be necessary to have a machine for mixing the materials. Suitable machines are on the market. A home-made mixer can be built by placing a tight barrel on an axis like a revolving cylinder, so that it can be turned with a crank. A small door which can be closed tightly is fitted into the side of the barrel. Boards are nailed inside the cylinder in such a manner that spiral flanges are formed for agitating the dust as the barrel revolves.

The dust is applied to the plants with a hand or machine duster, the same number of applications being made as are recommended for sprays. Every aerial part of the plant should be covered no matter how many nozzles are required for each row.

Copper sulfate solution.

The copper sulfate dissolved in water is frequently employed as a disinfectant. The concentration of the solution depends on the use for which it is meant. When boards around the seed-bed or cement floors and walls are to be scrubbed, one pound of the crystals may be dissolved in five gallons of water or less. If the solution is intended for a seed-bed after the seed is sown, it is diluted to one pound of the copper sulfate in twenty-five gallons of water.

Copper ammonia solution.

In Florida a solution made of two pounds of copper sulfate, three pints of ammonia, and fifty gallons of water has been used in combating the southern sclerotium wilt of tomato. The solution has long been known by the name *eau céleste.*

Ammoniacal copper carbonate is sometimes employed for spraying when ordinary bordeaux mixture cannot be applied because of the unsightly spotting which the latter produces on the sprayed plant parts. The copper carbonate leaves no stain. Six ounces of the copper carbonate are dissolved in three pints of ammonia, after which fifty gallons of water are added. This solution should be used at once, as it loses strength on standing exposed to the air.

Copper soap.

In treating spinach for mildew, a copper-soap solution has been tried successfully. One-half pound of copper sulfate is dissolved in twenty-five gallons of water. Five pounds of caustic-potash fish-oil soap also are dissolved in twenty-five gallons of water. The two solutions are then poured together, being stirred vigorously while being mixed.

A copper-soap dust has been tried on tomatoes with promising results. A saturated solution of copper sulfate in water is prepared. A hot aqueous solution of resin-fish-oil soap or caustic-potash fish-oil soap solution of about the consistency of sirup is poured into the copper sulfate solution, keeping both of them moderately hot. The precipitate which is formed by the union of the two solutions is dried in the air, then ground into a fine powder and used as a dust.

Mercuric chloride.

The most generally used chemical for seed treatment is mercuric chloride, or corrosive sublimate as it is usually known. It may be bought in tablet or powder form. One tablet ordinarily contains just the proper quantity of the poison to make a 1-1000 solution when dissolved in a pint of water. The tablets contain in most cases also a coloring matter, so that the solution will not be mistaken for drinking water, as the mercuric chloride is extremely toxic when taken into the stomach. It, however, may be used safely on open wounds on the hands or body. The white powder form of the poison is less expensive, and can now be bought in small four-ounce boxes which will supply enough of the chemical for thirty gallons of water, or one part of the poison in one thousand parts of water.

The principal objections to the use of mercuric chloride are that it is poisonous, that when being used the solution loses its strength rapidly, and that it can be placed only in wooden, earthenware or glass containers. When placed in metal or cement containers, the mercury is replaced by the metal with which it comes in contact. Cloth sacks and dirt also tend to weaken the solution.

The 1-1000 formula of the corrosive sublimate is used for nearly all purposes, except for wire-stem of plants when the solution is weakened to four ounces in forty gallons of water. The strength of the solution does not matter so much as does the temperature at which it is applied. There is little or no reaction at 40° F., but as the temperature is raised the fungicidal value of the poison increases rapidly. At 70° F. potatoes may be soaked one to two hours without injury, but if the solution is heated to 125° F., the tubers will be injured if treated only ten minutes. Many delicate vegetable seeds can be treated without injury when the temperature of the solution is 60° to 70° F. If the temperature is raised to 80° F., the same length of treatment may kill most of the seed. Consequently in treating seeds and tubers, the length of application should be reduced in inverse proportion to the increase of temperature of the solution.

Formerly the recommendation was made that the disinfecting solution be used three times and then discarded, but this entailed so much extra labor that growers failed to treat their vegetable seeds and potatoes. Cross has adapted a comparatively simple test for determining the strength of the corrosive sublimate solution, so that after using the poison once or twice, the test can be applied and the proper amount of bichloride added without throwing away any water. This saves time, labor, and expense. Crystals of potassium iodide, which may be procured at a drug store, are dissolved in water. It is not imperative to have any definite concentration, but a very convenient strength of solution is five grams (about one-sixth ounce) dissolved in 1000 cubic centimeters of water, or roughly, five grams in a quart of water. After the crystals are dissolved, one gram of copper sulfate which has been dissolved in a little of the quart of water is added. Various devices may be used for measuring the required solution when making the test. A small graduated glass cylinder may be obtained at the drug store, although a tall narrow bottle with marks filed on its side at quarter-inch intervals will serve the purpose.

The first test is given after the corrosive sublimate solution is made up and before any tubers or seeds are treated. Enough of the potassium iodide solution is poured into the glass measuring cylinder or into the bottle to reach to the first or second mark. Ten cubic centimeters of the iodide may conveniently be used in a graduated cylinder. The exact quantity is not important, except that the same amount of potassium iodide solution must be used each time that the test is made. Then the corrosive sublimate solution is added very slowly, shaking the cylinder, until an orange precipitate begins to form. The amount of corrosive sublimate solution needed to produce the precipitate is recorded. The tubers or seeds are then treated in the usual manner until the disinfecting solution is used twice, when another test is made. Before testing, however, enough water should be added to the corrosive solution to bring it up to its original volume of thirty gallons. Again, the same amount of the potassium iodide solution as was employed the first time is placed in the cylinder or bottle and the disinfecting solution added slowly. Since the latter is weaker, after being used, more will be required to form the orange-colored precipitate. The proportionately larger amount will indicate the quantity of corrosive sublimate which has been removed in the treating, and the amount of the poison to be added. For example, if two parts or measures of the bichloride solution are required to form the precipitate in the first test, and three parts or measures are necessary in the second, the solution is only two-thirds as strong as it was originally; therefore, one-third of the original four ounces of poison, or one and one-third ounces are to be added. In a similar manner the proportion can be figured out for any loss in strength that may have taken place in treating, and then it is very simple to weigh or measure out the necessary ounces of corrosive sublimate. In fact, the entire process of testing is much more simple than a written description would lead one to believe.

Formaldehyde.

This is one of the older fungicides, which in many cases has been replaced by corrosive sublimate. The formaldehyde gas is dissolved in water up to about 40 per cent, and is placed on the market as a colorless liquid. Its fumes are irritating to the eyes, nose, and throat. The formaldehyde has the advantage of retaining its strength and of not reacting with metals, so that it may be placed into any kind of containers.

The concentration of the formaldehyde is varied for nearly every different use. In sterilizing soil, it may be diluted one part in fifty of water or one part in two hundred of water. In its use for controlling onion smut, it is diluted one pint in sixty-four pints of water or less. The ordinary strength for seed treatment is usually one pint in thirty gallons of water.

The formaldehyde may also be used for disinfecting storage-houses after vegetables are removed. A gallon of the solution for each thousand cubic feet of space to be disinfected is placed in a large flaring receptacle. After all windows and cracks are closed tightly, thirty ounces of potassium permanganate for each gallon of formaldehyde are dropped into the solution, after which the operator leaves quickly. The fumes may be retained in such an inclosed place for several days or even a week, the longer the better. A storage-house cannot be disinfected with formaldehyde if vegetables or fruit are present, for any living plant parts are injured.

Hot water.

Water heated to a certain temperature is one of the most useful disinfectants known, yet is seldom employed by the average grower. The one disadvantage is the difficulty in keeping the temperature constant during a given operation. It also is expensive to heat large quantities of water. It is applied at a boiling temperature to the soil, or at temperatures of 90° to 125° F. to seeds.

Bleaching powder.

The bleaching powder, also known as calcium hypochlorite or chloride of lime, is usually sold in one-pound paper cartons. It is used extensively in laboratories, but is not well known by the growers. Two ounces of the powder are soaked in a quart of water for an hour or two, after which the water is decanted and employed for disinfecting seeds. In England the dry powder has been applied to the soil for disinfection, but similar experiments in America did not yield satisfactory results. The time or method of application may have been at fault.

References

Barker, B. T. P. and C. F. Gimingham. Further experiments on the Rhizoctonia disease of asparagus. Agr. Hort. Res. Sta. Univ. Bristol Ann. Rept. **1917**: 28-32. 1918.

Bourcart, E. Insecticides, fungicides, and weed-killers, pp. 1-431. Scott, Greenwood and Son, London. 1913.

Butler, O. Bordeaux mixture: I Physico-chemical studies. Phytopath. **4**: 125-180. 1914.

Butler, O. Bordeaux mixture: II Stimulatory action. N. H. Agr. Exp. Sta. Tech. Bull. 21: 1-50. 1922.

Duggar, B. M. and W. W. Bonns. The effect of bordeaux mixture on the rate of transpiration. Ann. Mo. Bot. Gard. **5**: 153-176. 1918.

Ewert, R. Weitere Studien über die physiologische und fungicide Wirkung der Kupfer-brühen bei kraütigen Gewächsen und der Johannisbeere. Ztschr. f. Pflanzenkr. **22**: 257-285. 1912.

Hawkins, L. A. Some factors influencing the efficiency of bordeaux mixture. U. S. Dept. Agr. Bur. Plant Ind. Bull. 265: 1-29. 1912.

Lodeman, E. G. Spraying of plants, pp. 1-399. Macmillan Company, New York. 1906.

Lutman, B. F. The covering power of the precipitation membranes of bordeaux mixture. Phytopath. **2**: 32-41. 1911.

Lutman, B. F. Some studies on bordeaux mixture. Vt. Agr. Exp. Sta. Bull. 196: 1-80. 1916.

Pritchard, F. J. and W. S. Porte. Use of copper-soap dust as a fungicide. Phytopath. **11**: 229-235. 1921.

Schrander, R. Ueber die physiologische Wirkung der Kupferkalkbrühe. Landw. Jahr. **33**: 517-584. 1904.

Smith, L. B. Control of spinach leaf-mold (downy mildew) by spraying. Va. State Crop and Pest Comm. Quart. Bull. **3**: 1-4. 1921.

Stewart, F. C. and W. O. Gloyer. The injurious effect of formaldehyde gas on potato tubers. N. Y. (Geneva) Agr. Exp. Sta. Bull. 369: 385-416. 1913.

Weimer, J. L. Reduction in strength of the mercuric chloride solution used for disinfecting sweet potatoes. Jour. Agr. Research **21**: 575-587. 1921.

Wilson, J. K. Calcium hypochlorite as a seed sterilizer. Am. Jour. Bot. **2**: 420-427. 1915.

GLOSSARY

Acervulus (acervuli). Open, saucer-shaped, asexual fruiting body, in which spores of some fungi are borne.

Aeciospore. One of the types of spores formed by rust fungi, and borne in an aecium or cluster-cup.

Aecium. A cup-like type of fruit-body in rusts.

Aplanobacter. A genus of bacteria and differentiated by having no flagellæ—not being motile.

Apothecium (apothecia). Open cúp or saucer-shaped sexual fruit-body, frequently borne on a slender stem.

Appressorium (appressoria). An organ of attachment developed by certain fungi.

Ascospore. Sexually formed spores which are borne within a sac known as an ascus.

Ascus (asci). Sac-like spore-bearing structures borne in apothecia, perithecia, or directly on the host.

Bacillus. A genus of bacteria and differentiated by having cilia or flagellæ over all parts of the body.

Bacteria (bacterium). Small microscopic plants each having one cell and often being motile. They lack the green coloring matter (chlorophyll) of higher plants.

Bacterium. A genus of bacteria and differentiated by having flagellæ only at one or both ends.

Basidiospore. Spores borne on basidia, which are club-shaped or thread-like stalks produced by fungi belonging to such groups as rusts, smuts, and mushrooms.

Calyx (calyces). The outer series of leaves of a flower, individually called sepals.

Cambium. In a woody stem or root the region of growth at which wood is formed on the inside and bark on the outside.

Canker. A definite dead area in the bark of woody plants or the cortex of herbaceous stems; smooth or roughened, sunken or raised.

Chlamydospore. A thick-walled resting spore developed directly from the mycelium.

Chlorophyll. The green coloring material produced in the leaves of the higher plants.

Cilium (cilia)). A thread-like organ by which the swarm-spores of fungi and some bacteria move.

Conidiophore. A stalk which bears conidia.

Conidium (conidia). An asexual spore.

Cortex. The tissue lying between the vascular bundles and the epidermis. In trees it is the outer bark.

Cuticle. A waxy layer which covers exposed plant surfaces, such as leaves, succulent stems, and fruits.

Enzyme. A chemical compound capable of assisting or hastening chemical transformation, but without itself entering into the final product.

Fruit-body. Any structure made of mycelium and in which the spores of fungi are borne (acervulus, pycnidium, apothecium, perithecium, aecium, etc.).

Fungus. A simple plant of low order lacking chlorophyll. The body is composed of mold threads which may be branched. Food is obtained by direct absorption, and reproduction is mainly by spores of various kinds.

Germ-tube. A tube developed from a spore on germination. It usually gives rise directly to mycelium.

Haustorium (haustoria). A special branch of the mycelium which is pushed into a cell to obtain food materials.

Heterœcious. Said of fungi (usually rusts) requiring more than one kind of host plant on which to complete the life history.

Host. Any plant on or in which a parasite grows.

Hypertrophy. An abnormal enlargement of plant parts.

Hypha (hyphæ). An individual thread of the vegetative part of the fungus.

Infect (infection). The act of a parasitic organism when it establishes parasitic relations with a host plant.

Inoculum. The transferable portion of a pathogene which may bring about infection. It usually consists of spores or mycelial fragments.

Lesion. Any definitely diseased place on the plant.

Middle lamella. The primary membrane between any two cells of higher plants.

Morphology. The science which treats of the form, size, and color of an organism.

Mycelium. The vegetative body of a fungus composed of threads (hyphæ).

Oospore. A special type of sexual spore such as is produced by fungi causing downy-mildews and white-rust.

Organism. An animal or plant.

Parasite. An organism which lives in or on another living organism during a portion or all of its life for the purpose of obtaining food and habitation.

Pathogene. A disease-producing organism.

Pedicel. A fruit-bearing or flower-bearing stalk.

Perennial. Lasting more than two years.

Perthecium (perthecia). A closed globose or flask-shaped sexual fruit-body containing asci and ascospores.

Protoplasm. The living substance within the cells of plants and animals.

Pseudomonas. Same as Bacterium.

Pycnidium. An inclosed globose or flask-shaped fruit-body containing asexual spores.

Pycnospore. An asexual spore borne in a pycnidium.

Rhizomorph. A compact bundle of mycelium arranged parallel to form a root-like structure.

Saprophyte. A living organism which obtains its food materials from dead organic matter.

Sclerotium (sclerotia). A compact structure of closely woven mycelial threads. Usually a resting body rich in stored food.

Seta (setæ).) A bristle or bristle-like body.

Sorus. A heap or collection of spores causing a dusty fruit-body.

Sporangium (sporangia). An asexual spore-containing sac. Not inclosed and usually borne on a stalk known as a **sporangiophore.**

Spore. A portion of the mycelium which is detached to serve as a propagative or reproductive body, corresponding to the seeds of higher plants. Spores may be formed sexually or asexually.

Stoma (stomata). A minute opening in the epidermis of stems, leaves, and fruits which admits free exchange of gases between the interior and exterior of the plant.

Stroma (stromata). A cushion-like growth of mycelium on which fruit-bodies are borne.

Swarm-spore. A spore which is able to move about in water by means of cilia or a cilium.

Teliospore. A type of spore formed by rust fungi.

Telium (telia). The shallow fruit-body in which teliospores are borne.

Urediniospore. A type of spore formed by the rust fungi—usually the summer type.

Uredinium (uredinia). A sorus in which urediniospores are borne.

INDEX

A

B

D

R

U

V

W

Z